Advanced Polymeric Materials
Synthesis and Applications

RIVER PUBLISHERS SERIES IN POLYMER SCIENCE

Series Editors

Indexing: All books published in this series are submitted to the Web of Science Book Citation Index (BkCI), to CrossRef and to Google Scholar.

The "River Publishers Series in Polymer Science" is a series of comprehensive academic and professional books which focus on theory and applications of Polymer Science. Polymer Science, or Macromolecular Science, is a subfield of materials science concerned with polymers, primarily synthetic polymers such as plastics and elastomers. The field of polymer science includes researchers in multiple disciplines including chemistry, physics, and engineering.

Books published in the series include research monographs, edited volumes, handbooks and textbooks. The books provide professionals, researchers, educators, and advanced students in the field with an invaluable insight into the latest research and developments.

Topics covered in the series include, but are by no means restricted to the following:

- Macromolecular Science
- Polymer Chemistry
- Polymer Physics
- Polymer Characterization

For a list of other books in this series, visit www.riverpublishers.com

Advanced Polymeric Materials
Synthesis and Applications

Editors

Didier Rouxel
Université de Lorraine
France

Sabu Thomas
Mahatma Gandhi University
India

Nandakumar Kalarikkal
Mahatma Gandhi University
India

Sajith T. Abdulrahman
Mahatma Gandhi University
India

Published 2018 by River Publishers
River Publishers
Alsbjergvej 10, 9260 Gistrup, Denmark
www.riverpublishers.com

Distributed exclusively by Routledge
4 Park Square, Milton Park, Abingdon, Oxon OX14 4RN
605 Third Avenue, New York, NY 10017, USA

Advanced Polymeric Materials Synthesis and Applications / by Didier Rouxel, Sabu Thomas, Nandakumar Kalarikkal, Sajith T. Abdulrahman.

Routledge is an imprint of the Taylor & Francis Group, an informa business

ISBN 978-87-93609-68-6 (print)

Contents

Preface

Recent advances in the polymers leads to the generation of high quality materials for various applications in day to day life. The synthesis of new functional monomers has shown strong potential in generating novel polymer material with improved properties. The progress in the blends, polymer nanocomposites and its various characterisation techniques are highlights of this book.

Chapter 1 discusses about the synthesis of polybenzoxazine containing bismaleimide as part of the benzoxazine molecule. The main objective of this study is to examine the polymerization and degradation properties of the blends of traditional benzoxazine based on Bisphenol-A with extended bismaleimide.

In Chapter 2, an investigation on curing and thermal stabilities of bispropargyl ether of bisphenol-A with 4,4'-bismaleimidodiphenyl methane BMIM and 4,4'-bismaleimidodiphenyl ether were performed. These studies reveals that different stuctural networks were formed through polymerisation of bispropargyl ether-bismaleimide blends.

Chapter 3 is an empirical study on solubility of 2-thiophenecarba benzhydrazone and oxovanadium(IV) complexes in different solvents through spectroscopic analyses. The complex was proposed to be square pyramidal in geometry. The molecular ion peaks were less intense suggesting complex possess lower stability in solution.

Chapter 4 considers sorption-desorption behaviour of various sorbents and different models dealing with oil-spill. Sorption abilities and release tendencies were quantified in terms of various absorption efficiencies and test-fluid release profiles. The analyses revealed that the polyurethane sorbents had good practical promise of a performance for oil-spill applications.

Chapter 5 disribes the synthesis of Hemiaminal from 4,4'-methylene dianiline, paraformaldehyde, and N-methylpyrrolidone and its thermal studies. The DSC and TG studies indicated the loss of N-methylpyrrolidone and water during the formation of polyhexahydrotriazine.

In Chapter 6, the effect of Melamine Formaldehyde and PolyCarboxylate Ether with CNT composites on various behavioral properties of cement and concrete were studied. The PCE nanocomposites improved strength of the cement while the MF nanocomposites has very little effect.

Chapter 7 discuss the effect of disphenols on blending with 4,4'-bisisomaleimidodiphenyl methane through thermal polymerisation. The disphenols Resorcinol, Quinol, 4,4'-Dihydroxybiphenyl and Bisphenol-A in 1:1 mole ratio for these studies. All the polymers from the blends show considerably lower char yield indicating lower thermal stability which may be attributed to the structure of the resulting copolymer from the blend.

Chapter 8 is a review which discuss about the processing, characterisation and applications of Natural fibre reinforced polymer composites. The overview on development in the area of Natural fibre reinforced polymer composites in terms of description and classification of the composites, fabrication methods, characterization of composites and applications are addressed here.

Chapter 9 is an investigation on factors effection the tribological performance of various polymer composites. The role of nature and size of the fillers on the mating surfaces of the polymer composites were reviewed. Various preparation methods have greater importance in the final performance of the composites.

Chapter 10 gives a brief account of advancement in molecular imprinting technology for enantioselective sensing system (MIT) in accordance with the theoretical and computational aspects were discussed. The molecular imprinting technique for the preparation of polymers with predetermined selectivity, specificity and high affinity which involves arrangement of polymerizable functional monomers around a print molecule.

Chapter 11 is a review which discuss the properties and specialties of transition metal dichalcogenide materials and how ultrafast process could help to characterize them, which will be directed to the evolution of Nanoelectronics to another level and futuristic applications.

List of Contributors

Aravind Krishnan, *Research and Post Graduate Department of Chemistry, St. Berchmans College, Changanassery, India*

Arun Kumar Chakraborty, *Associate Professor, Indian Institute of EngineeringScience and Technology, Howrah, India*

Ashok N. Bhaskarwar, *Department of Chemical Engineering, Indian Institute of Technology Delhi, New Delhi, India*

Beena Mathew, *School of Chemical Sciences, Mahatma Gandhi University, Kottayam, India*

Bilahari Aryat, *International Interuniversity centre for Nanoscience and Nanotechnology, Kerala, India*

C. T. Vijayakumar, *Department of Polymer Technology, Kamaraj College of Engineering and Technology, Virudhunagar, India*

J. Dhanalakshmi, *Department of Chemistry, Kamaraj College of Engineering and Technology, Virudhunagar, India*

Jyothy G. Vijayan, *Department of Chemistry, Christ University, Bengaluru, India*

K. G. Sudhamani, *Department of Polymer Technology, Kamaraj College of Engineering and Technology, Virudhunagar, India*

M. K. Gupta, *Department of Mechanical Engineering, Motilal Nehru National Institute of Technology, Allahabad, India*

M. M. Thahajjathul Kamila, *Department of Polymer Technology, Kamaraj College of Engineering and Technology, Virudhunagar, India*

Mainak Ghosal, *1. Adjunct Assistant Professor, JIS College of Engineering, Kalyani, India*
2. Research Scholar, Indian Institute of Engineering Science and Technology, Howrah, India

Neelakandan M. S., *International Interuniversity centre for Nanoscience and Nanotechnology, Kerala, India*

Neha Bhardwaj, *Department of Chemical Engineering, Indian Institute of Technology Delhi, New Delhi, India*

Nitish Paul Tharakan, *Department of Polymer Technology, Kamaraj College of Engineering and Technology, Virudhunagar, India*

Parvathy Prasad, *International Interuniversity centre for Nanoscience and Nanotechnology, Kerala, India*

R. K. Srivastava, *Department of Mechanical Engineering, Motilal Nehru National Institute of Technology, Allahabad, India*

R. Mahalakshmy, *Postgraduate and Research Department of Chemistry, Thiagarajar College, Madurai, India*

R. Surender, *Department of Polymer Technology, Kamaraj College of Engineering and Technology, Virudhunagar, India*

Raghvendra Kumar Mishra, *1. Director, BSM Solar and Environmental Solution, A-348, Awas Vikas Colony, Unnao, UP, 261001, India*
2. International Interuniversity centre for Nanoscience and Nanotechnology, Kerala, India
3. Indian Institute of Space Science and Technology, India

S. Shamim Rishwana, *Department of Chemistry, Kamaraj College of Engineering and Technology, Virudhunagar, India*

Sabu Thomas, *School of Chemical Sciences, Mahatma Gandhi University, Kottayam, India*

Sajith T. Abdulrahman, *International Interuniversity centre for Nanoscience and Nanotechnology, Kerala, India*

T. Sajini, *Research and Post Graduate Department of Chemistry, St. Berchmans College, Changanassery, India*

V. Sarannya, *Postgraduate and Research Department of Chemistry, Thiagarajar College, Madurai, India*

Yadu Nath V. K., *International and Inter University Centre for Nanoscience and Nanotechnology, Mahatma Gandhi University, Kottayam, India*

List of Figures

List of Tables

List of Abbreviations

α	Extent of reaction
β	Heating rate
Δ	Heat
ΔG	Gibbs free energy
ΔHc	Enthalpy of curing
ΔHf	Enthalpy of fusion
ΔS	Entropy change
β	Heating rate
μ	Magnetic moment
%	Percentage
°C	Degree centigrade
°C/min	Degree centigrade per minute
2Na-a	2, 2'-*p*-phenylene-bis-(2,3-dihydro-1Hnapth [1, 2-e]-m-oxazine)
3-APMI	N-(3-acetylenephenyl) maleimide
3-D	Three dimentional
4VP	4-Vinyl Pyridine
AAS	Atomic absorption spectra
ACN	Acetonitrile
Acac	Acetyl acetonate
Anal	Analysis
ACNC	Alkynylated cellulose nanocrystals
ANDAD	4-Amino-N, N-dimethyl aniline dihydrochloride
AT	Acetylene terminated
ATBN	Amine-terminated butadiene acrylonitrilerubber
ATR	Acetylene-terminated resin
ATRP	Atom transfer radical polymerization
B.M	Bohr magneton
BA-af	Bis(3-furfuryl-3,4-dihydro-2H-1,3-benzoxazinyl) isopropane

BAB	Bis(3,4-dihydro-2H-3-phenyl-1,3-benzoxazinyl) isopropane
bACE	Acrylic Polymer
B-ala	Bis(3-allyl-3,4-dihydro-2H-1,3-benzoxazinyl) isopropane
BAX	Bisamic acid
BI	N,N'-Biisomaleimide
BMI	Bismaleimide
BMIE	4,4'-Bismaleimidodiphenyl ether
BMIM	4,4'-Bismaleimidodiphenyl methane
B.P	Boiling point
BPEBPA	Bispropargyl ether of bisphenol-A
BT-resins	Bismaleimidetriazine-resins
bvMD	Molecular Dynamics
BZ	Benzoxazine
Calc.	Calculated
CAPRO	6-Amino caproic acid
CB	Carbon black
CB	Conduction Band
CE	Cyanate ester
CFs	Carbon fibers
cm	Centimeter
CNT	Carbon Nanotubes
C-S-H	Calcium Silicate Hydrate
CVD	Chemical vapor deposition
COSY	Correlated spectroscopy
d	Doublet
DABA	2,2'-Diallyl bisphenol-A
DA-BPAPCNB	2,2-Bis(4-aminophenoxy phenyl)propane
DCC	Dicyclohexycarbodiimide
DCD	Dicyandiamide
DFT	Density functional theory
DGEBA	Diglycidyl ether of bisphenol-A
DMA	Dynamic mechanical analysis
DMF	Dimethyl formamide
DMSO	Dimethyl sulphoxide
DN-BPAPCNB	2,2-Bis(4-nitrophenoxyphenyl)propane
DODEC	Dodecylamine
DRS	Diffuse reflectance spectra

DSC	Differential scanning calorimetric
DTG	Differential thermogravimetric analysis
Ea	Apparent activation energy
Ea-D	Apparent activation energy for the thermal degradation
EBPEBPA	BPEBPA-BMIE blend
EDMA	Ethylene Di Methacrylate
Epa	Anodic peak potential
Epc	Cathodic peak potential
EPR	Electron paramagnetic resonance
EtOH	Ethanol
EXBMI	2,2-Bis[4-(4-maleimido-phenoxy phenyl)]propane
FCB	Surface functionalized carbon black
FET	Field-effect transistor
FTIR	Fourier transform infrared spectroscopy
g	Gram
G	Gauss (EPR)
GAP	Glycidyl azide polymer
GC-MS	Gas chromatography–mass spectrometry
GNS	Graphite nanosheets
GPa	GigaPascal
h	Hour
HA	Hemiaminal
HA-MDA	Hemiaminal from 4,4'-methylene dianiline
HPM	Hydroxyphenylmaleimide
HPM-Ba	[N-(4-hydroxyphenyl)maleimide]-benzoxazine
HTBD	Hydroxyl-terminated polybutadiene
HSQC	Heteronuclear single quantum correlation
Ipa	Anodic peak current
Ipc	Cathodic peak current
Jg^{-1}	Joule per gram
$Jg^{-1}deg^{-1}$	Joule per gram per degree
LLDPE	Linear low density polyethylene
m	Multiplet (NMR)
m/z	Mass/Charge
MAA	Methacrylic acid
MALDI	Matrix assisted laser desorption -ionization
MBPEBPA	BPEBPA-BMIM blend
MDA	4,4'-methylene dianiline

MF	Melamine Formaldehyde
mg	Milligram
M.P	Melting point
MIB	Maleimido benzoxazine
min	Minutes
MIPs	Molecularly Imprinted Polymers
mL	Millilitre
mL min^{-1}	Millilitre per minute
mL/min	Millilitre per minute
mmol	Millimole
MMT	Montmorillonite
mol	mole
MoS_2	Molybdenum disulfide
$MoSe_2$	Molybdenum diselenide
$MoTe_2$	Molybdenum ditelluride
MP	Melting point
Na–MMT	Sodium montmorillonite
NMP	N-methylpyrrolidone
NMR	Nuclear magnetic resonance
NOB	Norbornene benzoxazine
OAPS	Octa(aminophenyl) silsesquioxane
OLEDs	Organic light-emitting diode
OMMT	Organically modified montmorillonite
ONPS	Octa(nitrophenyl)-silsesquioxane
Pa.s	Pascal-second
P-af	3-Furfuryl-3,4-dihydro-2H-1,3-benzoxazine
P-ala	3-Allyl-3,4-dihydro-2H-1,3-benzoxazine
P-alp	Allyl amine-based benzoxazine
PBE	Perdew-Burke-Ernzerhof
PBMIE	Polymerized BMIE
PBMIM	Polymerized BMIM
PBPEBPA	Polymerized BPEBPA
PBZ	Polybenzoxazines
PCE	PolyCarboxylate Ether
PCL	Polycaprolactone
PCL-NZ	Polycaprolactone-naphthoxazine
PE	Polyethylene
PEBPEBPA	Polymerized EBPEBPA
PEEK	Polyether ether ketone

PES	Polyethersulfone
PF	Paraformaldehyde
pH	Potential of hydrogen
PHEN	*p*-Phenetidine
PHT-MDA	Polyhexahydrotriazine formed from HA-MDA
PHTs	Polyhexahydrotriazines
PL	Photoluminescence
PMBPEBPA	Polymerized MBPEBPA
PMMA	Poly(methyl methacrylate)
PP	Polypropylene
PPOA	Polypropylene oxides
PPO-CouBenz	Poly [propylene oxide(s)] bearing coumarinebenzoxazine
PT	Propargyl terminated
PTFE	Polytetrafluoroethylene
PTR	Propargyl-terminated resin
PU	Polyurethane
RMC	Ready Mixed Concrete
T_i	Onset temperature
T_{max}	Temperature maxima
T_{10}	10% Weight loss temperature
T_{25}	25% Weight loss temperature
T_5	5% Weight loss temperature
T_{50}	50% Weight loss temperature
TBAH	Tetrabutylammonium hydrogen sulphate
Te	Endset
TEM	Transmission electron microscopy
Tg	Glass transition temperature
TG	Thermogravimetric
TGA	Thermogravimetric analysis
Tm	Melting temperature
TMAN	2,4,6-trimethylaniline
Tmax	Maximum temperature
TMDC	Transition metal dichalcogenide
Ts	Onset
UF	Urea formaldehyde
UHMWPE	Ultra-high-molecular-weight polyethylene
UV	Ultra violet
VB	Valance Band

VS	4,4'-Bisisomaleimidodiphenyl methane
VS 1	VS + Resorcinol
VS 2	VS + Quinol
VS 3	VS + 4,4'-Dihydroxybiphenyl
VS 6	VS + Bisphenol-A
WS_2	Tungsten disulfide
WSe_2	Tungsten diselenide

1

Bisbenzoxazine–Bismaleimide Blends: Thermal Studies

S. Shamim Rishwana[1], M. M. Thahajjathul Kamila[2] and C. T. Vijayakumar[2]

[1]Department of Chemistry, Kamaraj College of Engineering and Technology, Virudhunagar, India
[2]Department of Polymer Technology, Kamaraj College of Engineering and Technology, Virudhunagar, India

Abstract

The compounds bis (3,4-dihydro-2H-3-phenyl-1,3-benzoxazinyl) isopropane (BAB) and 2,2-bis[4-(4-maleimido-phenoxy phenyl)]propane (EXBMI) were synthesized. The blend of these two compounds was prepared in the 1:1 mole ratio. The structures of the synthesized compounds were confirmed by FTIR. The thermal curing behaviour of the monomers and their blend was studied using Differential Scanning Calorimeter (DSC). Blending bismaleimide with bisbenzoxazine considerably reduced the cure exotherm and the amount of heat liberated during thermal curing. The polymer from the blend shows higher char yield than the polymers from pure monomers. The differences in the thermal stabilities of these materials may be attributed to the structural aspects of the materials investigated.

1.1 Introduction

Due to the higher potential of the high-performance materials, they are found to be used in the automotive, aerospace, electrical, and construction industries [1]. Mostly the thermosetting polymers are used as high-performance materials. These high-performance materials are prepared by the chemical reaction of the reactive groups present in the matrix resin along with the appropriate fiber and/or particulate filler reinforcements. Therefore the properties of these materials depend on the chemical reactions during the process. Further, the rate of curing material and the amount of heat liberated during the curing process are also important factors [2].

Polybisbenzoxazine is a novel class of phenolics and it can be used as the matrix of high-performance composites because of its superior mechanical properties and high temperature stability [3–5]. Although benzoxazines possess many advantages over other state-of-the-art thermosetting resins, their brittleness, long cure time, and higher cure temperature point to the need for modifications on the matrix. The polybenzoxazine properties could be improved or modified via the formation of blends, alloys, copolymers, and organic/inorganic nanocomposites.

Benzoxazine (BZ) is a single benzene ring fused to another six-membered heterocycle containing one oxygen atom and a single nitrogen atom. There are a number of possible isomeric benzoxazines depending upon the relative position of the two hetero atoms and the degrees of oxidation of this oxazine ring system. 3,4-Dihydro-2H-1,3-benzoxazine is one kind of hydrogenated derivative of benzoxazine. When the benzene ring is replaced by naphthalene, the corresponding oxazine becomes naphthoxazine

Interest in oxazines and their benzo derivatives dates back well into the early part of the last century because some derivatives exhibit color. Monocyclic oxazines function as a base and as very useful synthetic intermediates, particularly for the construction of carbonyl derivatives and as starting materials for many complex heterocyclic systems. Some benzoxazines and their salts have been shown to possess interesting pharmacological properties, such as having antidepressant activity and being useful as insecticides and antiarrhythmics. As a specific kind of hydrogenated benzoxazine, 3-substituted-3,4-dihydro-2H-1,3-benzoxazine was studied mainly as an alternative product of the Mannich reaction in the middle of this century. Few applications of this compound had been reported in the polymer area until it was identified as an intermediate of the amine catalyzed phenolic resin.

Aromatic oxazines were first synthesized in 1944 by Holly and Cope [6] through Mannich reaction from phenols, formaldehyde and amines. They first proposed the 3,4-dihydro-2H-1,3-benzoxazine structure from the condensation of formaldehyde with *o*-hydroxybenzylamine. From the1950s to the 1960s, Burke et al. synthesized many benzoxazines and naphthoxazines [7] for the purpose of antitumor activity test. Reiss et al. investigated the polymerization of monofunctional benzoxazines with and without phenol as an initiator, resulting in linear polymers under 4000 molecular weight [8].

The precursors of polybenzoxazine are formed from phenols and formaldehyde in the presence of aliphatic or aromatic amines. The choice of phenol and amine permits design flexibility and tailoring of polymer properties. The most-investigated polybenzoxazine is derived from bisphenol-A and aniline. This benzoxazine is treated as a "benchmark" among polybenzoxazine and the properties of the new polybenzoxazine are compared with it. The material property balance of polybenzoxazines is excellent.

Good thermal, chemical, electrical, mechanical, and physical properties make them attractive alternatives to existing applications. Additional new applications can be developed by utilizing unique properties of polybenzoxazines that have not been often observed by other well-known polymers. Those properties include near-zero shrinkage, very high char yield, fast development of mechanical properties as a function of conversion, much higher glass transitions than curing temperatures, excellent electrical properties, and low water uptake despite having many hydrophilic groups. They can be synthesized from inexpensive raw materials and polymerized by a ring opening addition reaction, yielding no reaction by-product. The superb molecular design flexibility of the polybenzoxazines allows the properties of the polymerized material to be tailored to a wide range of properties for the specific requirements of individual applications. The synthesized mixture consists of monomer and oligomers that contain phenolic groups. For practical applications, the mixture is sufficient, but for controlled structure and properties, the monomer is freed of the oligomers.

Polybenzoxazines (PBZ) that can overcome many shortcomings associated with traditional phenolic resins have been synthesized and characterized extensively by Ishida et al. [3]. They possess excellent resistance to chemicals and UV light, and amazingly high glass transition temperature (Tg) values. A wide variety of BZ monomers can be readily prepared by varying the precursor components by the Mannich condensation reaction. They are therefore of great interest in synthetic fields.

3-Allyl-3,4-dihydro-2H-1,3-benzoxazine (P-ala), bis (3-allyl-3,4-dihydro-2H-1,3-benzoxazinyl) isopropane (B-ala), and 3-phenyl-3,4-dihydro-8-allyl-2H-1,3-benzoxazine (also called bisphenol-A and allyl amine-based benzoxazine; P-alp) were the first allylbenzoxazines reported. Of these, P-alp showed difficulty in polymerising because the active ortho sites for polymerization were blocked, so the exotherm was observed at a higher temperature.

The P-ala exhibited two exotherms, of which the first was attributed to allyl polymerization and the second to oxazine polymerization.

The material B-ala showed a broader exotherm with higher exothermic features than normal bisphenol-A and aniline based BZ (BAB). This was attributed to the polymerization of allyl and BZ groups. The polymers of P-ala and B-ala showed Tg of 297°C and 322°C, respectively, whereas P-alp had poor thermal properties with a Tg of 107°C.

A series of novel hybrid monomers of benzoxazine and maleimide moieties were made because the maleimide group can serve as an additional polymerizable group to produce high-performance polybenzoxazine resins. A BZ monomer with a maleimide pendant [*N*-(4-hydroxyphenyl) maleimide]-benzoxazine (HPM-Ba) has shown attractive processing and thermal properties. The monomer HPM-Ba could be polymerized to produce a fusible polymer with benzoxazine pendants (PHPM-Ba-I). The further polymerization of PHPM-Ba-I to PHPM-Ba-II via crosslinking of the pendant benzoxazine groups enhanced the Tg to 204°C.

The relatively high thermal stability of PHPM-Ba-II was due to the restriction of initial-stage thermal degradation by the formation of benzoxazine cross-linked networks, reflecting its good flame retardant characteristics. Another series of maleimides containing BZ were prepared from hydroxyphenyl maleimide and various amines (e.g., aniline, allylamine and amino phenyl propargyl ether). Mono-functional benzoxazine containing norbornene functionalities have also been reported. Maleimido benzoxazine (MIB) and norbornene benzoxazine (NOB) showed benzoxazine polymerization exotherm at 213°C and 261°C, respectively. The MIB has a maleimide group, so it could be further polymerized by the free radical mechanism by thermal activation or with an initiator. The polymerization features of the mixtures of BZ and *N*-phenyl maleimides have been investigated. Three *N*-phenyl maleimides, which have hydroxyl, carboxylic acid, and hydrogen moieties, respectively, are attached on the phenyl group.

The propargyl functional group is interesting because they can polymerize independently and impart thermal stability to the polybenzoxazine structure. Hence, monofunctional (phenol and propargyl amine based, P-appe) and difunctional (bisphenol-A and propargyl amine based, B-appe) propargyl BZ monomers were synthesized and their polymerization investigated. The furan containing BZ monomers, 3-furfuryl-3,4-dihydro-2H-1,3-benzoxazine (P-af) and bis (3-furfuryl-3,4-dihydro-2H-1,3-benzoxazinyl) isopropane (BA-af), were prepared using furfuryl amine as a raw material. The activation energies of polymerization were 96 and 98 kJ mol^{-1} for P-af and BA-af, respectively, by the Kissinger method. This implied that functionality does not have a bearing on reactivity. High char-yielding PBZ were obtained from acetylene-functional BZ monomers. Polymerization of acetylene-functional groups (in addition to oxazine polymerization) contributes to thermal stability. The reaction exotherm of acetylene polymerization broadly overlaps with the reaction exotherm of BZ ring opening polymerization.

A significant change in char yield was found for these PBZ in air and an inert atmosphere. These BZ polymerized in air, resulting in a higher char yield and thermal stability than those polymerized under an inert atmosphere. It is due to the different concentrations and structures of the newly-formed polyene chains by acetylene group polymerization, and due to the different extent of ring opening polymerization. These polymers provide many desirable properties, such as resistance to solvents and moisture, and good physical properties as well as high thermal stability. These characteristics make them good candidates for matrix resins for advanced composite materials. The acetylenic group can also react under cationic, coordination, free radical, photolytic, and thermal inducement.

Phthalonitrile PBZ showed a lower cure temperature (250°C) and high char yield (68%) at 800°C as a result of the polymerization of the nitrile groups at a higher temperature. The reactive phthalonitrile terminal group contributed to the formation of highly thermally stable cross-linked structures. A polymerization temperature of 250°C was sufficient to achieve material of high thermal stability and a Tg of 275–300°C. Similarly, the phenylnitrile crosslinking sites containing polymers showed excellent thermal properties and improved mechanical properties.

Several new synthetic techniques have been introduced for preparing diverse BZ, including polymers with BZ units in the main chain, which enables BZ technology to be much more versatile to tailor desired properties in the final product. These polymers can be processed into self-supporting flexible films. During this phase of processing, they behave as thermoplastics.

A linear PBZ molecule with oxazine rings in the main chain was synthesized with a molecular weight of ~10 kDa. The resultant polymer had a moderately broad polydispersity index. However, the insolubility of the products due to extreme rigidity resulted in a low molecular weight and broad polydispersity. To overcome this difficulty, flexible and thus more soluble segments (i.e., an aliphatic amine) were made.

High molecular weight polyether esters containing BZ units showed thermal stability more or less close to the stability of BA-a, but they exhibited better film properties. The resultant polymer showed better toughness induced by the soft ether ester group, but the aliphatic ether ester group molecules had some drawbacks, such as high water absorption and low thermal stability. Linear main chain PBZ with a molecular weight in the range 20–40 kDa was also prepared by the "click chemistry approach."

Azide containing BZ monomers produce PBZ with a triazole unit in the chain, enhancing the Tg up to 278°C with low moisture absorption. A polycaprolactone-naphthoxazine (PCL-NZ) containing poly-caprolactone (PCL) has been reported. This naphthoxazine macro monomer does not exhibit the exotherms usually observed with low molecular weight BZ due to the polymeric nature of the macro monomers. Thermally curable naphthoxazine functionalized polymers were synthesized by the reaction of linear (diamines) and branched (triamines) polypropylene oxides (PPOA) of various molecular weights with para-formaldehyde and 2-naphthol to prepare thin films. The DSC results showed that the maximum cure temperature increased with an increase in the molecular weight of PPOA and that it reduced the exothermicity. The films of polybenzoxazine had a water-contact angle of 67°. The water-contact angles of the BA-a incorporated system increased with the increase in the amount of monomer in the film and by curing.

Zhang et al. [9] studied the synthesis and characterization of pyridine-based benzoxazines and their carbons. The thermogravimetric analysis and elemental analysis results demonstrated that the cured benzoxazines had a high char yield and their carbides had high nitrogen content. The nitrogen configurations of the carbons obtained from pyridine-based benzoxazines were pyridinic N, pyrrolic N and graphitic N, which were the active centers in nitrogen-doped carbons. Zhang et al. believe that these pyridine based benzoxazines will be applied as the precursors for functional carbon materials in many fields such as hydrogen storage, fuel cell and super capacitor.

Kiskan and Yagchi [10] studied the self-healing of poly (propylene oxide)-polybenzoxazine thermosets by photo-induced coumarine dimerization. A self-healing strategy for poly [propylene oxide(s)] bearing coumarine-benzoxazine units (PPO-CouBenz) s based on light-induced coumarine dimerization reactions is described. Four different types of poly (propylene oxide) amines with molecular weights ranging from 440 to 5000 Da were reacted with formaldehyde and 4-methyl-7-hydroxycoumarin to yield desired (PPO-CouBenz) s. The cross-linked polymer films were prepared by solvent casting of various compositions of PPO-CouBenzs in chloroform, followed by the thermal ring opening reaction of benzoxazine groups at 210–240°C. The thermal curing and thermal stability of the initial PPOs and final products were investigated. Using allyl benzoxazine in the formulation, it was demonstrated that the toughness of the films was improved. The photo-induced healing of coumarin based cured PPO-CouBenz polymer films was investigated.

Selvi et al. [11] studied the carbon black–polybenzoxazine nano composites for high K dielectric applications. This work involves the development of surface functionalized carbon black (FCB) reinforced polybenzoxazine (PBZ) nano composites, using a benzoxazine monomer and varying weight percentages of functionalized carbon black (0.5, 1.0, and 1.5 wt%) through ring opening polymerization via thermal cure. Data from the morphological studies indicate that the nano sized FCB particles are homogeneously distributed in the polybenzoxazine matrix. The incorporation of CB improves the dielectric constant, electrical conductivity and the reduction in resistivity of the resulting FCB–PBZ nano composites when compared to neat PBZ. A significant improvement is observed in the glass transition temperature and thermal stability of the resulting FCB–PBZ nano composites, when compared to neat PBZ.

Rajput et al. [12] studied the preparation and characterization of flexible polybenzoxazine-linear low density polyethylene (LLDPE) composites. It has been observed that the thermal stability of the composites increased with the increasing amount of polybenzoxazine in the composition. A mechanical property of the composites was investigated by using tensile and flexural tests. The presence of 5 wt% compatibilizer in the composition showed very good mechanical properties. Pure polybenzoxazine possessed high tensile strength and less elongation at break whereas LLDPE showed low tensile strength and significantly more elongation at break. Composites consist of LLDPE and polybenzoxazine with additional LLDPE-g-MA, exhibited higher tensile strength than pure LLDPE and more elongation at break than pure polybenzoxazine. Prepared composites possess very good mechanical flexibility, which makes them good candidates to prepare of complex structures and coating.

Ishida et al. have used carbon fiber, glass fiber and natural fiber to develop high-performance fiber reinforced polybenzoxazine composites and reported their properties [13, 14]. They also investigated the use of $CaCO_3$ as filler [15]. The preparation of titania polybenzoxazine as an organic–inorganic hybrid material by using sol gel process was reported by Agag et al. [16].

A unique synthetic route was reported by Kiskan et al. [17] for the synthesis of a macromonomer for which the benzoxazine ring was anchored to the polystyrene polymer. Dibromophenyl terminated polystyrene was synthesized using Atom Transfer Radical Polymerization (ATRP), which was then followed by a Suzuki coupling reaction to prepare amino functional polymers. These amino functional polymers when reacted with phenol and paraformaldehyde at 110°C for 2 h produced benzoxazine functionalized polystyrene macromonomer.

Kimura et al. prepared the samples containing 50 mol% B-a and 50 mol% diglycidyl ether of bisphenol-A (DGEBA) and cured in a mold in the oven using the curing condition of 150°C/1h+170°C/1h+190°C/2h+200°C/2h+ 220°C/2h [18]. As it is reported that terpene diphenols formaldehyde resin possesses superior heat resistance, water resistance, and mechanical properties, terpene diphenol-based benzoxazine monomers were synthesized.

Cured blend samples containing 50 mol% DGEBA and 50 mol% benzoxazine monomers were prepared employing the above-mentioned cure conditions [19]. The molding compounds were prepared by hot roll kneading of a mixture of 50 phr (per hot roll kneading) Ya, 50 phr OCNE (*o*-Cresol novolac type epoxy resin). Rao et al. prepared copolymers of chain extended epoxy (40 mol%) with benzoxazine (bisphenol-A and aniline based) (60 mol%) by using the solution mixing method in acetone and investigated the effects of molecular weight of the added epoxy resins [20] wax and 100 phr fused silica.

In the first approach, a novel phosphorous containing dopotriolbenzoxazine was copolymerized with a commercial benzoxazine [6,6-bis (3-phenyl-3,4-dihydro-2*H*-1,3-benzoxazineyl) methane (F-a)] or diglycidyl ether of bisphenol-A (DGEBA). In the second case, the element phosphorus was incorporated into benzoxazine via the curing reaction of dopotriol and F-a. In the third approach, dopo reacted with benzoxazine to incorporate the element phosphorus.

A nano composite composition comprising clay and benzoxazine monomer, oligomer and/or polymer was first developed by Ishida et al. The presence of benzoxazine in the clay resulted in about at least a 5% increase in the spacing between platelets of the clay. In another study, Agag and Takeichi [21] prepared the polybenzoxazine–clay hybrid nano composites from a polybenzoxazine precursor (B-a) and organically modified montmorillonite (OMMT), as layered silicates. The OMMTs were prepared by surface treatment of montmorillonite (MMT) by octyl, dodecyl or stearyl ammonium chloride. A melt of B-aand OMMT was mixed by a mechanical stirrer at 100°C, with the addition of a small amount of methylene chloride to achieve better dispersion. The mixture was then heated at 120°C for 2 h to remove solvents, followed by film casting on glass plates. Then the film was cured by stepwise increase of heating up to 230°C.

The amines used as the modifying agent were dodecylamine (DODEC), 6-amino caproic acid (CAPRO), 4-amino-*N*, *N*-dimethyl aniline dihydrochloride (ANDAD), *p*-phenetidine (PHEN) and 2,4,6-trimethylaniline (TMAN). Mixtures of 3 wt% OMMT with benzoxazine monomers were prepared using solvent, binary solvent or non-solvent systems. All samples were cast on aluminium foil surface and solvents were allowed to evaporate and then cured at 230°C for 90 min.

The major disadvantages of the typical polybenzoxazines are their brittleness and the high cure temperature needed for the ring opening polymerization. Two major approaches are generally considered: (1) by preparing specially designed novel monomers or (2) by blending with a high performance polymer or filler and fiber. In the first approach, the introduction of ethynyl or phenyl ethynyl, nitrile, etc. groups can offer additional cross-linking site during polymerization and were found to be an acceptable choice for that purpose. According to the second approach, the mechanical and thermal properties of polybenzoxazines can be improved by the preparation of copolymers, polymer alloys, composites and polymer clay nano composites (vide infra).

The main advantage of the allyl group is not only that it provides additional cross-linkable sites, but that it can easily be cured at a temperature lower than that needed for acetylene groups. Allyl-containing monomers have attracted much attention because they are used as reactive diluents of bismaleimides to improve the toughness of the cured resin.

Because of the absence of an activated ortho position to the phenolic hydroxyl group, these ally phenol-based benzoxazine monomers, however, are considered to be difficult to polymerize through ring opening and are not good candidates for preparing high-performance polybenzoxazines.

The synthesis of a series of allyl groups containing mono functional benzoxazine monomers, where the allyl group is attached with nitrogen and derived from cresol and allyl amine by a solventless method has been reported and the effect of these ally groups on polymerization reaction and the performance enhancement of the cured polymers at high temperature has been reported.

The synthesis of easily processable benzoxazine monomers with acetylene functionality has been reported by Ishida et al. It has been observed that the high thermal stability of the polybenzoxazines derived from this class of monomers is a combined result of the polymerization of the acetylene terminal functional group and oxazine ring opening polymerization. Agag and Takeichi have prepared novel benzoxazine monomers containing a propargyl ether group as the cross-linkable functional group and obtained novel polybenzoxazines with attractive thermal properties [21]. The ring opening polymerization of oxazine ring and the cross-linking of propargyl ether group occurred at almost the same temperature range, at 230°C for mono functional and 249°C for bifunctional monomers.

The synthesis of adamantyl modified benzoxazine monomers has been reported using 4-(1-adamantyl)-phenol, formaldehyde, and aniline (or methylamine) in dioxane. It was expected that the rigid structure of the adamantane would hinder the chain mobility (boat anchor effect) and substantially enhance the thermal properties of the resulting polymer, including the glass transition temperature and decomposition temperature, especially for poly (6-adamantyl-3-methyl-3,4-dihydro-2*H*-1,3-benzoxazine). In the poly (6-adamantyl-3-phenyl-3,4-dihydro-2H-1,3-benzoxazine) system, however, the opposite result for the glass transition temperature, was observed and explained by lowering of crosslinking density.

A monomer, 4-methyl-9-*p*-tolyl-9,10-dihydrochromeno[8,7-e][1,3] oxazin-2(8H)-one, possessing both benzoxazine and coumarin rings in its structure was synthesized by the reaction of 4-methyl-7-hydroxycoumarin, paraformaldehyde, and *p*-toluidine in 1,4-dioxane. As the coumarin dimerizes, the level of unsaturation decreases because of the formation of the cyclobutane ring.

Andreu et al. reported the synthesis and polymerization of glycidylic derivatives of benzoxazines obtained from aniline and 4-hydroxy benzoic acid, and from phenol and 4-amino benzoic acid [22]. By introducing epoxy groups into the molecular structure of benzoxazine, another attractive way of improving the thermal stability and glass transition temperatures of the resulting polybenzoxazines was achieved.

On replacement of the benzene ring by the naphthalene, the corresponding oxazine becomes a naphthoxazine. Naphthoxazines were synthesized employing a similar strategy, i.e., reaction of napthol, formaldehyde, and primary amines. But along with it, alkylaminomethyl-2-napthol also formed as a by-product.

Solvent, temperature, and basicity of amine play important roles upon the yield of the corresponding naphthoxazine monomer formation. Difunctional amines like *p*-phenylene diamine when reacted with formaldehyde and napthol (1:4:2 molar ratio), produced 2, 2'-*p*-phenylene-bis-(2,3-dihydro-1H napth [1, 2-e]-*m*-oxazine) (2Na-a). Several other difunctional naphthoxazines were also synthesized from dihydroxynapthalene, formaldehyde, and primary amines. Apart from naphthoxazine, some fluorinated benzoxazine, and furan containing benzoxazine have also been reported in the literature.

Ishida et al. used hydroxyl-terminated polybutadiene (HTBD) rubber, having various epoxy content, as the toughening modifier. The epoxidized polybutadiene rubber can undergo copolymerization with the hydroxyl groups produced upon ring opening of benzoxazine and thus can be chemically grafted into the matrix network [23]; a toughened composite with a higher compatibility was obtained. A melt mixing method was used to obtain rubber-modified polybenzoxazines.

Agag and Takeichi [21] reported the preparation of hydroxyphenyl maleimide (HPM) and ATBN-modified polybenzoxazine by mixing benzoxazine monomer (Bz), HPM and amine-terminated butadiene acrylonitrile rubber (ATBN) in melting state, followed by film casting and curing.

According to Ishida and Lee [24], the driving forces that result in the miscibility of the poly carbonate (PC)/benzoxazine blend in the entire composition range is the interaction between the hydroxyl groups of polybenzoxazine and the carbonyl groups of the PC. A solution blending method was employed for the preparation of all the blend samples. Solutions of the purified benzoxazine monomer based on *p*-cresol and aniline, 3-benzyl-3,4-dihydro-6-methyl-2H-1,3-benzoxazine (abbreviated as p-Ca), and PC were blended at room temperature to form a homogeneous mixture with the aid of chloroform to obtain a transparent yellow solution.

The unique property of poly (ε-caprolactone) (PCL) makes it a potential candidate for blending with polybenzoxazine to achieve easy processibility and improved thermal properties. Apart from that, as intermolecular hydrogen bonding between hydroxyl groups of polybenzoxazine main chain and the carbonyl groups of PCL may form, it can enhance the miscibility of PCL with polybenzoxazine.

The preparation and characterizations of PCL polybenzoxazine (PBz) blends by a melt blending process was reported by Ishida and Lee [24]. Different concentrations of PCL were added to Bz at 120°C. After thorough mixing, a clear homogeneous mixture was obtained. This mixture was then step-cured in a compression molder after degassing.

Octa(aminophenyl) silsesquioxane (OAPS) was prepared from octa (nitrophenyl)-silsesquioxane (ONPS) by using the reducing agent hydrazine hydrate of around 86% yield. The decomposition temperature and char yield of resulting materials increased with increasing OAPS loading. The DSC data showed that the glass transition temperature of nanocomposite was enhanced by 36°C with the content of OAPS around 15 wt%.

Bismaleimide-modified novolac resin/clay nanocomposites were synthesized by Wang et al. [25] using *p*-phenylene diamine as a swelling agent. Sodium montmorillonite (Na–MMT) was successfully swollen with protonated *p*-phenylene diamine, and it was converted to maleamic acid. Exfoliation of the organoclay took place easily in the melt of the bismaleimide-modified allylated novolac resin. Major improvement in thermal stability, flexural strength and modulus was achieved by incorporation of the exfoliated clay layers.

Nanocomposites of bismaleimide (BMIM) with different proportions of nanometer SiC were prepared by a high shear dispersion process and casting method at elevated temperature. The results indicate that the nanocomposites exhibited lower friction coefficient and wear loss as well as higher bending and impact strength than BMI resin under the same testing conditions. The lowest wear rate was obtained with the nanocomposite containing 6.0 wt% SiC, while the highest mechanical properties were obtained with the nanocomposite containing 2.0 wt% SiC.

The physical and mechanical properties of butyl rubber have been improved by the mixing of organo clay cloisite 15a. The nanoclay was mixed with rubber using melt mixing methods in different ratios such as 3, 7, 11, and 15 phr. The effects of organo-clay content and structure on mechanical and rheological properties of nanocomposites as well as permeability of CO_2 gas through their films were evaluated. It was shown that as small an amount as 3 phr montmorillonite organo-clay reinforces butyl rubber to a great extent, affects its elastic and viscous behavior in dynamic conditions, and increases its barrier properties to CO_2 gas. The rate of improvement in properties slows down at organo-clay contents higher than 3 phr.

Rheology, morphology, and thermal behavior of HDPE/clay nanocomposites were studied by Rezanavaz et al. [26]. The grafting experiments were carried out in an internal mixer (BrabenderW50EHT, Duisburg, Germany) and two sets of compatibilizers were used in the preparation of nanocomposites. The first set was used without any purification after the grafting process, and the second set was used after removal of ungrafted maleic anhydride. Nanocomposites were prepared using the same internal mixer at 170°C with a rotor speed of 60 rpm. The degree of maleation played a major role in determining the extent of exfoliation and/or intercalation of clay in the nanocomposites. Incorporation of clay decreased the onset temperature of degradation due to the Hofmann elimination reaction but increased remarkably the mid-point of the degradation temperature. Intercalated nanocomposite displayed higher thermal stability than exfoliated samples, due to the barrier effect of clay layers to oxygen and volatile products.

The bismaleimide (BMI) systems dominate over the other polymer matrices primarily due to their high performance-to-cost ratio and relatively high temperature resistance. They have superior thermal and oxidative stability, low propensity for moisture absorption, and good flame retardance. They offer excellent thermomechanical properties and withstand high stress at high temperatures at which typical phenolics and epoxies as well as most high-performance plastics are unstable. However, their process ability and fracture toughness are not promising. Attempts to reduce brittleness by way of reduction of cross-link density through structural modification, toughening etc. adversely affect the high-temperature performance. There are a few reports on polybenzoxazine containing imide units as part of the molecule. Preliminary investigations on the co-curing of polybenzoxazine containing BMI as part of the benzoxazine molecule have been reported. The objective of the present work is to examine the polymerization and degradation properties of the blends of traditional benzoxazine based on bisphenol-A with extended bismaleimide.

1.2 Experimental

1.2.1 Materials

Bisphenol-A and paraformaldehyde were purchased from SISCO Research Laboratory Pvt. Ltd., Mumbai, India. Sodium hydroxide, ethyl alchol, hydrazine hydrate and 1,4-dioxane were purchased from Lobachemie Pvt. Ltd., Mumbai, India. Aniline and diethyl ether were purchased from MERCK Specialist Pvt. Ltd., Mumbai, India and aniline was used after distillation. Maleic anhydride and *p*-chloronitrobenzene were supplied by S. D. fine-chem Ltd., Mumbai, India.

1.2.2 Synthesis of Bis(3,4-Dihydro-2H-3-Phenyl-1, 3-Benzoxazinyl) Isopropane (BAB)

Stoichiometric amounts of bisphenol-A (0.06 mol), paraformaldehyde (0.24 mol) and aniline (0.12 mol) were mixed together at room temperature and then placed in a three-necked flask equipped with a distillation setup in order to collect the water evolved during the condensation reaction. The reaction mixture was heated for 1 h in an oil bath maintained at a constant temperature of 120°C with constant stirring. The resultant product was

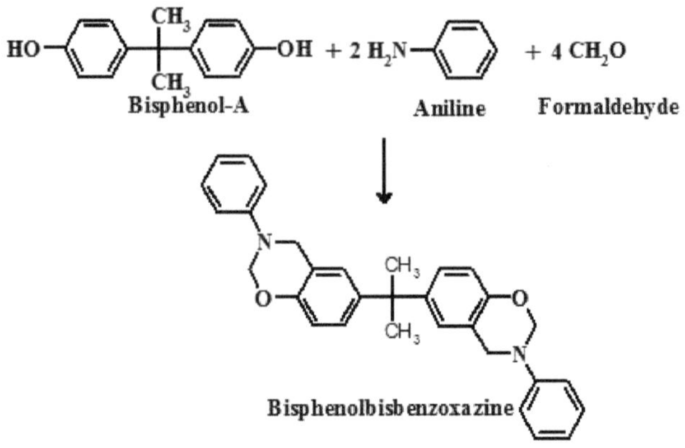

Figure 1.1 Preparation of BAB.

dissolved in diethyl ether and washed with 2N NaOH solution and finally with copious amount of distilled water. The yield was 75% (Figure 1.1).

1.2.3 Synthesis of 2,2-Bis(4-Nitrophenoxyphenyl) Propane (DN-BPAPCNB)

Bisphenol-A (22.8 g), *p*-chloronitrobenzene (34.6 g), and anhydrous potassium carbonate (30.4 g) were taken in a round bottom flask containing 125 mL of *N,N'*-dimethylformamide. After a few minutes, the orange-colored reaction mixture changed its color to blood red and potassium chloride gets precipitated. The resultant reaction mixture was refluxed for 12 h. After this period, the precipitated potassium chloride was filtered off. The filtrate was poured into a copious amount of crushed ice with effective stirring. The separated yellow-colored 2,2-bis(4-nitrophenoxy phenyl)propane, DN-BPAPCNB, was filtered, washed with ice-cold water, and dried at 50°C for 24 h in a hot air oven. The yield was found to be 90%. Elemental analysis: C% = 67.93; H% = 4.65; O% = 20.41; N% = 5.88.

1.2.4 Preparation of 2,2-Bis(4-Aminophenoxy Phenyl) Propane (DA-BPAPCNB)

Dinitro compound (DN-BPAPCNB) (28.2 g) and 0.15 g of 10% Pd/C dispersed in 180 mL of ethanol were taken in a round-bottom flask. About 60 mL of hydrazine hydrate was added dropwise and the resultant reaction mixture was refluxed at 85°C for 12.5 h. The hot black-colored solution was

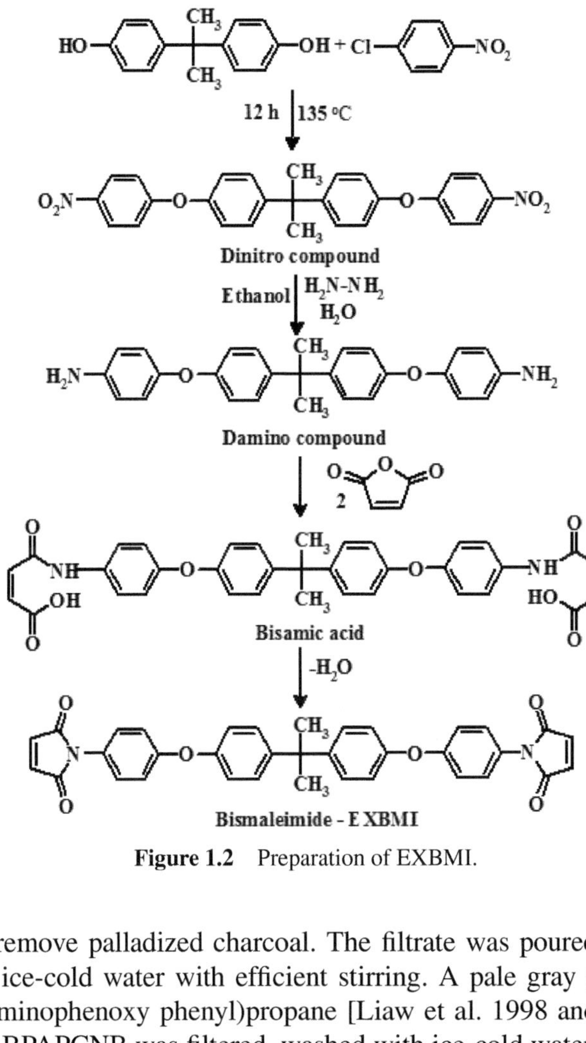

Figure 1.2 Preparation of EXBMI.

filtered to remove palladized charcoal. The filtrate was poured into a large amount of ice-cold water with efficient stirring. A pale gray precipitate of 2,2-bis(4-aminophenoxy phenyl)propane [Liaw et al. 1998 and Hsiao et al. 1994], DA-BPAPCNB was filtered, washed with ice-cold water, and dried at room temperature. The yield of the diamino compound was 85%. Elemental analysis: C% = 78.86; H% = 6.32; O% = 7.72; N% = 6.81 (Figure 1.2).

1.2.5 Preparation of Bisamic Acid (BAX)

Exactly 12.3 g of diamine (DA-BPAPCNB) was dissolved in 155 mL of acetone with constant stirring at room temperature. To this solution, 6.5 g of powdered maleic anhydride was added in portions. A yellow precipitate was formed and it was stirred continuously for half an hour. It was filtered and washed with ice-cold acetone to remove the acetone-soluble materials and dried. The yield was 86%. Elemental analysis: C% = 69.24; H% = 4.86; O% = 21.02; N% = 4.58.

1.2.6 Preparation of 2,2-Bis[4-(4-Maleimidophenoxy Phenyl)]propane (EXBMI)

The yellow bisamic acid (BAX) was dispersed in 160 mL of dry acetone kept in a 500 mL round-bottom flask. Anhydrous sodium acetate (3.44 g) and 40 mL of acetic anhydride were added into the reaction mixture and refluxed for 3 h. The brown-colored solution was poured into a copious amount of crushed ice and 2,2-bis[4-(4-maleimidophenoxy phenyl)propane] was obtained as yellow precipitate. The material was filtered, washed with ice-cold water, and dried in vacuum. The yield was 90%. Elemental analysis: C% = 73.59; H% = 4.46; O% = 16.73; N% = 4.82 (Figure 1.2).

1.2.7 Blending of the Materials

The dry bismaleimide (EXBMI) and the dry bisbenzoxazine (BAB) were taken in amortar in the mol ratio 1:1 and ground with the pestle thoroughly for intimate mixing.

1.2.8 Polymerization of the Materials

The synthesized monomers and the blend were taken in separate micro test-tubes and flushed with dry oxygen-free-nitrogen and thermally polymerized. After the polymerization, the materials were allowed to reach room temperature and the samples were removed from the micro test tubes, ground to coarse powder, packed, and stored for further analyses.

1.2.9 FTIR Studies

The FTIR spectrum of the material was run on a Fourier transform infrared-8400S spectrophotometer, Shimadzu, Japan using KBr disc technique. About 500 mg of potassium bromide was taken in a mortar. Approximately 10 mg of the sample was added and ground well with a pestle. Approximately 100 mg of the above mixture was made into a transparent disc using pelletizer. The prepared disc was placed in the pellet holder and the IR spectrum was recorded using 16 scans with 4 cm^{-1} resolution.

1.2.10 Differential Scanning Calorimetric (DSC) Studies

Differential scanning calorimetric curves were recorded on TA Instruments DSC Q20. The sample (2–3 mg) was weighed, placed in the non-hermetic aluminium pan, and sealed with the aluminium lid. The samples were heated from ambient temperature to 350°C at a heating rate of 20°C min^{-1} in nitrogen (Flow rate = 50 mL min^{-1}) atmosphere. The obtained DSC curves were analyzed using the Universal Analysis 2000 Software provided by TA instruments.

1.2.11 Thermogravimetric (TG) Studies

Thermogravimetric analyses were performed on a TA Instruments TG Q50 thermogravimetric analyzer. To avoid the secondary reaction of evolved gases in thermogravimetric analysis such as thermal cracking, recondensation, and repolymerization, the nitrogen flow maintained at balance and sample were 50 mL min^{-1} and 60 mL min^{-1} respectively. The sample (3–4 mg) was weighed into a platinum crucible and was heated from ambient to 800°C at a heating rate of 10°C min^{-1}. The obtained TG and DTG curves were analyzed using the Universal Analysis 2000 Software provided by TA instruments.

1.3 Results and Discussion

1.3.1 FTIR Studies

The FTIR spectra of all the polymers are shown in Figure 1.3. The absorption bands noted are given in Table 1.1. The presence of an intense band at 3,361 cm^{-1} in PBAB indicates the formation of phenolic groups during the ring opening polymerization of bisbenzoxazine, BAB. The complete disappearance of the band at 948 cm^{-1} specific for the oxazine group in the FTIR spectrum of the polymer obtained by thermally polymerizing BAB, indicates the near-completion of the thermal polymerization of BAB. The structure of the polymerized BAB is shown in the Figure 1.4.

The thermally cured EXBMI shows bands at 1,157, 1,396, and 1,713 cm^{-1} which correspond to the stretching of -C-O-C- group, C=C bonds in the phenyl nucleus and the cyclic imide ring ($-N-(CO-)_2$) respectively and this confirms the structure of the thermally cured EXBMI. The absence of a band at 1,635 cm^{-1} (C=C maleimide double bond) [5] confirms the complete polymerization of EXBMI. The structure of the polymerized compound is shown in the Figure 1.5.

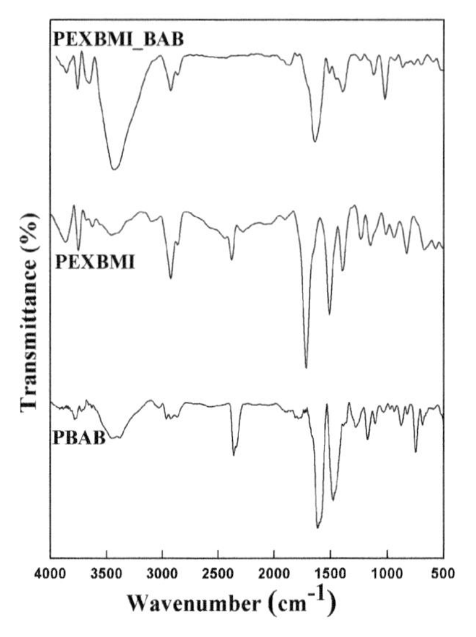

Figure 1.3 FTIR spectra of polymerized compounds.

Table 1.1 FTIR studies of PBMIX-BAB and related materials

Band	BMIX	PBMIX	PBAB	PBMIX-BAB
948 (oxazine)			Absent	Absent
1383			Gem dimethyl	Gem dimethyl
3361			Phenolic OH	
1149, 1157	-C-O-C stretching	-C-O-C stretching		
1396	C=C stretching (Phenyl nucleus)	C=C stretching (Phenyl nucleus)		
1635	C=C stretching (imide ring)	Absent		
1712	Cyclic imide –OC-N-CO stretching	Cyclic imide –OC-N-CO stretching		Absent
1121				C-O stretching
1169				C-O-C
1622				Ester carbonyl C=O stretching
1655				Amide C=O stretching
1512				N-H bending
3424				N-H stretching

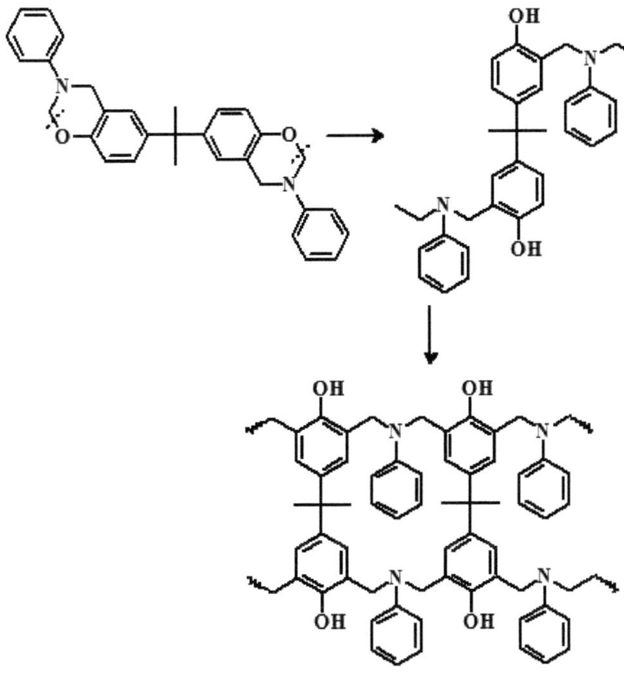

Figure 1.4 Structure of polymerized BAB.

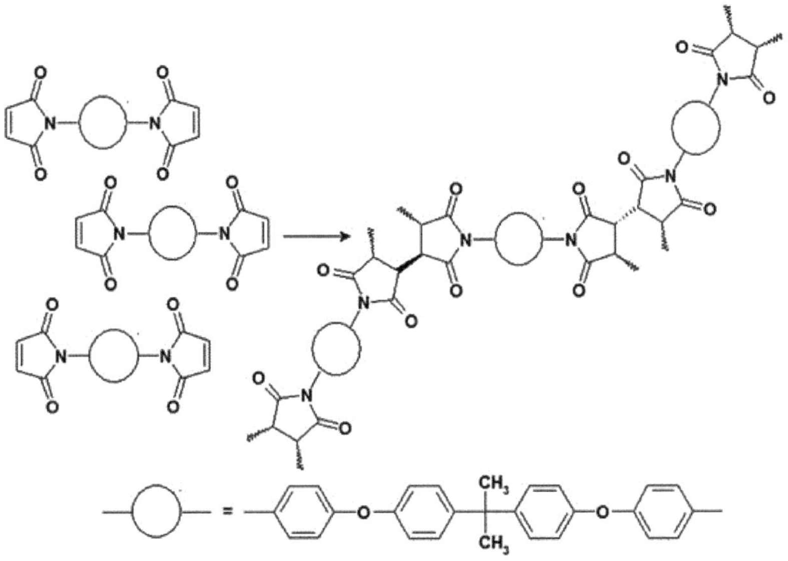

Figure 1.5 Structure of polymerized EXBMI.

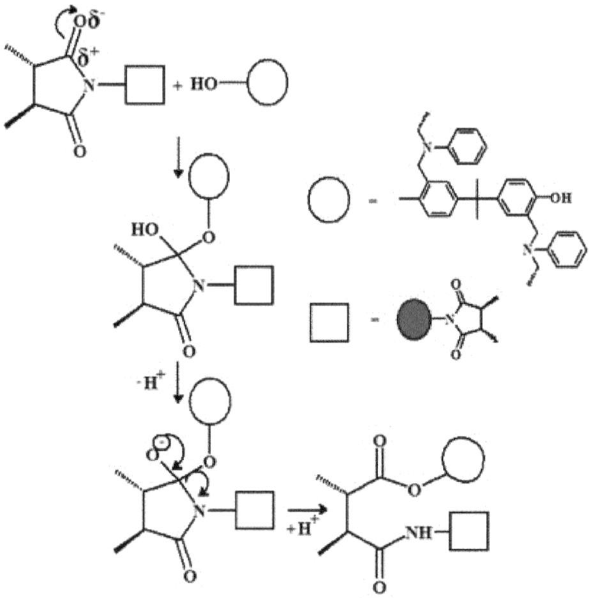

Figure 1.6 Structure of the polymer from the blend BMIX-BAB.

The presence of an intense band at 3,361 cm^{-1} in PEXBMI_BAB indicates the formation of phenolic groups during the thermal polymerization of the blend BAB/EXBMI. The absence of a band at 1,635 cm^{-1} indicates the utilization of maleimide double bond (C=C) for the formation of the polymer. This confirms that the blend BAB/EXBMI undergoes polymerization both by the ring opening polymerization of the oxazine rings in BAB and the C=C polymerization of the maleimide double bond present in EXBMI during the thermal curing process (Figure 1.6).

1.3.2 DSC Studies

The DSC curves for the monomers and blend are shown in Figure 1.7. The different parameters obtained from the DSC curves are given in Table 1.2. The bisbenzoxazine BAB shows the onset of the curing around 163°C; curing attained a maximum at 229°C and ends around 304°C. The value of enthalpy of thermal curing (ΔHC) was 273 J/g and the temperature region of the curing window was 141°C. The bismaleimide EXBMI showed a sharp melting point (T_m) at 82°C. The onset of curing was noted at around 195°C; curing attained a maximum at 289°C and ended at 379°C. The value of ΔHC was

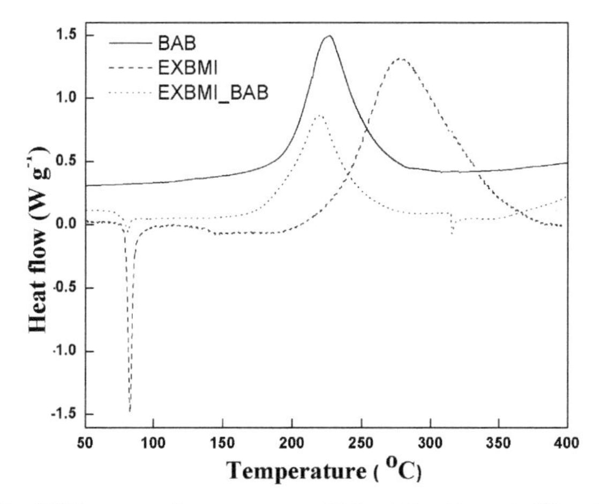

Figure 1.7 DSC curves of monomers and blend [heating rate (β) = 10°C/min].

Table 1.2 DSC studies: Curing characteristics of monomers and its blend (β = 10°C/min)

Sample	*MP* (°C)	T_i (°C)	T_{max} (°C)	T_e (°C)	$T_{max} - T_i$ (°C)	$T_e - T_i$ (°C)	ΔH_C (J/g)	$\Delta H_C/T_e - T_i$ (J/g°C)
BAB	–	163	229	304	66	141	273	1.94
EXBMI	82	195	289	379	94	184	169	0.92
EXBMI-BAB	81	136	221	287	85	151	202	1.34

169 J/g and the temperature region of the curing window was 184°C. The uniform curing curves indicate that both BAB and EXBMI are undergoing thermal curing smoothly without much complication. Further, BAB cures at considerably lower temperatures than EXBMI. Blending EXBMI with BAB does not have much influence over the melting temperature of EXBMI. There is no characteristic change in the melting point of the blend EXBMI_BAB. Blending of BAB with EXBMI shifts the curing onset and the cure maxima to the lower region. This indicates that the ring opening of benzoxazine is much favored in the presence of the maleimide entities of EXBMI. The double bonds present in the maleimide rings help the earlier curing of BAB to form the copolymer. The high enthalpy of curing (ΔH_C) of BAB is considerably lowered by the addition of EXBMI (Table 1.1). The amount of heat released per every degree rise in temperature is also decreased by the addition EXBMI to BAB. Thus a slight increase in the cure window (T_e – T_i), the lowering of the curing parameters (T_i and T_{max}), and the decrease in the quantity, $\Delta H_C/T_e$ – T_i, will facilitate the composite fabricator towards getting void-free thermal stress-free laminates of good thermal stability.

Table 1.3 TG studies: The degradation parameters for the polymers (β = 10°C/min)

		Temperature (°C)				T_{max}(°C)		Char yield % at 800°C
Sample	T_5	T_{10}	T_{25}	T_i	T_e	T_{M1}	T_{M2}	–
PBAB	281	307	359	261	571	359	437	27
PEXBMI	396	414	452	324	602	423	486	43
PEXBMI-BAB	311	360	424	273	616	400	463	48

1.3.3 TG and DTG Studies

The TG and DTG curves of the polymers (heating rate = 10°C/min) are shown in Figures 1.8 and 1.9, respectively. The onset, maximum, and end-set temperatures for the degradation and the char residue obtained at 800°C for all the samples are tabulated in Table 1.2. The T_{max} values presented in Table 1.3 are obtained from the DTG curves. The initial weight-loss for the compounds PBAB and PEXBMI are found to be at 261 and 324°C respectively. From the DTG curve of PEXBMI, it is obvious that the degradation is proceeding in two different stages and the maximum weight loss is noted at the first stage of degradation. Compared to PEXBMI, the thermal degradation of PBAB is much more complex involving several overlapping degradations and covers broad temperature region. For the polymer PEXBMI_BAB the initial degradation temperature is found to be at 273°C and it follows multiple degradations of which two degradation stages are more conspicuous. The first degradation stage shows a maximum at 400°C and the maximum for the

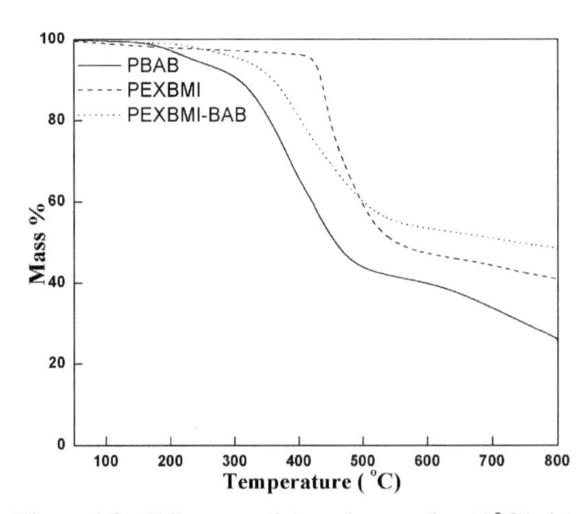

Figure 1.8 TG curves of the polymers (β = 10°C/min).

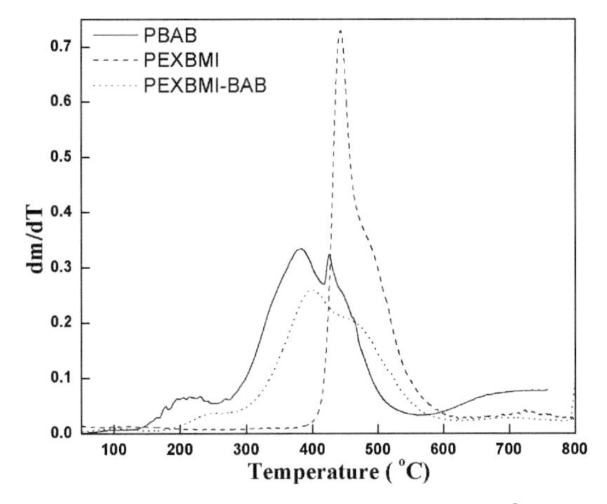

Figure 1.9 DTG curves of the polymers (β =10°C/min).

second degradation stage is seen at 463°C. The polymer, PEXBMI_BAB shows the highest char value compared to PBAB and PEXBMI. The char value of the polymer from the blend is slightly higher than the polymer PEXBMI and much higher than PBAB. If the char value is calculated on the basis of the amount of different monomers in the feed, then the expected char is around 35% only. But the observed char for the polymer from the blend is 48%, indicating the thermoset resulting from the thermal polymerization of a blend of BAB and EXBMI has sufficient thermally stable structural units.

1.4 Conclusions

The compounds BAB and EXBMI were prepared and a 1:1 mol ratio blend of this monomer were also prepared and thermally polymerized. The FTIR studies of the polymers indicate the ring opening polymerization of benzoxazine units leading to phenolic structures. Whether in pure or in blended form, the BMI polymerizes through the maleimide double bonds. Reduction in curing temperature, reduction in the heat liberated during thermal curing, and the broadening of the curing temperature window are the major consequences of blending bismaleimide with bisbenzoxazine. Based on the char residue, the polymer from the blend is comparatively more stable than polymers from pure monomers. Although PEXBMI is thermally more stable on the basis of the thermal degradation temperature, the rate of evolution of volatiles from the PEXBMI-BAB is slow.

Acknowledgements

The authors would like to acknowledge the Management, the Principal, and the Dean of Kamaraj College of Engineering and Technology, Virudhunagar, India for providing all the facilities to do this work.

References

[1] Agag, T., and Takeichi, T. (2001). Novel benzoxazine monomers containing *p*-phenyl propargyl ether:? Polymerization of monomers and properties of polybenzoxazines. *Macromolecules* 34, 7257–7263.

[2] Ishida, H. (1996). Process for preparation of benzoxazine compounds in solventless systems. U.S. Patent No 5,543,516.

[3] Ishida, H., and Sanders, D. P. (2000). Improved thermal and mechanical properties of polybenzoxazines based on alkyl substituted aromatic amines. *J. Polym. Sci. B* 38, 3289–3301.

[4] Vijayakumar, C. T., Surender, R., Rajakumar, K., and Alam, S. (2010). Synthesis and thermal studies of bisphenol-A based bismaleimide: Effect of nanoclays. *J. Therm. Anal. Calorim.* 103, 693–699.

[5] Surender, R., Vijayakumar, C. T., Thamaraichelvan, A., and Alam, S. (2013). Curing studies of bisphenol-A based bismaleimide and cloisite 15a nanoclay blends using DSC – Model free kinetics *J. Appl. Polym. Sci.* 128, 712–724.

[6] Holly, F. W., and Cope, A. C. (1994). Condensation products of aldehydes and ketones with o-Aminobenzyl Alcohol ando-Hydroxybenzylamine. *J. Am. Chem. Soc.* 66, 1875–1879.

[7] Burke, W. J., Nasutavicus, W. A., and Weatherbee, C. (1964). Synthesis and study of mannich bases from 2-naphthol and primary amines. *J. Org. Chem.* 29, 407–410.

[8] Reiss, G., Schwob, J. M., Guth, G., Roche, M., and Lande, B. (1985). *Advances in Polymer Synthesis*, eds B. M. Culbertson, J. E. McGrath. New York, NY: Plenum, 27–49.

[9] Zhang, H., Gu, W., Ran, Q., and Gu, Y. (2014). Synthesis and characterization of pyridine-based benzoxazines and their carbons. *J. Macromol. Sci. A* 51, 783–787.

[10] Kiskan, B., and Yagci, Y. (2014). Self-healing of poly (propylene oxide)-polybenzoxazine thermosets by photoinduced coumarine dimerization. *J. Polym. Sci. A* 52, 2911–2918.

[11] Selvi, M., Devaraju, S., Sethuraman, K., and Alagar, M. (2014). Carbon black–polybenzoxazine nanocomposites for high K dielectric applications. *Polym. Compos.* 35, 2121–2128.
[12] Rajput, A. B., Rahaman, S. J., Sarkhel, G., and Ghosh, N. N. (2012). Preparation and characterization of flexible polybenzoxazine-LLDPE composites. *Des. Monomers Polym.* 16, 177–184.
[13] Ishida, H., and Lee, Y.H. (2001). Study of hydrogen bonding and thermal properties of polybenzoxazine and poly-(epsilon-caprolactone) blends. *J. Polym. Sci. B* 39, 736–749.
[14] Dansiri, N., Yanumet, N., Ellis, J. W., and Ishida, H. (2002). Resin transfer molding of natural fiber reinforced polybenzoxazine composites *Polym. Compos.* 23, 352–360.
[15] Suprapakorn, N., Dhamrongvaraporn, S., and Ishida, H. (1998). Effect of CaCO3 on the mechanical and rheological properties of a ring-opening phenolic resin: polybenzoxazine. *Polym. Compos.* 19, 126–132.
[16] Agag, T., Tsuchiya, H., and Takeichi, T. (2004). Novel organic–inorganic hybrids prepared from polybenzoxazine and titania using sol–gel process. *Polymer* 45, 7903–7910.
[17] Kiskan, B., Colak, D., Muftuoglu, A. E., Cianga, I., and Yagci, Y. (2005). Synthesis and characterization of thermally curable benzoxazine-functionalized polystyrene macromonomers. *Macromol. Rapid Commun.* 26, 819–824.
[18] Kimura, H., Matsumoto, A., Hasegawa, K., Ohtsuka, K., and Fukuda, A. (1998). Epoxy resin cured by bisphenol A based benzoxazine. *J. Appl. Polym. Sci.* 68, 1903–1910.
[19] Kimura, H., Murata, Y., Matsumoto, A., Hasegawa, K., Ohtsuka, K., and Fukuda, A. (1999). New thermosetting resin from terpenediphenol-based benzoxazine and epoxy resin. *J. Appl. Polym. Sci.* 74, 2266–2273.
[20] Rao, B. S., Reddy, K. R., Pathak, S. K., and Pasala, A. (2005). Benzoxazine-epoxy copolymers: effect of molecular weight and crosslinking on thermal and viscoelastic properties *Polym. Int.* 54, 1371–1376.
[21] Agag, T., and Takeichi, T. (2001). Effect of hydroxyphenylmaleimide on the curing behaviour and thermomechanical properties of rubber-modified polybenzoxazine. *High Perform. Polym.* 13, S327–S342.
[22] Andreu, R., Espinosa, M. A., Galia, M., Cadiz, V., Ronda, J. C., and Reina, J. A. (2006). Synthesis of novel benzoxazines containing glycidyl groups: a study of the crosslinking behavior. *J. Polym. Sci. A* 44, 1529–1540.

[23] Ishida, H., and Allen, D. J. (1996). Physical and mechanical characterization of near-zero shrinkage polybenzoxazines. *J. Polym. Sci. B* 34, 1019–1030.

[24] Ishida, H., and Lee, Y. H. (2001). Study of hydrogen bonding and thermal properties of polybenzoxazine and poly-(epsilon-caprolactone) blends. *J. Polym. Sci. B* 39, 736–749.

[25] Wang, H., Zhao, T., and Yu, Y. (2005). Synthesis of bismaleimide-modified novolac resin/clay nanocomposites using p-phenylenediamine as swelling agent. *J. Appl. Polym. Sci.* 96, 466–474.

[26] Rezanavaz, R., RazaviAghjeh, M. K., and Babaluo, A. A. (2010). Rheology, morphology, and thermal behavior of HDPE/clay nanocomposites. *Polym. Compos.* 31, 1028–1036.

2

Studies on Thermosetting Resin Blends: Bispropargyl Ether-Bismaleimide

J. Dhanalakshmi[1], K. G. Sudhamani[2] and C. T. Vijayakumar[2]

[1]Department of Chemistry, Kamaraj College of Engineering and Technology, Virudhunagar, India
[2]Department of Polymer Technology, Kamaraj College of Engineering and Technology, Virudhunagar, India

Abstract

The compound bispropargyl ether of bisphenol-A (BPEBPA) is prepared using a phase-transfer catalyst. The synthesized BPEBPA is separately blended with different bismaleimides (BMIs) [4,4'-bismaleimidodiphenyl methane (BMIM) and 4,4'-bismaleimidodiphenyl ether (BMIE)] in the mole ratio 0.5:0.5. Curing behaviour of bispropargyl ether, BMIs and bispropargyl ether-BMI blends are studied using a differential scanning calorimeter (DSC) and the polymers from these materials are studied using a thermogravimetric analyzer (TGA). Bispropargyl ether-BMI blends show considerable differences in the thermal curing behaviour as evidenced by the DSC studies. Among the materials investigated, the BMIM blended bispropargyl ether show higher-end set degradation temperature. From this, one can conclude that the polymerization of bispropargyl ether-BMI blends leads to structurally different network structures.

2.1 Introduction

2.1.1 Thermosetting Resins

The technological revolution in the field of engineering and structural materials has required timely advances in material performance. With the development in the areas of polymer science and technology, polymers started to be used in various applications as replacements for conventional materials. Polymers are therefore comparatively new materials and have proved to be suitable substitutes for conventional materials in diverse applications. The driving force for replacement of conventional materials by polymers is their obvious advantages: light weight, low cost, ease of processing, and wide scope of modifications to tailor the desired properties. These polymers are classified into thermoplastic and thermoset resins.

Thermosetting resins are reactive low molecular weight oligomers or high molecular weight polymers having reactive functional groups for cross-linking reactions. After thermosetting process these resins become infusible and insoluble materials due to the cross-linking reactions and are irreversible. Cross-linking is responsible for providing hardness, strength, brittleness and better dimensional stability. Thermosets generally have better mechanical, thermal and chemical properties. They also have excellent electrical resistance and dimensional stability compared to thermoplastics. The most important thermosetting resins are formaldehyde condensation products with phenol (phenolic resins) or with urea or melamine (amino resins).

2.1.2 High Performance Thermosets

In the early 1960s, there was a great need and demand for materials that survive at high temperatures to be applied in the field of aerospace and automobile industries. With advancement in technology, polymers with better performance—known as high-performance high-temperature thermoset—became a better alternative. These high-performance thermosets are defined as polymers having extraordinary stability upon exposure to the environment and their properties outdo those of conventional polymers, i.e., high thermal decomposition temperatures, low weight-loss rates at high temperatures, and long-term durability.

Progress in the field of polymers were summarized [1] in the 1960s and pointed out three aspects wherein efforts had been made to synthesize polymer systems capable of withstanding prolonged exposure to elevated temperatures: (i) improving existing polymers by introducing structural modifications; (ii) devising new organic systems tailored to resist the efforts of heat; (iii) synthesizing an entirely new class of inorganic and organic–inorganic (semiorganic) polymers that are thermally stable. The three approaches developed in earlier days to gain high temperature-resistant polymers are still applied to current studies, particularly the first two approaches, used in developing thermosetting polymers, which are used for composite materials manufacturing.

High performance thermosets have high dimensional stability at elevated temperatures, excellent thermal and thermo-oxidative resistance, low water absorption, good chemical resistance, high mechanical strength, excellent stiffness, and high compression strength. High performance thermosets have found a broad range of applications from structural materials in fiber reinforced composites to thin films for use in electronics packaging and in emerging technologies. In structural applications, fiber-reinforced high-temperature polymer matrix composites can offer significant advantages over other materials because of their low density and high specific strength. These composites are quite attractive for use in aerospace structural applications, such as aircraft engines, airframe, missiles, and rockets [2].

The processability of high-performance polymers can only be achieved by the reduction of their molecular weight. However, this approach results in a significant decrease of their final physical properties. The solution to this is to reduce the size of the polymer molecule for fabrication and to incorporate reactive functional moieties on the molecule for further reaction. The reactive groups eventually polymerize upon curing and give cross-links to further improve material performance. The curing reaction should provide not only crosslinking but also additional rigid ring structures. This becomes the base concept for high-performance thermosets [3].

The various high performance oligomeric segments [4] such as imide, quinoline, quinoxaline, amide, amide–imide, ester, sulfone, and ether ketone have been synthesized from a number of different monomers to meet performance requirements. Because of the molecular weight reduction, the properties of high performance thermosets become significantly dependent on the type of reactive functional groups.

The end capping groups such as maleimide, nadi-imide, acetylene, benzocyclobutene, cyanate ester, arylethynyl, styrenyl, phthalonitrile, biphenylene, epoxy, *etc.,* are used to prepare the usual high-performance thermosets such as bismaleimides (BMIs), bisnadimides, acetylene-terminated resins (ATRs), benzocyclobutene resins, cyanate esters, and arylethynyl resins [5].

2.1.3 Bismaleimide

Condensation polyimides are typically based on high molecular weight polyamic acid precursors that release void-forming volatile by-products during the thermal imidization process. Addition polyimides are usually low molecular weight resins containing unsaturated moieties such as maleimido or nadimido or ethynyl end capping groups for subsequent cross-linking reactions.

Among addition imides, BMIs are used for advanced material applications due to its high performance-to-cost ratio, to provide fatigue resistance, and for prevention of fracture propagation. Owing to the low mobility of atoms in cured BMIs, they have in general a high glass transition temperature. In a small molecular unit there are more double bonds. There lies the chance for all the double bonds to be cross-linked. Due to this, the cross-linking densities of BMIs are large. When the cross-linking density increases, the brittleness of the cured BMI resin also increases. This reduces the toughness of the component and so the component will fail even when small loads are applied. Brittleness of cured BMIs resin is of major concern for structural applications. Various approaches such as basic structure change of BMI resins, chain extensions, copolymerization, alkenyl phenyl compounds as co-reactants, and rubber-toughened BMI have been developed to improve the toughness of a cured BMI resin [3].

Generally BMI resins are prepared by reacting diamines with maleic anhydride in two reaction steps: (a) bismaleamic acid formation, (b) imidization (similar to synthesis of condensation polyimides) (Figure 2.1). The formation of the bismaleamic acid is a fast exothermic reaction and the imidization is carried out thermally or chemically. The standard synthesis of an aromatic BMI involves the chemical imidization using acetic anhydride in the presence of a catalytic amount of anhydrous sodium acetate at temperatures below 80°C. Some addition of amine to BMI double bonds (Michael

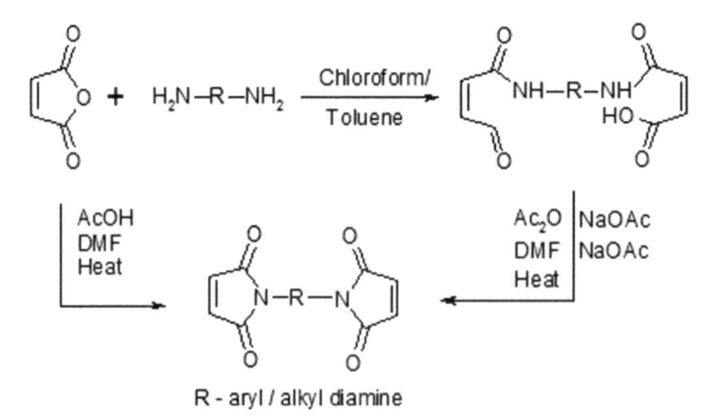

Figure 2.1 Preparation of bismaleimide.

addition) also takes place when the reaction is carried out at 140°C for a long duration [6].

The BMI resins, low molecular weight compounds, and oligomers require significant molecular weight increase and structure modification to become useful. The BMI resins are reactive towards various reactive species under mild conditions. A few curing reactions such as thermal polymerization, Michael additions, ene reaction, and Diels–Alder reactions have been developed to increase the application performance of BMI resins. The most important curing reaction of BMI resins has been their homo polymerization at elevated temperatures. The reaction produces no volatile by-products and provides void-free thermosets. The BMI resins on curing give rigid, solvent-resistant, highly cross-linked thermosets. Aromatic BMI resins usually are high melting solids. They are usually dissolved in solvents to improve their processability.

The BMI resins contain reactive unsaturated olefinic groups that are further activated by two strong electron withdrawing carbonyl groups and are known to undergo Michael additions with hydrogen active moieties such as phenols, thiols, and amines. The reactions have been used to prepare not only various high molecular weight linear polymers but also high-performance thermosets. During imidization, maleamic acid intermediates prepared from maleic anhydride and amines undergo reversible reactions to regenerate amines that eventually add to unsaturated olefinic bonds. Hence the properties of a BMI resin strongly depend on the synthetic conditions [6].

2.1.4 Acetylene-terminated Resins (ATRs)

Acetylene-terminated (AT) monomers and polymers are gaining importance in view of the increased need for easily processable thermally stable polymers. ATRs have become very popular and are commercially employed in composite resin structures such as in aircraft and aerospace structures where high strength, lightweight materials capable of withstanding high temperatures are required.

Acetylene-terminated resins was developed with various functionalized backbones which have greater advantage of processability and curability upon heating without releasing void-causing by-products. These developed ATRs are used in various applications where properties such as excellent thermal and heat resistance, high service temperature, reduced defects, low moisture uptake, and high strength are important [7].

The synthetic procedure having the potential to prepare reasonably large quantities of AT oligomers is the dehydrohalogenation of the corresponding bis(1,2-dihaloethyl) compound reported by Hay [8]. Introduction of an acetylenic group onto a phenyl group can be done by Stephens-Castro coupling reaction [9] between an aryl iodide and a protected copper(I) acetylide in pyridine at reflux. This reaction provides aryl acetylenes in high yields.

2.1.5 Propargyl-terminated Resins (PTRs)

High-performance propargyl resins are yet another class of attractive ATRs. These are a novel class of thermally stable composite matrix resins. These high-performance propargyl resins are thermosetting compounds bridged with thermal and heat resistant segments, and terminated with propargyl groups.

The high-performance thermosets, specifically propargyl-terminated resins (PTRs) [10], was developed for the fabrication of composite materials which can have continuous service at high temperatures. These PTRs were hydrophobic thermosetting materials. They act as a substitute for epoxy resin in advanced composites, electronics, adhesives and coating. The majority of known thermosets produced industrially, such as epoxy, BMIs, etc., are hydrophilic. This hydrophilic nature has an adverse effect on physico-mechanical properties. Another problem of hydrophilic thermoset is the easy delamination of their composites. Water concentrates at the polymer matrix/fiber interface and can cause easy delamination and composite failure. Adhesives and coatings exhibit a similar defect. As a result of these hydrophilic thermosets, there is a strong need for hydrophobic material. Hence a new type of ATR based on propargyl derivatives of bisphenols was developed.

The aromatic PTRs from bisphenol-A (Figure 2.2) was prepared in the presence of phase transfer catalyst (tetrabutylammonium bromide). The final product is isolated simply by filtration of dipropargylether of bisphenol which crystallizes from the initial aqueous solution. The yield is again quantitative with a purity of greater than 99% without the formation of rearranged products [11].

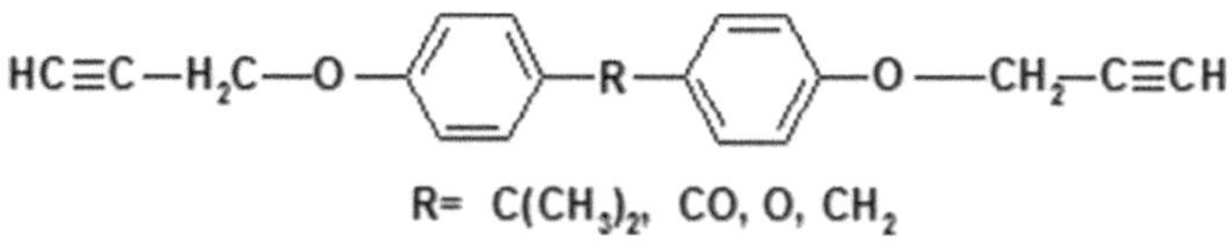

$HC{\equiv}C{-}H_2C{-}O{-}C_6H_4{-}R{-}C_6H_4{-}O{-}CH_2{-}C{\equiv}CH$

R= $C(CH_3)_2$, CO, O, CH_2

Figure 2.2 Structure of aromatic propargyl terminated resin.

The major advantages of PTRs are long shelf life, good thermal stability, low moisture absorption, excellent adhesion, low dielectric constant and acceptable physico-mechanical properties. The formation of void-free materials without the application of high pressures is an additional advantage of PTRs. The polymeric product formed after curing of the aromatic propargyl ethers exhibit an unexpected substantial improvement in the flexural modulus and flexural strength, compared to other resins [12].

In general, PT resins are prepared by a one-step condensation reaction using cost-effective industrial raw materials such as polyphenols and propargyl halides. A PT resin is initially prepared by refluxing the solution of polyphenols and propargyl bromide in acetone for 72 h in the presence of potassium carbonate [13]. The reaction can be much faster in other polar solvents. For example, the reaction of bisphenol-A with propargyl bromide in *N,N*-dimethylacetamide in the presence of potassium carbonate at room temperature produces a quantitative yield of bispropargyl ether of bisphenol-A (BPEBPA) in 2 h [14]. In the presence of a phase-transfer catalyst such as tetraalkylammonium salt, cost-effective propargyl chloride can be employed for preparing propargyl ethers of polyphenols at room temperature [15].

The phenolic materials containing propargyl groups were prepared in short reaction periods by the reaction of a polyhydric phenolic material with propargyl bromide, the reaction being conducted in an aqueous sodium hydroxide solution [16]. This method provides the advantage of a shorter reaction time and the utilization of an aqueous medium. However, Claisen rearrangement occurs during the preparation and results in a large amount of 3,3'-propargyl substituted phenols.

The various PTR monomers were synthesized from numerous polyphenols such as bisphenol-A, 4,4'-(hexafluoroisopro-pylidenyl)diphenol, 4,4'-dihydroxybenzophenone, 4,4'-sulfonyl-diphenol, 4,4'-diphenylsulfide, resorcinol and novalac [14]. They all have excellent shelf life and are water insensitive.

Propargyl-terminated resin undergoes a thermal rearrangement reaction to form a chromene ring from an arylpropargyl ether structure. The final thermosetting reaction is the polymerization of the chromene rings. The formation of the chromene ring is very important for the cross-linking of PT resin, whereas the final cross linking mechanism after the formation of the chromene ring has not been determined. The thermal curing mechanism of bispropargyl ethers resin is still obscure because of the numerous ways in which polymerization can proceed; combined with the fact that only a few characterization techniques are available to study these complex thermoset structures.

2.1.6 Property Enhancement in PT Resins

Propargyl ether resins have a set of properties including physico-mechanical and dielectric properties which are superior to that of other thermosetting materials. They appear to be a very attractive low-cost substitute for the expensive ATRs potential substitute for the hydrophilic epoxy resin and BMIs as a hydrophobic matrix in advanced composites, electronic, adhesives and coatings.

There has been growing interest in propargyl ether resin since it possesses excellent thermal properties. Many studies have reported on its thermal rearrangement and it was studied that the high cure temperature of these resins makes it unsuitable to be a matrix to fabricate high-performance fiber-reinforced composites. Furthermore, the mechanical properties of the cured resin and its composites are poor. Therefore, enhancing the properties of propargyl ether is a key factor for the advancement of these materials. In general, an effective way to improve the properties of the resin is required. Since the thermal properties of PT resins can be enhanced, many investigators focused their attention in developing high thermally stable PT resins.

The thermosetting resin with terminal propargyl groups has recently been studied as high-performance polymers. The propargylic resins are good candidates as epoxy resin substitutes. But their thermal stability is marginally inferior to that of BMIs. One attainable method to derive both thermal stability and processability is to make a blend by mixing a BMI resin and a propargylic resin. Such a blend could possibly benefit from the good physicochemical attributes of the two resins. Both the BMI resin and the PTR contain unsaturated terminal groups that can polymerize without forming volatiles.

Allyl compounds are very stable at room temperature. At high temperatures or with peroxide initiators the homopolymerization of allyl functionality is difficult to obtain. This is because the free radicals produced are very stable due to conjugation and thus more unlikely to propagate. At room temperature, *o,o*'-diallyl bisphenol-A is an amber liquid with viscosity of 12∼20 Pascal-second (Pa.s) and at temperatures above 100°C, it can dissolve BMIs by stirring. Then, at temperatures above 150°C, it is thought to react with BMIs through the Diels–Alder mechanism rather than the free radical copolymerization mechanism.

Blends of BMI and allyl derivatives of phenols [17] were extensively studied. The results indicated that the blends showed better toughness and high heat resistance. The curing and thermal properties of a diallyl bisphenol-A novolac epoxy system cocured with bisphenol-A BMI was investigated [18]. They reported that blending of BMI improved the overall thermal stability and the modulus of the resultant composites. The increase in BMI concentration in the system resulted in enhanced glass transition temperature with a consequent improvement in high temperature performance.

One important class of high-performance thermosets is derived from BMI resins and allylphenyl oligomers. Upon thermal curing, the mixtures copolymerize to produce polymers with high glass transition temperature. For example, the mixture of 3,3'-diallylbisphenol-A and bis(4-maleimidophenyl)methane displays low and high temperature exotherms occurring at approximately at 130°C and 255°C, respectively. On the basis of the nuclear magnetic resonance spectrum (NMR) results, the low-temperature exothermic reaction was concluded to be an ene reaction between allyl and maleimide groups (Figure 2.3). The high-temperature exothermic reaction is related to a Diels–Alder reaction of the functional groups produced in the ene reaction [19].

Diels–Alder polymerization requires a reacting diene monomer and a monomeric dienophile, which is normally activated by electron withdrawing substituents. During the polymerization, a stiff, heat-resistant ring structure is formed by every Diels–Alder reaction unit. Therefore it is an excellent approach to increase the polymer performance as in the reaction of difuran end capped monomers with BMI resins [20]. The maleimide group of BMI is capable of reacting with the allyl group through the addition reaction; many studies have been devoted to blending BMI with allyl derivates of phenols, such as 2,2'-diallyl bisphenol-A (DABA), allyl ether novolac, etc. The results indicated that the blends exhibited better toughness and high heat resistance. Owing to the similar chemical property of allyl with propargyl, we expected to improve the processability and thermal properties of bispropargyl ethers by blending it with bis(4-maleimidophenyl)methane.

2.1.7 Literature

The chemistry of resins with terminal acetylenic groups has undergone a very rapid development in the past two decades. The reason that much effort is being made in this direction is that these resins cure with no volatile byproducts during the processing stage. The reaction can proceed without

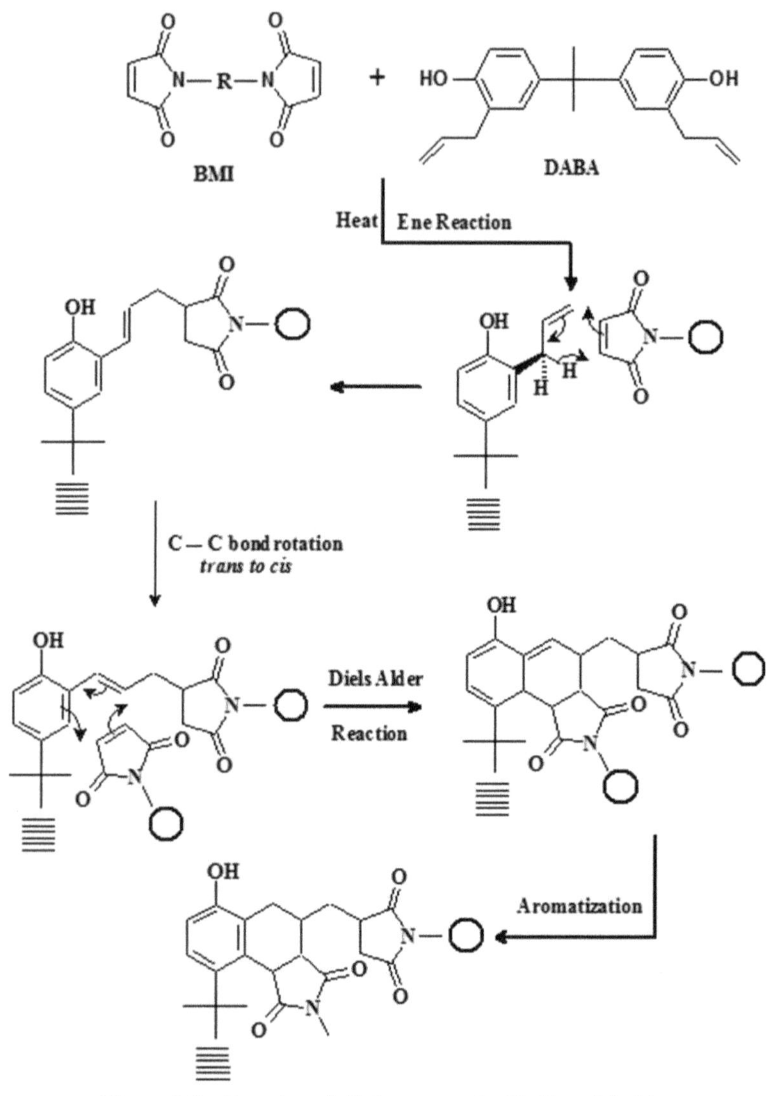

Figure 2.3 Reaction of allyl compound with bismaleimide.

catalyst which is an advantage because catalysts often contribute to long-term instability in the cured resin. In addition, resins afford cure products with high thermal stability and good mechanical properties. So, they are currently being developed to replace epoxies used under hot wet conditions. Highly cross-linked polymers have desirable properties such as excellent dimensional stability with low creep rates, high resistance to solvents, and higher softening temperatures. Oligomers and polymers containing acetylenic groups have excellent shelf life and when cured, provide resins which are resistant to solvents and have a favorable combination of physical and mechanical properties.

Acetylene-terminated resins fall into two main categories; those terminated by aryl acetylene and those terminated by aryl prop-2-ynyl ether groups. The uncatalyzed curing mechanism of the former type involves a free radical reaction which results in a network with linear conjugated polyenes. However, it has recently been discovered that uncatalyzed curing takes place via sigmatropic thermal rearrangement of the aryl prop-2-ynyl ether groups to form 2H-1-benzopyran rings, which then undergo polymerization.

A polymeric matrix makes the largest contribution to the heat resistance of composite material. Phenolic resins release volatiles in the curing process. This causes the formation of gas bubbles in the final materials and increased pressure is required to avoid it. Also cured resins usually have low thermo oxidation resistance. The way to solve the problems is through a chemical modification of the resins to change the mechanism of polymerization from condensation to addition type. This can be achieved by the chemical modification of resin, by introduction of allyl and propargyl groups.

Aromatic propargyl ether compounds are believed to represent the most robust resin chemistry currently available to meet the many performance requirements associated with under fill applications. Propargyl ether resins are hydrophobic, hydrolytically stable, low-toxicity monomers that can be cured to thermally stable thermosets having high T_g, thermally stable thermosets. Liquid propargyl ether monomers have been found and/or described in the literature which can be used alone or in combination to yield diluent-free underfill compositions. Alternatively, mixtures of two or more propargyl ether monomers (wherein one or more of these monomers may be solids at room temperature) can be used to create diluent-free, room temperature stable, eutectic liquid resin compositions.

Further, when aryl propargyl ether-terminated monomers are heated to around 220°C, they first undergo a thermal sigmatropic Claisen rearrangement to form 2H-chromene or 2H-1-benzopyran. These intermediates possess reactive double bonds that can subsequently undergo thermal polymerization, generating polymers. Hay [13] studied photosensitive acetylenic polymers. He reported that polymeric compositions containing acetylenic groups and a photosensitizing agent, which may be part of the polymer molecule or an additive to the polymer and which upon absorbing actinic radiation accelerates the cross linking of the polymer, are rendered insoluble when exposed to ultraviolet radiation, thereby making them suitable as photoresists in the graphic arts.

Harfenist and Thom [22] studied the cyclization of monofunctional propargyl ethers to monofunctional chromenes in yields of 36 to 76%, by heating in a solvent such as dimethyl aniline, trimethylene glycol or dichlorobenzene. Balasubramanian and Venugopal [23] reported the formation of bischromenes derived from the bispropargyl ethers of 2,7-dihydroxynaphthalene and hydroquinone by heating the latter compounds in *N,N*-diethyl aniline. Pomeranz and Schmid [24] reported that propargylphenyl ethers rearrange at about 200°C to 2H-chromenes. They also reported that rearrangement occurs in benzene or chloroform at lower temperatures (20–80°C) in the presence of catalysts such as silver tetrafluoroborate (or trifluoracetate).

The thermal cyclization of substituted arylpropargyl ethers was carried out by Anderson and Voie [25]. Specifically, they reported that the cyclization of simple 3-aryloxypropynes to the corresponding chromenes in approximately 60% yield, using the solvent diethylaniline at temperatures between 210°C and 215°C. Balasubramanian et al. [26] reported that the mercuric oxide and concentrated sulfuric acid catalyzed rearrangement of bispropargyl ether leads to bischromenes. They obtained the bischromene in 61% yield as crude material which had to be purified by column chromatography.

Picklesimer [16] prepared phenolic materials containing propargyl groups in short reaction periods by the reaction of a polyhydric phenolic material with propargyl bromide, the reaction being conducted in an aqueous sodium hydroxide solution. The process suffers from the disadvantage of providing both *O*-propargylated (desired) and *C*-propargylated (undesired) materials (Figure 2.4). For example, bisphenol-A is claimed to provide a 45.4% yield of the desired bispropargyl ether and a 43.6% yield of the undesired C-propargylated bisphenol. Additionally, the process employs rather vigorous conditions, such as reflux conditions of 100°C, for 1–3 h. A further drawback of the process is that propargyl bromide is used rather than propargyl chloride. The bromide is a relatively expensive inaccessible compound on a commercial scale and it is shock-sensitive.

Greenwood et al. [27] prepared a series of high molecular weight aromatic polyamide copolymers derived from 4,4’-bis(methylamino)-diphenylmethane, 4,4’-bis(propargylamino)diphenylmethane, and isophthaloyl dichloride as potential candidates for use as matrix resins in Kevlar

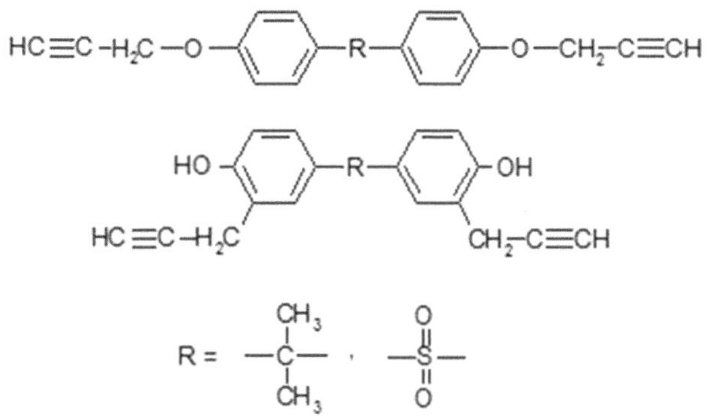

Figure 2.4 Structure of *O*-propargylated (desired) and *C*-propargylated (undesired) materials.

fiber composites. These polyamides contain pendant propargyl groups that underwent facile thermal crosslinking at 280°C, as evidenced by a dramatic increase in their glass transition temperatures and the accompanying loss of solubility. Other attempts to affect cross-linking by exposure to ultraviolet light, electron beam, or gamma radiation were unsuccessful.

Usha and Balasubramanian [28] studied the Claisen rearrangement of aryl propargyl ethers in poly (ethylene glycol) at 200°C. Aryl propargyl ethers containing electron donating groups yield (2H)-benzopyrans and those containing electron withdrawing groups yield 2-methylbenzofurans (Figure 2.5).

Dirlikov and Feng [14] prepared dipropargylether of bisphenol-A in quantitative yield by the reaction of bisphenol-A with propargyl bromide in *N,N*-dimethylacetamide in the presence of potassium carbonate at room temperature in 2 h. Various PTR monomers have been synthesized from numerous polyphenols such as bisphenol-A, 4,4'-(hexafluroisopropylidenyl) diphenol, 4,4'-dihydroxybenzophenone, 4,4'-sulfonyldiphenol, 4,4'-dihydrox ydiphenyl sulfide, resorcinol, and novolac.

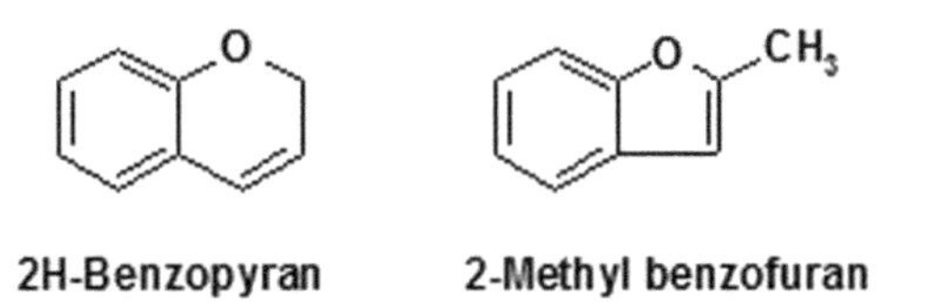

Figure 2.5 Structure of (2H)-benzopyran and 2-methyl benzofuran.

Inbasekaran and Dirlikov [11] synthesized an aromatic propargyl ether, preferably bispropargyl ether from phenolic compounds comprising vigorously stirring a propargyl halide, preferably propargyl chloride, with a phenolic compound in an aqueous sodium hydroxide solution at a temperature of from about 0°C to 100°C in the presence of a phase-transfer catalyst such as tetrabutylammonium bromide. The mixture is then filtered to recover a solid product which is 98% pure.

Dirlikov [10] developed a new type of ATR based on propargyl derivatives of bisphenols and phenol formaldehyde resin (PTRs) without ATR limitations. These PTR monomers polymerize in the presence of copper chloride with the formation of linear polymers with diacetylene linkages between the monomer units. Their two conjugated triple bonds require very low energy emission for radiation cure and easily cross-link. These linear polymers appear very suitable for preparation of radiation and/or thermally curable coatings.

Lucotte and Delfort [15] synthesized the bispropargyl ether, 1,2-bis (3-propargyloxyphenoxy)ethane (Figure 2.6) by phase transfer catalyzed [tetrabutylammonium hydrogen sulphate (TBAH)] etherification. And they studied the polymerization using copper catalyst. The copper catalyst shifts the polymerization temperature of 1,2-bis(3-propargyloxyphenoxy) ethane to a lower temperature. The catalytic amount of such salts was efficient to modify the thermal behaviour of the reaction.

Kolb et al. [12] prepared different aromatic bispropargyl ether systems from different bisphenols using propargyl chloride, sodium hydroxide and tetrabutylammoniumbromide and studied the formation of chromene containing mixtures from propargyl ether systems. They reported that the conversion of aromatic propargyl ethers to a preformed low-viscosity

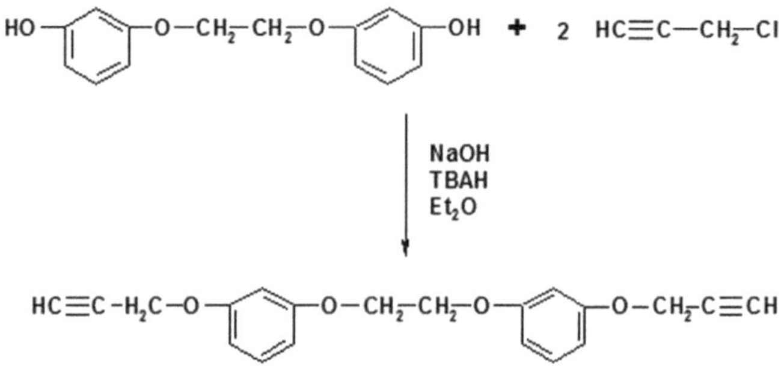

Figure 2.6 Synthesis of 1,2-bis(3-propargyloxyphenoxy)ethane.

chromene-containing mixture prior to polymerization affords two principle advantages. First the low viscosity character of the inventive chromene containing mixtures facilitates processability. If desired, a higher viscosity product may be attained by B-staging the chromene-containing mixture. Second, the partial conversion of the aromatic propargyl ether moiety into chromene moiety reduces the heat of polymerization. Milder heats of polymerization translate to substantially void-free parts exhibiting minimal shrinkage. The polymers are moisture insensitive and have a wide range of end uses, e.g., in high-temperature composites and electrical laminates.

Loustalot and Sanglar [29] studied the kinetic and polymerization mechanism of a series of monofunctional propargylic compounds. The results obtained have shown that the reaction starts by intramolecular ring formation (Claisen rearrangement). Based on a set of consistent physicochemical data, a reaction mechanism for these systems and a kinetic model of polymerization in the molten state can be proposed, corresponding to the formation of the first species before the gel point.

Bindu et al. [30] synthesized bispropargyl ethers of bisphenol-A (BPA), bisphenol ketone (BPK) and bisphenol sulfone (BPS) (Figure 2.7). These monomers were thermally polymerized to the corresponding poly bischromenes. The non isothermal kinetic analysis of the cure reaction using four integral methods such as Coats-Redfern, MacCallum-Tanner, Horowitz-Metzger, and Madhusudhanan-Krishnan-Ninan revealed that the presence of the electron withdrawing group did not favor the cyclisation reaction leading to formation of chromene, which precedes the polymerization, and this is in confirmation with the proposed mechanism of polymerization. The substituent group bridging the two phenyl rings also influenced the thermal stability of the resultant polymers. Thus sulphone and ketone containing polymers were more stable than the isopropylidene containing one.

Reghunadhan Nair et al. [31] compared the thermal and pyrolysis characteristics of four different types of addition cure phenolic resins as a function of their structure. Whereas the propargyl ether resins and phenyl azo functional phenolics underwent easy curing, the phenyl ethynyl and maleimide functional ones required higher thermal activation to achieve cure. All addition cure phenolics exhibited improved thermal stability and char-yielding property in comparison to conventional phenolic resole resin. The maleimide functional resins exhibited the lowest thermal stability and those cross-linked via ethynyl phenyl azo groups were the most thermally stable systems. Propargylated novolac and phenyl ethynyl functional phenolics showed intermediate thermal stability. The maximum char yield was also

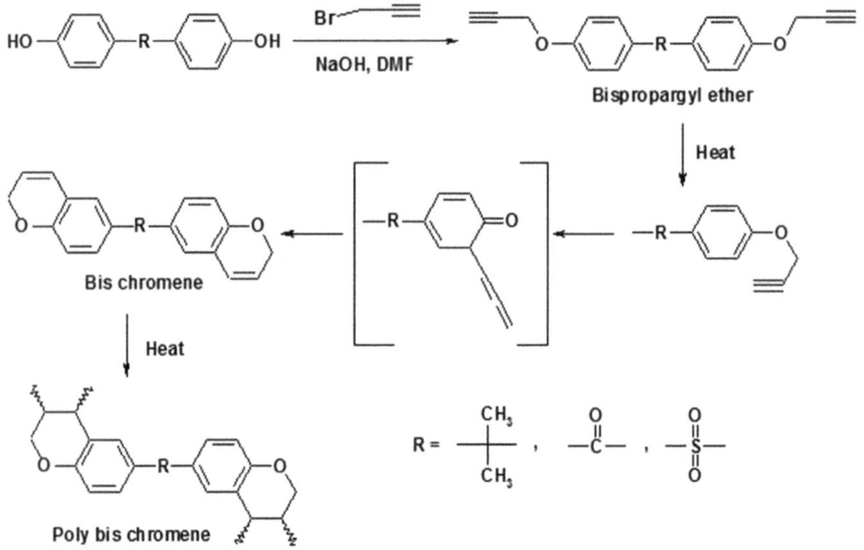

Figure 2.7 Synthesis and thermal polymerization of bispropargyl ethers.

given by ethynyl phenyl azo system. Non-isothermal kinetic analysis of the degradation reaction implied that all the polymers undergo degradation in two steps, except in the case of ethynyl phenyl azo resin, which showed apparent single-step degradation. The very low pre-exponential factor common to all polymers implied the significance of the volatilization process in the kinetics of degradation. Isothermal pyrolysis studies led to the conclusion that in the case of a nitrogen-containing polymer, the pyrolysis occurs via loss of nitrogenous products, which is conducive for enhancing the carbon content of the resultant char. The FTIR spectra of the pyrolysed samples confirmed the presence of C-O groups in the char.

Badarau and Wang [32] synthesized a series of diacetylene-containing polymers by oxidative coupling polymerization of 9,9'-bis(4-propargyloxyph enyl)fluorine and dipropargyl 4,4'-(hexafluoro-isopropylidene)diphthalimide. All the polymers showed typically amorphous diffraction patterns, good solubility in organic solvents, and good film-forming capability. The polymers were readily cross-linkable upon exposure to UV light at ambient temperature and could also be thermally cured. Upon UV irradiation with a low- or medium-pressure mercury lamp, the diacetylene moieties within the crystalline domains of the polymers undergo intermolecular 1,4-addition reaction to yield highly cross-linked networks. The refractive indices and birefringence of a series of diacetylene polymers can be adjusted and controlled through copolymerization of two different monomers.

Liu et al. [33] investigated the curing kinetics of bispropargyl ether bisphenol-A (BPEBPA) and 4,4'-bismaleimido diphenylmethane blends. They synthesized BPEBPA via Williamson reaction and blended with 4,4'-bismaleimido diphenylmethane at different molar ratios. The results indicated that the onset cure temperatures of the blend resins were about 20–30°C lower than that of pure BPEBPA and the exothermic enthalpy of curing of the resins are significantly reduced from 1,320 (BPEBPA) to 493 Jg^{-1}. The activation energy (Ea) values of blends are lower than the pure monomers which lead to high thermal stability resin system. The thermal stability of the resins improved markedly with the increase in 4,4'-bismaleimidediphenylmethane content. Finally they concluded that the blended resins presented better processability and thermal properties. These blends have a potential application as matrix for advanced heat-resistant composite materials.

Tsvetanka et al. [34] studied dipropargyl ethers of high aspect ratio bisphenols. They prepared difunctional propargyl monomers in a one-step reaction of propargyl bromide with corresponding bisphenol in an aqueous solution of sodium hydroxide in the presence of a phase-transfer catalyst (tetrabutylammonium bromide) for several hours at room temperature

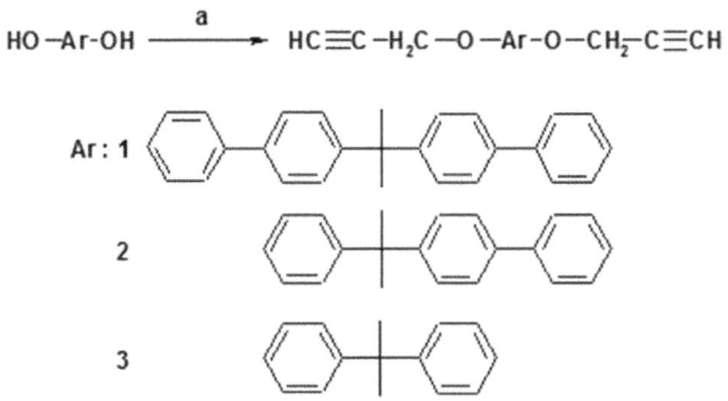

$$HO{-}Ar{-}OH \xrightarrow{a} HC{\equiv}C{-}H_2C{-}O{-}Ar{-}O{-}CH_2{-}C{\equiv}CH$$

Ar: 1

2

3

where **a**: $HC{\equiv}CCH_2Br$, $NaOH/H_2O$, $[CH_3(CH_2)_3]_4NBr$, 8 h at room temperature for **1** and **2**; $HC{\equiv}CCH_2Cl$, $NaOH/H_2O$, $[CH_3(CH_2)_3]_4NBr$, 16 h at room temperature for **3**

Figure 2.8 Synthesis of propargyl monomers.

(Figure 2.8). They further concluded that high aspect ratio monomers with additional phenyl rings over and above those of bisphenol-A are easily synthesized and elaborated to new proparylated monomers for resins. These materials were thermally polymerized in high yield to solid resins without the addition of catalyst.

Luo et al. [35] reported that the *N*-(3-acetylenephenyl) maleimide (3-APMI) was synthesized by the traditional two-steps method. The resultant monomer presents a series of excellent properties. The DSC studies show that the monomer possesses excellent cure reactivity. The peak temperature of cure reaction is 197.9°C and postcuring at 200°C for 4 h can drive the cure reaction to approach completion. The thermal properties of the cured monomer were determined by dynamic mechanical analysis (DMA) and the results show that glass transition temperature (represented by the onset temperature of storage modulus) is high, up to 460°C. The results of thermogravimetric analysis (TGA) reveal that the cured monomer possessed excellent thermal stability, whose 10% weight loss temperature ($T_{10\%}$) is 515.6°C and char yield at 800°C is 59.1%. All these characteristics make the 3-APMI monomer an ideal candidate for the matrix of thermo-resistant composites.

Rong et al. [36] developed a kind of modified BMI resin, with good heat resistance and processing properties for advanced composites. The modifier, BPEBPA, was prepared by a phase-transfer catalyzing procedure and used to modify bis(4-maleimidodiphenyl) methane. The results of DMA showed that the cured BPEBPA-modified BMIM resins had a glass transition temperature higher than 320°C. The carbon fiber-reinforced composites showed excellent flexural properties at ambient temperature and at 250°C.

Vingayagamoorthi et al. [37] investigated the BMI [4,4'-bismaleimido diphenyl ether (BMIE)] and BPEBPA blend. The DSC results indicated that the BPEBPA-BMIE blend has low ΔH cure values for the thermal polymerization and the whole temperature window for the exothermic curing reaction is shifted to a lower temperature than BPEBPA. The thermally polymerized BMI (BMIE), BPEBPA, and their blends were investigated using off-line pyrolysis GC/MS technique and the volatile products evolved during the pyrolysis were listed. The detailed analysis of the degradation products by GC-MS revealed the formation of phenols and several substituted phenols. This findings hint that the competitive C-C and C-O scissions of the chromene ring units formed via the Claisen rearrangement of the aryl propargyl ether system present in BPEBPA is operative. The proposed thermal degradation scheme for the polymer from BPEBPA is given in Figure 2.9.

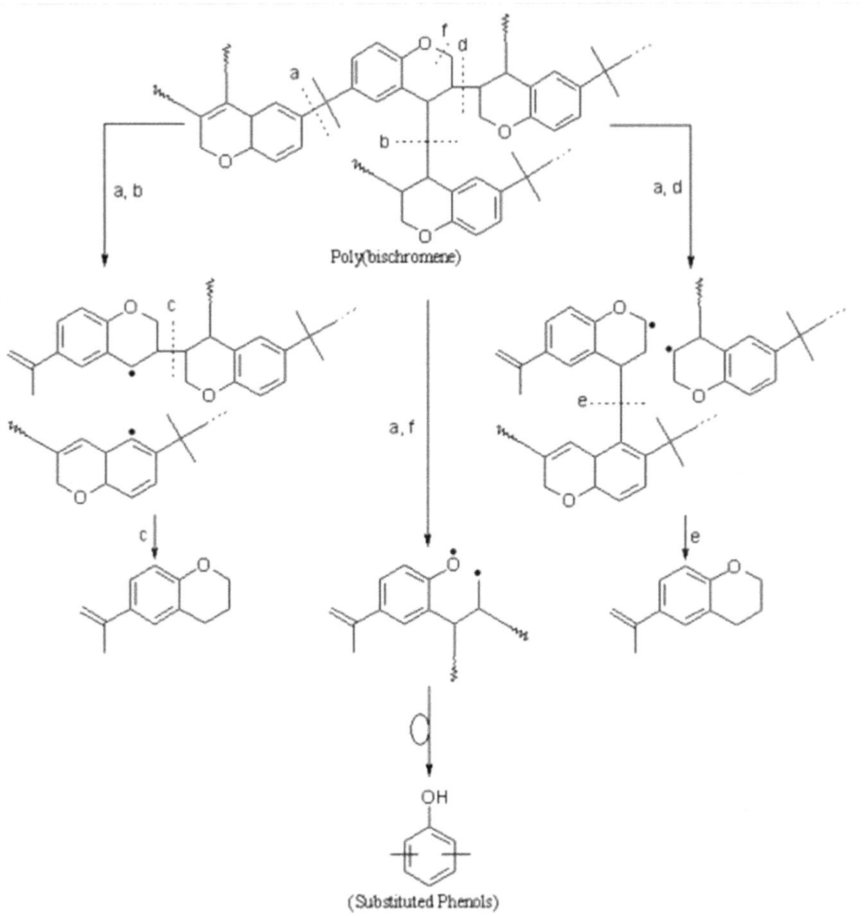

Figure 2.9 Thermal degradation of poly bispropargyl ether of bisphenol-A.

Liu et al. [38] synthesized a novel series of propargyloxyphenyl maleimides such as *N*-(2-propargyloxyphenyl) maleimide, *N*-(3-propargyloxy phenyl) maleimide, and *N*-(4-propargyloxyphenyl) maleimide. The cure behaviour of these monomers indicated that they had a broader processing window than normal BMI resins. The 5% mass loss temperatures of the cured above said monomers was high (~405°C) and this indicates that the cured resins possessed excellent thermal stability. Also, the char yields at 900°C for cured monomers were much higher than those of cured blended BMI resins.

The higher char yields suggest that it is possible to use these monomers for high-performance composite matrices.

Chen et al. [39] synthesised a series of reactive GAP (glycidyl azide polymer)/PTPB (propargyl-terminated polybutadiene) nanocomposites reinforced by alkynylated cellulose nanocrystals (ACNC) by Huisgen (Rostovtsev et al., 2002) click chemistry. In comparison with the neat GAP/PTPB (GP2) material, the tensile strength, Young's modulus, and the elongation at break of the GP2/ACNC-1.0 nanocomposite (containing only 1.0 wt% ACNC) were increased by 103.3%, 100.0%, and 12.4%, respectively. This study is a promising attempt to develop advanced polymeric composites reinforced with biomass-based nanoparticles with the simultaneous improvement of strength, modulus, and toughness.

The literature review shows that several possible routes have been investigated in order to modify the physico-chemical properties of polymers and their composites. Works pertaining to bispropargyl ethers and blends of bispropargyl ethers with BMI are scarce. Among the BMI 4,4'-bismaleimidodiphenyl methane (BMIM) and BMIE were important thermoset BMIs and industrially available. Due to these reasons, in the present investigation BMIM is chosen for blending with bispropargyl ethers. The blending of two high-temperature matrix resins will enhance the properties of both the resins and reduce the cost of the material.

Hence in the present investigation, the blends of BPEBPA with BMIM and BMIE in 0.5:0.5 mol ratios were made. The thermal curing behaviour of BPEBPA, BMIM, BMIE, and bispropargyl ether-BMI blends (MBPEBPA and EBPEBPA) were investigated using a differential scanning calorimeter (DSC). The pure BPEBPA, BMIM, BMIE, and BMI-blended BPEBPA were cured thermally. The thermal stabilities of the polymeric materials were investigated using TGA and all the results are presented and discussed.

2.2 Experimental

2.2.1 Preparation of BPEBPA, BMIM, and BMIE

The synthetic procedures for the preparation of BPEBPA and BMIE were reported by Vijayakumar et al. [37] BMIM was obtained from ABR Organic Ltd., Hyderabad, India. The structure of BPEBPA, BMIM, and BMIE are shown in Figure 2.10.

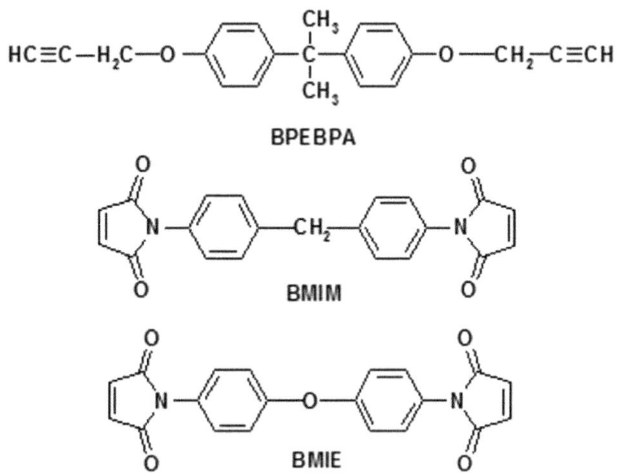

HC≡C–H$_2$C–O–C$_6$H$_4$–C(CH$_3$)$_2$–C$_6$H$_4$–O–CH$_2$–C≡CH

BPEBPA

BMIM

BMIE

Figure 2.10 Structures of BPEBPA, BMIM and BMIE.

2.2.2 Blending of Bispropargyl Ether of Bisphenol-A with BMIM and BMIE

The BPEBPA was separately blended with BMIM and BMIE (0.5:0.5 mol) in an agate mortar and the mixture was ground repeatedly to have effective mixing. The mixture was then dried in a vacuum oven and preserved for polymerization. The following blends were made: 0.5 BPEBPA + 0.5 BMIM (MBPEBPA), and 0.5 BPEBPA + 0.5 BMIE (EBPEBPA).

2.2.3 Thermal Curing of the Materials

The materials BPEBPA, BMIE, and the blends MBPEBPA and EBPEBPA were taken in separate micro test-tubes and flushed with dry oxygen-free nitrogen. The material was thermally polymerized at 250°C for 6 h. Similarly the material BMIM was thermally polymerized (240°C) for 6 h. After the polymerization, test tubes were cooled and the samples were removed from the micro test-tubes, ground to coarse powder, packed and stored in a vacuum desiccator for further analysis.

2.2.4 Methods

2.2.4.1 FTIR analysis

The Fourier transform infrared (FTIR) spectra of the materials were recorded on a SHIMADZU-8400S infrared spectrophotometer using the KBr pellet technique. The absorption bands in FTIR spectra were used to identify the functional groups in the materials investigated. About 500 mg of potassium bromide was taken in a mortar. Approximately 10 mg of the sample was added and ground well with a pestle. Approximately 100 mg of the above mixture was made into a transparent disc with a pellet maker. The above prepared disc was placed in the infrared radiation path using a pellet holder and the IR spectrum was recorded by plotting percentage of transmittance as a function of wave number in cm^{-1}.

2.2.4.2 DSC analysis

The differential scanning calorimetric (DSC) curves for materials were recorded in a TA Instruments DSC Q20. The sample (2–3 mg) was weighed, placed in the non-hermatic aluminium pan and sealed with aluminium lid. The samples were heated from ambient temperature to 350°C at a rate of 20°C min^{-1} in dry nitrogen (Flow rate = 50 mL min^{-1}) atmosphere. The obtained DSC curves were analyzed using the universal analysis 2000 software provided by TA instruments. The programme Microsoft Office Excel was used for kinetic analysis.

2.2.4.3 TG analysis

Thermogravimetric (TG) analyses were performed on a TA Instruments TGA Q50 TGA. To avoid the secondary reaction of evolved gases in the TGA such as thermal cracking, re-condensation, and repolymerization reactions, the nitrogen flow was maintained at balance and the samples were 50 mL min^{-1} and 60 mL min^{-1} respectively. The sample (3–4 mg) was weighed into platinum crucible and was heated from ambient temperature to 800°C at a rate of 20°C min^{-1}.

2.3 Results and Discussion

2.3.1 FTIR Studies

The monomer BPEBPA shows the characteristic absorption owing to ≡C-H band of the propargyl groups at 3271 cm^{-1} in its FTIR spectrum. The absorption due to the C≡C appeared as a weak band at 2,121 cm^{-1}. The band pertaining to the -OH group of the precursor bisphenol-A was absent at 3,500 cm^{-1} and this confirms the structure of BPEBPA. The strong peak noted at 1,234 cm^{-1} confirms the presence of the aryl ether (-Ar-O-CH_2-) group.

The FTIR spectrum of the thermally cured BPEBPA shows the band at 1,504 cm^{-1} for the phenyl nucleus stretching. The appearance of a new broad band at 3,430 cm^{-1} indicates the formation of phenol entities during the curing process and the appearance of a band at 1,604 cm^{-1} indicates the propargyl group has been converted into chromene structure, and the disappearance of ≡C–H stretching band at 3,263 cm^{-1}, C≡C stretching at 2,121 cm^{-1}, and ≡C–H deformation at 696 cm^{-1} confirm the polymerization of the propargyl groups in BPEBPA.

The presence of absorption bands for phenyl nucleus stretching (1,598 and 1,501 cm^{-1}), cyclic imide ring stretching (1,711 cm^{-1}), and Ar-O-Ar stretching (830 cm^{-1}) in the FTIR spectrum of BMIE confirm the structure of BMIE. The FTIR spectrum of the thermally cured BMIE show the absorption bands for phenyl nucleus stretching at 1,598 and 1,501 cm^{-1}, cyclic imide ring stretching at 1,711 cm^{-1} and presence of Ar-O-Ar stretching at 830 cm^{-1}. The identification of these bands and the absence of the maleimide double bond absorption band in the FTIR spectrum of thermally cured BMIE confirm the polymerization of BMIE.

The FTIR spectrum of thermally cured BMIM shows the absence of maleimide ≡CH band at 3,095 cm^{-1} and the presence of aliphatic C-H band at 2,931 cm^{-1}, which indicate the participation of the maleimide double bonds in the thermal polymerization of BMIM. Thermally polymerized material show an absorption band at 1,504 cm^{-1} and is attributed to the phenyl nucleus stretching. The presence of a band at around 1,705 cm^{-1} confirms the presence of the cyclic imide ring structure in the cured material.

The FTIR spectra of the thermally cured bispropargyl ether-BMIM and bispropargyl ether-BMIE blends show the absorption bands corresponding to the cured pure bispropargyl ether and pure BMI. From the above observations, it is concluded that the blending of propargyl ether with BMIs does not affect the basic polymerization mechanism of the cured pure materials, *i.e.*, the Claisen rearrangement of the propargyl groups, the conversion to the chromene structure followed by the addition polymerization of the double bond present in the chromene ring.

2.3.2 DSC Studies

The DSC curves for BPEBPA, BMIM, BMIE, and the blends MBPEBPA and EBPEBPA were recorded at a heating rate of 20°C min^{-1} and are shown in Figure 2.11. The parameters obtained from the DSC curves such as melting temperature (T_m), enthalpy of fusion (ΔH_f), onset of curing (T_s),

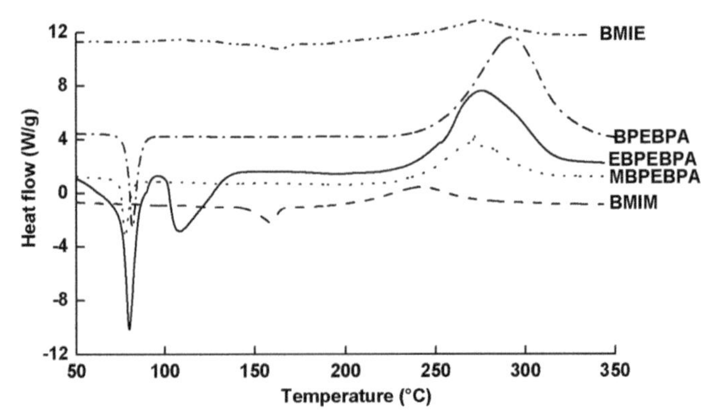

Figure 2.11 DSC curves recorded at 20°C min^{-1} for BPEBPA, BMIM, BMIE and BPEBPA blends with bismaleimides.

curing maximum ($T_{\max}$), endset of curing (T_e), and total enthalpy of cure reaction (ΔH_c) for the materials investigated (BPEBPA, BMIM, BMIE, and bispropargyl ether-BMI blends: MBPEBPA and EBPEBPA) are tabulated in Table 2.1.

The compound BPEBPA shows a sharp melting point at 82°C with an enthalpy of fusion of 117 Jg^{-1}. The thermal curing of BPEBPA starts at 218°C and attains maximum at 292°C and ends at 340°C, and the total enthalpy of the curing process is 964 Jg^{-1}. The DSC trace of BMIM shows a broad melting region at around 159°C with an enthalpy of fusion of 38 Jg^{-1} and the thermal curing of BMIM starts at 184°C, attains maximum at 243°C and ends at 328°C. The total enthalpy associated with the curing exotherm is 202 Jg^{-1}. Similarly the DSC curve of BMIE shows an endotherm at 164°C which is attributed to the melting of BMIE and it is immediately followed by an exothermic peak. The curing exotherm of BMIE starts at 189°C and attains maximum at 275°C and ends at 335°C. The enthalpy of curing for BMIE is 232 Jg^{-1}. From Table 2.1, it is obvious that BPEBPA and the BMI-blended BPEBPA (MBPEBPA and EBPEBPA) show sharp melting points. The monomer BPEBPA on blending with BMIE during heating shows two melting regions, one characteristic for BPEBPA and the other for BMIE. The presence of BPEBPA in BMIE considerably reduces the melting point of BMIE. Thus, the melting of one of the monomers is drastically influencing the melting character of the other monomer in the case of EBPEBPA. Such an effect is not seen in MBPEBPA.

Table 2.1 DSC studies: Parameters derived from DSC curves of BPEBPA, BMIM, BMIE and their blends [heating rate (β) = 20°C min^{-1}]

Samples	T_m-1 (°C)	T_m-2 (°C)	ΔH_f-1 (Jg^{-1})	ΔH_f-2 (Jg^{-1})	T_s (°C)	T_{max} (°C)	T_e (°C)	$T_e - T_s$ (°C)	ΔH_c (Jg^{-1})	ΔS_f (Jg^{-1}deg^{-1})
BMIM	–	159	–	38	184	243	328	144	202	0.2389
MBPEBPA	78	–	–	56	220	271	317	97	332	0.7179
BPEBPA	82	–	–	117	218	292	340	122	964	1.4268
EBPEBPA	80	108	324	245	219	275	324	122	752	–
BMIE	–	164	–	18	189	275	335	146	232	0.1097

Inspection of the ΔH_f values (Table 2.1) indicates a significant decrease in the ΔH_f values when one compares the values of the BPEBPA with that of the blends of BPEBPA with BMIM. This provides a good evidence for the interactions set in between BPEBPA and BMIM in the molten state. The cure window ($T_e - T_s$) of MBPEBPA is less than BPEBPA but there is no characteristic change in EBPEBPA (Table 2.1). It reveals that the co-cure reaction between BPEBPA and BMIM takes place more easily than the self-cure reaction of BPEBPA and BMIM.

There was no characteristic change in the onset temperature of curing for BPEBPA when blended with BMIM and BMIE, but lowers the T_{max} and T_e temperatures by around 20°C. From this observation in these systems, one can reasonably state that the curing mechanism of BPEBPA is affected in a uniform fashion when blended with BMIM and BMIE. The observed ΔH_c values of MBPEBPA and EBPEBPA are sufficiently less than the ΔH_c value for BPEBPA. Within the materials investigated, the maximum reduction in the parameter ΔH_c is noted in MBPEBPA and the value amounts to over 632 Jg^{-1}. Thus, blending of BMIM and/or BMIE in BPEBPA will help in reducing both the processing temperature and the amount of heat energy released during the thermal cure of pure bispropargyl ether monomer. Both these factors will help the processor to obtain void-free composites.

Melting is an equilibrium phenomenon and hence the change in free energy associated with melting is zero and is expressed as: $\Delta G_f = \Delta H_f - T_m \times \Delta S_f$ where ΔG_f is the change in free energy associated with melting, ΔH_f is the change in enthalpy associated with melting, T_m is the melting temperature and ΔS_f is the change in entropy associated with the melting phenomenon. The change in the entropy values (ΔS_f) calculated from the ΔH_f and T_m obtained from the DSC studies of the materials is presented in Table 2.1. Comparison of the ΔS_f values for BPEBPA, BMIM, and MBPEBPA, clearly indicate that BPEBPA shows much disorderliness compared to BMIM. Blending these two materials leads to a value in between these two limits. This may be one of the possible reasons for the ΔH_c value noted for MBPEBPA.

2.3.3 TG and DTG Studies

The TG and DTG curves for the polymers PBPEBPA, PBMIM, PBMIE, PMBPEBPA, and PEBPEBPA recorded at a heating rate of 20°C min^{-1} in nitrogen atmosphere are shown in Figures 2.12 and 2.13 respectively. For clarity both the TG and DTG curves are shifted uniformly in the Y axis. The

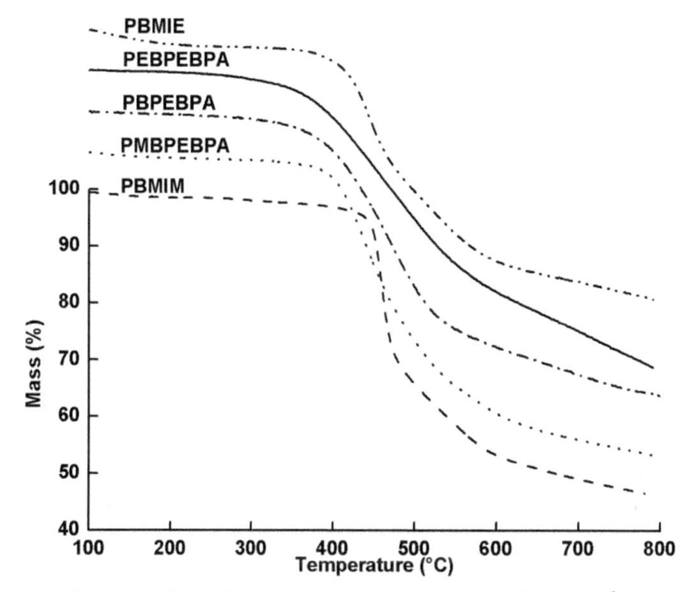

Figure 2.12 TG curves of the polymers (β = 20°C min^{-1}).

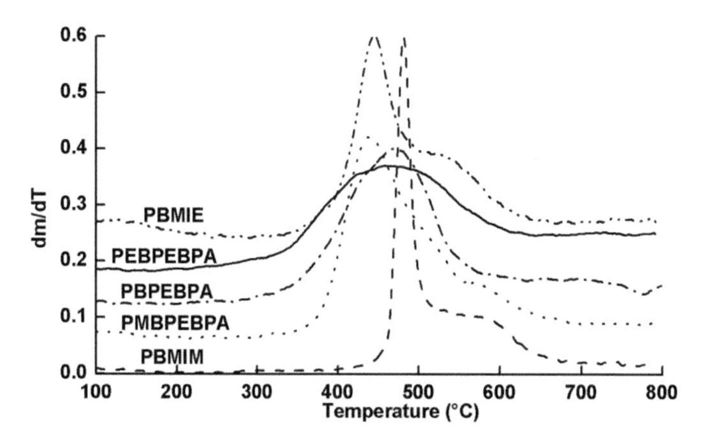

Figure 2.13 DTG curves of the polymers (β = 20°C min^{-1}).

thermal degradation onset (T_S), maximum (T_{max}), endset (T_E) temperatures, and the char residue obtained at 700°C for all the samples are tabulated in Table 2.2. The detailed observations made from the thermogravimetric data are discussed below.

The thermal degradation of the thermoset PBPEBPA starts at 260°C, attains maximum at 472°C and ends at 630°C. The char residue for

Table 2.2 TG studies: The degradation parameters for the polymers (β = 20°C min^{-1})

Samples	T_S (°C)	T_{max} (°C)	T_E (°C)	$T_E - T_S$ (°C)	Char residue at 700°C (%)
PBMIM	430	470	532	102	49
PMBPEBPA	300	440	725	425	48
PBPEBPA	260	472	630	370	52
PEBPEBPA	212	453	687	475	54
PBMIE	365	446	617	252	54

PBPEBPA noted at 700°C is 52%. This material shows nearly a smooth degradation behaviour and the degradation is slower, covering a wide temperature region. The polymer PBMIM shows the onset degradation temperature at 430°C and degradation reaches maximum at 470°C and the process is over at 532°C. The char residue for PBMIM noted at 700°C is 49%. From the DTG curve for PBMIM it is easy to state that the polymer degrades very fast at the initial degradation stage followed by a much slower degradation and the thermal degradation is bimodal. The initial degradation of thermally cured BMIE starts at 365°C, attains maximum at 446°C and ends at 617°C. The char residue for PBMIE noted at 700°C is 54%. This material also shows a bimodal thermal degradation.

Thermally cured MBPEBPA shows the onset degradation temperature at 300°C and the degradation was found to have maxima at 440°C and 572°C, respectively. The thermal degradation was nearly over by 725°C and the char residue at 700°C is 48%. The thermally cured EBPEBPA shows the onset degradation temperature at 212°C, attains maximum at 453 and ends at 687°C. The char residue for the PEBPEBPA noted at 700°C is 54%.

From the DTG curves (Figure 2.13) of PMBPEBPA and PEBPEBPA it is obvious that these materials show thermal degradation very similar to the thermal degradation of PBPEBPA. The degradation is a composite of several degradations overlapping over a wide temperature region. If the initial degradation temperature is taken as the main criterion for thermal stability, and from Table 2.2, one can easily identify that the thermally cured MBPEBPA showed higher onset degradation temperature than thermally cured pure BPEBPA. The above result shows BMIM-blended BPEBPA showed better thermal stability than thermally cured pure BPEBPA. But the situation is totally different in the case of EBPEBPA. Hence the structure of the BMI plays a definite role in determining the thermal stability of the blends.

It is well known that BMI polymerizes through the maleimide double bonds but the polymerization of propargyl ethers is highly complex due to the formation of various isomers resulting from Claisen rearrangement. Further it has been established that the propargyl group undergoes cyclization reaction leading to the formation of chromenes during the thermal polymerization of propargyl ethers. It is further complicated by the cyclic trimerization of the acetylene unit present in the propargyl moiety. Therefore in bispropargyl ethers depending on the structure of the monomer, these reactions are taking place at varying degrees and are dictated by the curing temperature. Hence addition of BMI to bispropargyl ethers definitely affects the curing behavior of bispropargyl ethers to a distinct level and the level being controlled by the structure of the BMI.

Among the cured materials (PBPEBPA, PBMIM, PBMIE, PMBPEBPA, and PEBPEBPA) investigated, both PMBPEBPA and PEBPEBPA show higher end set degradation temperature. The thermally cured BMIE blended BPEBPA exhibit higher char residue (~54%) at 700°C. From the TG and DTG results it is explicit that the thermosetting polymer resulting from the thermal polymerization of the blend BPEBPA-BMIM shows better thermal stability than the cured pure bispropargyl ether. From this one can conclude that the polymerization of bispropargyl ether-BMI blends leads to structurally different network structures.

2.4 Conclusions

The two thermosetting monomers BPEBPA and BMIE were prepared. The synthesized BPEBPA was separately blended with BMIM and BMIE in the 0.5:0.5 mole ratios. The structural characterization of these materials was carried out using FTIR spectrometer. Curing characteristics of these materials are investigated using DSC. Differential scanning calorimetric studies show that the difference in melting characteristic is caused by the incorporation of BMI in BPEBPA. Addition of BMIs to bispropargyl ether reduces the energy released during the thermal curing. The materials are thermally polymerized and the structural characterisation and the thermal properties of these crosslinked materials are investigated using FTIR and TGA. The DTG curves of both PBMIM and PBMIE show a bimodal degradation pattern. All the other materials show broad overlapping thermal degradations. The polymers PBMIE and PEBPEBPA show higher char residue values among the cured materials. Thus incorporation of BMI in BPEBPA alters the degradation as evidenced by the thermogravimetric studies.

Acknowledgements

The authors would like to acknowledge the Management, the Principal, and the Dean of Kamaraj College of Engineering and Technology, Virudhunagar, India for providing all the facilities to do this work.

References

[1] Frazer, A. H. (1968). *High Temperature Resistant Polymers*. Hoboken, NJ: John Wiley & Sons.
[2] Hergenrother, P. M. (2003). The use, design, synthesis and properties of high performance/high temperature polymers: an overview. *High. Perform. Polym.* 15, 3–45.
[3] Lin, S. C., and Pearce, E. M. (1994). *High Performance Thermosets: Chemistry, Properties, Applications*. New York, NY: Carl Hanser Verlag.
[4] Hergenrother, P. M. (1980). Acetylene-containing precursor polymers, *J. Macromol. Sci. Polym. Rev.* 19, 1–34.
[5] Fink, J. K. (2005). *Reactive Polymers Fundamentals and Applications – A Concise Guide to Industrial Polymers*. New York, NY: William Andrew Publishing.
[6] Varma, I. K., and Tiwari, R. (1987). Thermal characterization of bismaleimide blends. *J. Therm. Anal. Calorim.* 32, 1023–1037.
[7] Kovar, R. F., Ehlers, G. F. L., and Arnold, F. E. (1977). Thermosetting acetylene-terminated polyphenylquinoxalines. *J. Polym. Sci. Polym. Chem. Ed.* 15, 1081–1095.
[8] Hay, A. (1960). Notes-preparation of m-and p-diethynylbenzene. *J. Org. Chem.* 25, 637–638.
[9] Stephens, R. D. V., and Castro, C. E. (1963). Castro-Stephens reaction. *J. Org. Chem.* 28, 3313–3315.
[10] Dirlikov, S. K., and Feng, Y. (1988). Propargyl terminated resins (PTR): preparation and thermostability. *Polym. Mater.* 59, 990–993.
[11] Inbasekaran, M. N., and Dirlikov, S. K. (1989). Process for Making Propargyl Ethers of Bisphenols. U. S. Patent 4,885,403.
[12] Kolb, G. C., Scheck, D. M., Dirlikov, S. K., Inbasekaran, M. and Godschalx, J. P. M. (1992). Polymer resulting from the cure of a preformed chromene-containing mixture. U. S. Patent 5,155,196.
[13] Hay, A. S. (1971). Photosensitive acetylenic polymers. U. S. Patent 3,594,175.

[14] Dirlikov, S. K., and Feng, Y. (1988). Propargyl terminated resins (PTR): preparation and thermostability. *Polym. Mater. Sci. Eng.* 59, 990–993.

[15] Lucotte, G., and Delfort, B. (1991). 1,2-bis(3-propargyloxyphenoxy) ethane synthesis and polymerization. *Polym. Bull.*26, 1–6.

[16] Picklesimer, L. G. (1980). Synthesis of acetylene-terminated compounds. U. S. Patent 4,226,800.

[17] Gouri, C., Nair, C. P. R., Ramasamy, R., and Ninan, K. N. (2002). Thermal decomposition characteristics of alder-ene adduct of diallyl bisphenol-A novolac molar mass and bismaleimide structure, *Eur. Polym. J.* 38, 503–510.

[18] Ambika devi, K., John, B., Nair, C. P. R., and Ninan, K. N. (2007). Syntactic foam composites of epoxy-allyl phenol-bismaleimide ternary blend-processing and properties. *J. Appl. Polym. Sci.* 105, 3715–3722.

[19] Abraham, T. (1988). New bismaleimide resin systems: Decreased moisture absorption and increased impact strength, *J. Polym. Sci. C Polym. Lett.* 26, 521–528.

[20] Jones, R. J., Cassey, H. N., and Green, H. E. (1976). Polyimide sealant compositions. U. S. Patent 3,951,902.

[21] Liang, G. Z., and Gu, A. J. (1997). New bismaleimide resin with improved tack and drape properties for advanced composites. *J. Appl. Polym. Sci.* 64, 273–279.

[22] Harfenist, M., and Thom, E. (1972). The Infuence on the rate of thermal rearrangement of aryl propargyl ethers to the chromenes. The gem-dimethyl effect. *J. Org. Chem.* 37, 841–848.

[23] Balasubramanian, K. K., and Venugopalan, B. (1973). Studies in claisen rearrangements claisen rearrangement of bispropargyl ethers. *Tetrahedron Lett.* 14, 2707–2710.

[24] Pomeranz, K. U., Hansen, H. J., and Schmid, H. (1973). Die durch silberionen katalysierte umlagerung von propargyl-phenylether. *Helv. Chim. Acta* 56, 2981–3004.

[25] Anderson, W., and Voie, E. L. (1973). Thermal cyclization of substituted aryl propargyl ethers. The scope and regioselectivity of the reaction in the synthesis of substituted 3-chromenes. *J. Org. Chem.* 38, 3832–3835.

[26] Balasubramanian, K. K., and Venugopalan, B. Studies in claisen rearrangements claisen rearrangement of bispropargyl ethers. *Tetrahedron Lett.* 14, 2707–2710.

[27] Greenwood, T. D., Armistead, D. M., and Wolfe, J. F. (1982). N-propargyl-substituted aromatic polyamides: preparation and thermal crosslinking. *Polymer* 23, 621–625.

[28] Usha, R., and Balasubramanian, K. K. (1983). Claisen rearrangement of aryl propargyl ethers in poly (ethylene glycol)- a remarkable substituents and solvent effect. *Tetrahedron Lett.* 24, 5023–5024.
[29] Loustalot, M. F. G., and Sanglar, C. (1997). Prepolymers with propargylic terminal residues-I. Simulation of reaction mechanisms and kinetics of monofunctional models. *Eur. Polym. J.* 33, 1125–1134.
[30] Bindu, R. L., Nair, C. P. R., Krishnan, K., and Ninan, K. N. (1999). Bis propargyl ether resins: synthesis and structure-thermal property correlations. *Eur. Polym. J.* 35, 235–246.
[31] Reghunadhan Nair, C. P., Bindu, R. L., and Ninan, K. N. (2001). Thermal characteristics of addition-cure phenolic resins, *Polym. Degrad. Stab.* 73, 251–257.
[32] Badarau, C., and Wang, Z. Y. (2004). Synthesis and optical properties of thermally and photochemically cross-linkable diacetylene-containing polymers. *Macromolecules* 37, 147–153.
[33] Liu, F., Li, W., Wei, L., and Zhao, T. (2006). Bismaleimide modified bispropargyl ether bisphenol-A resin: Synthesis, cure and thermal properties. *J. Appl. Polym. Sci.* 102, 3610–3615.
[34] Tsvetanka, S. F., Josiah, T. R., and David, A. B. (2007). Dipropargyl ethers of high aspect ratio bisphenols. *Polym. Prepr.* 48, 388–389.
[35] Luo, Z., Wei, L., Liu, F., and Zhao, T. (2007). Study on thermal cure and heat-resistant properties of N-(3-acetylenephenyl)maleimide monomer. *Eur. Polym. J.* 43, 3461–3470.
[36] Rong, Z., Huang, F., Shen. X., and Huang, F. (2008). Preparation and properties of dipropargyl ether of bisphenol A-modified bismaleimide resins and composites. *Polym. Composite* 29, 483–488.
[37] Vinayagamoorthi, S., Vijayakumar, C. T., Alam, S., and Nanjundan, S. (2009). Structural aspects of high temperature thermosets Bismaleimide/propargyl terminated resin system-polymerization and degradation studies. *Eur. Polym. J.* 45, 1217–1231.
[38] Liu, F., Liu, J., and Zhao, T. (2010). Synthesis of a novel series of propargyloxyphenyl maleimides and their characterization as thermal-resistance resins. *J. Appl. Polym. Sci.* 115, 3103–3109.
[39] Chen, J., Lin, N., Huang, J., and Dufresne, A. (2015). Highly alkynyl-functionalization of cellulose nanocrystals and advanced nanocomposites thereof via click chemistry. *Polym. Chem.* 6, 4385–4395.

3

Synthesis, Characterization, Magnetic, Thermal and Electrochemical Studies of Oxovanadium(IV) Complex of 2-thiophenecarba Benzhydrazone

Jyothy G. Vijayan

Department of Chemistry, Christ University, Bengaluru, India

Abstract

The hydrazone ligand obtained from 2-thiophene carboxaldehyde and benzhydrazide react with an equimolar mixture of vanadyl acetyl acetonate in methanol to yield oxovanadium(IV) complex of 2-thiophenecarba benzhydrazone. The prepared compound shows effective solubility in organic solvents like acetonitrile, DMF and DMSO. Molar conductivity data of oxovanadium(IV) complex of 2-thiophenecarba benzhydrazone revealed its non-electrolytic behavior in DMF and DMSO. EPR spectra of 2-thiophenecarba benzhydrazonato oxovanadium(IV) was recorded in DMF at LNT and g and A values were calculated. The complex was proposed to be square pyramidal in geometry. Cyclic voltammograms of the complex in DMF were studied by changing the scan rates 50, 100, and 200 mV/s. ΔE values of the complex showed the reversible criterion and ipc/ipa values which were close to 1 indicating the redox couple as reversible. Thermograms of the complex were recorded to find the weight loss at different temperature ranges. Matrix-assisted laser desorption ionization time-of-flight (MALDI-TOF) mass spectra showed mass number of the molecular ions.

3.1 Introduction

Acid hydrazone complexes of oxovanadium, dioxovanadium, oxomolybdenum, ruthenium, and palladium are used as potential catalysts in organic reactions. Katsuki et al. have reviewed the unique asymmetric catalysis of metal complexes of salen and related ligands [1]. Researchers have reviewed the catalytic activities of transition metal complexes—both simple and polymer-anchored [2].

Recent studies on the non-linear optical properties exhibited by metal complexes have received considerable attention since they serve as opto-electronic devices. Many scholars have proposed the relevance of hydrazones for non-linear optics by the measurements of the molecular hyper-polarizabilities of some phenyl hydrazone derivatives [3]. Hydrazones on coordination to metal center through the iminol form improves the conjugation and enhances (boosts) the non-linearity [4–6].

There have been extensive studies on the physicochemical properties and structures of metal complexes containing hydrazone derivatives. In view of their chelating capability and pharmacological applications hydrazones have attracted considerable attention [7–11]. The design, synthesis and characterization of hydrazones and their metal complexes have come from their ease of syntheses, easily tunable steric and electronic properties, and good stability in common organic solvents [12–15].

3.2 Experimental

3.2.1 Physical Measurements

Microanalysis of carbon, hydrogen, nitrogen, and sulfur in acid hydrazone and in its metal complex was carried out on an elemental model vario EL III CHNS analyzer. Molar conductivity of the complex in DMF and DMSO (10^{-3}M) were measured at room temperature using a direct reading digital conductivity meter. The magnetic susceptibility measurement on the complex was carried out at room temperature, using VSM method. Infrared spectra of the ligand and the complex were recorded on a Thermo Nicolet AVATAR 370 DTGS model FT-IR spectrophotometer as KBr pellets. The photoluminescence emission spectra were recorded at room temperature on JY Flourolog-FL3-II. ^{1}H NMR, ^{13}C NMR, ^{1}H-^{1}H COSY, and ^{1}H-^{13}C HSQC of the ligand was recorded using a Bruker AMX 400 FT-NMR spectrometer with DMSO-d_6 as solvent and TMS as the internal standard.

The EPR spectrum of the complex at liquid nitrogen temperature (LNT) in DMF was recorded on the X-band JES-FA 200 ESR spectrometer. Cyclic voltammetry of the complex was carried out using a suitable reference, working and counter electrodes in the presence of a supporting electrolyte. The three-electrode system consisted of a glassy carbon (working), platinum wire (counter), and Ag/AgCl (reference) electrodes. Measurements were made in DMF (10^{-3}M) containing 0.1 M tetrabutyl ammonium fluoroborate as supporting electrolytes. Cyclic voltammograms were recorded on an electrochemical analyzer. A TG-DTG analysis of the complex was carried out in the heating range of 15–900°C by using a Perkin Elmer Diamond TG/DTG Analyzer. Mass spectra of the complex was obtained from Ultra flextreme Maldi TOF spectrometer (Bruker Daltonics).

3.2.2 Materials

2-Thiophene carboxaldehyde, benzhydrazide, methanol, glacial acetic acid, and vanadyl acetylacetonate were procured from Sigma Aldrich and were used without further purification.

3.2.3 Synthesis of Ligand

3.2.3.1 Synthesis of 2-thiophenecarba benzhydrazone

Benzhydrazide (0.14 g, 1 mmol) was dissolved in methanol (30 ml) and to this was added 2-thiophene carbaldehyde (0.11 ml, 1 mmol) followed by two drops of glacial acetic acid. The mixture was refluxed for 6 h and kept aside for cooling when colorless 2-thiophenecarba benzhydrazone crystals separated out. The mixture was filtered and the crystals were washed with methanol. The crystals were dried over P_4O_{10} under vacuo. Yield, 73%.

3.2.4 Synthesis of Complex

3.2.4.1 Preparation of 2-thiophenecarba benzhydrazonato oxovanadium(IV)

To a solution of 2-thiophenecarba benzhydrazone (0.23 g, 1 mmol L) in methanol (20 ml), vanadyl acetyl acetonate (0.27 g,1 mmol) dissolved in methanol (20 ml) was added. The reaction mixture was stirred at room temperature with a magnetic stirrer for 5 h. The resulting solution was allowed to stand at room temperature for slow evaporation, when a greenish brown

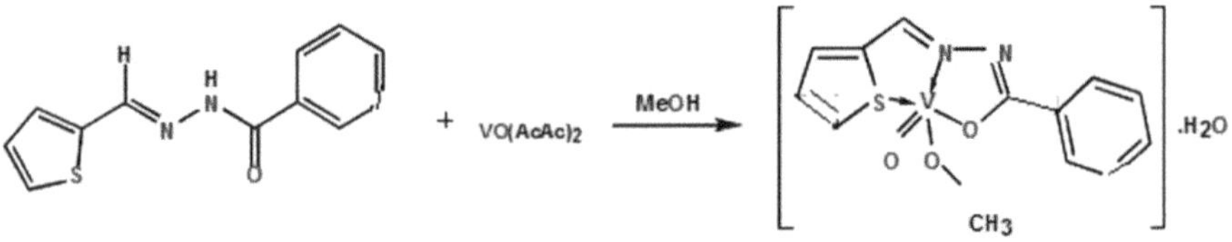

Scheme 3.1 Preparation of 2-thiophenecarba benzhydrazonato oxovanadium(IV).

colored precipitate separated out. It was filtered, washed with methanol, and dried under vacuo. Yield, 71% (Scheme 3.1).

3.3 Results and Discussion

3.3.1 Characterization of the Ligand (2-Thiophenecarba Benzhydrazone)

The colorless crystalline product was isolated in excellent yield (>80%). The ligand was characterized by ^{1}H,^{13}C, IR, UV-Visible, and elemental analysis. Elemental analyses of the ligand suggest the formulae proposed (Table 3.1). The strong bands at 1642 and 3253 cm^{-1} in the spectrum of 2-thiophenecarba benzhydrazone are due to ν(C=O) and ν(N-H) respectively. The C=N stretching band in 2-thiophenecarba nicotinic hydrazone is at 1558 $cm^{-1}$and in 2-thiophenecarba benzhydrazone it is at 1549 cm^{-1}. This indicates the condensation between carbonyl group and hydrazide. The electronic spectra of substituted hydrazone was recorded in acetonitrile solution (10^{-4}M). The UV-Vis spectra of the ligand displayed bands around 400 and 290 nm which are assigned to n-π* and π-π* transitions. The proton NMR spectrum of 2-thiophenecarba benzhydrazone recorded in DMSO-d_6 (Figure 3.1) exhibited a singlet at δ, 11.8 ppm and this is ascribed to NH proton at position 8. A signal at δ, 8.8 ppm is attributed to the proton on azomethine carbon at position 6. The resonances due to protons on the thiophene and benzene rings are in the range 7.1 to 8.0 ppm. The ortho protons

Table 3.1 Analytical data of acid hydrazone

Compound	Color	Melting Point (°C)	Analytical Data Found (Calc.)
2-Thiophenecarba benzhydrazone ($C_{12}H_{10}N_2OS$)	Colorless	204–206	C 62.90 (62.59) H 4.19 (4.38) N 12.17 (12.16) S 14.05 (13.92)

Figure 3.1 ^{1}H NMR spectrum of 2-thiophenecarba benzhydrazone.

Table 3.2 Elemental analytical data of the complex

Compound	Color	Melting Point (°C)	Analytical Data Found (Calc.)
2-Thiophenecarba Benzhydrazonato oxovanadium(IV)	Greenish Brown	>300	C 45.3 (45.04) H 3.9 (4.33) N 8.4 (8.11) S 9.7 (9.24)
[VOL(OMe)]H_2O			**V 14.49 (14.73)**

of the benzene ring at 11, 15 showed a doublet at 7.9 ppm. The meta and para protons exhibited triplets at δ, 7.6, and 7.7 ppm respectively owing to vicinal coupling.

The ^{13}C NMR spectrum of 2-thiophenecarba benzhydrazone was recorded in DMSO-d_6 (Figure 3.2). The carbonyl carbon at position 9 is the most deshielded because of the π bond and the electronegative oxygen associated with it. It resonated at δ, 162.9 ppm. Carbon 1 in the thiophene ring has displayed a signal at δ, 142.9 ppm implying its deshielded environment.

3.3.2 Characterization of the Complex

Elemental analyses of the ligand suggest the formulae proposed (Table 3.2). The complex is soluble in acetonitrile, DMF, and DMSO but insoluble in other organic solvents. The solutions are non-conducting and the complex

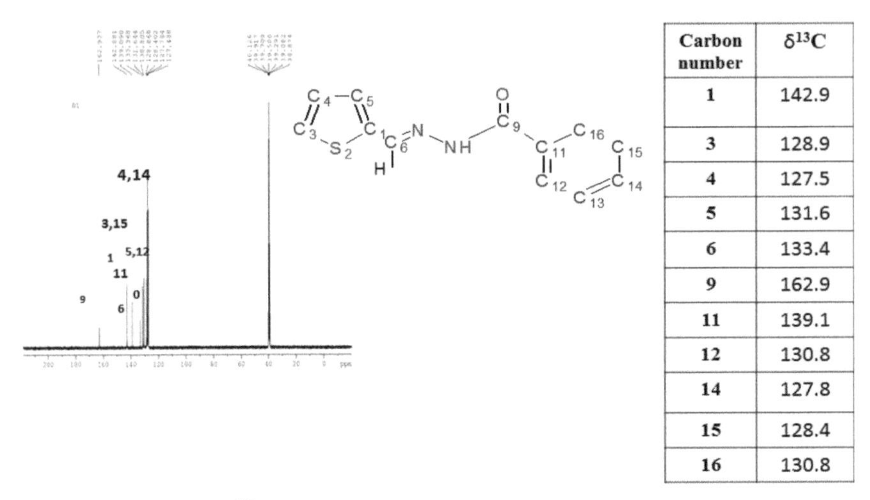

Carbon number	δ¹³C
1	142.9
3	128.9
4	127.5
5	131.6
6	133.4
9	162.9
11	139.1
12	130.8
14	127.8
15	128.4
16	130.8

Figure 3.2 ^{13}C NMR spectrum of 2-thiophenecarba benzhydrazone.

is non-electrolytic in nature. The magnetic moments of the complex were determined by VSM at room temperature. The value 1.83 is close to the spin-only value of 1.73 BM, which indicates that the complex is paramagnetic in nature, having one unpaired electron [16–19]. The IR spectrum of 2-thiophenecarba benzhydrazonato oxovanadium(IV) does not show ν(C=O) and ν(N-H) at 1642 and 3253 cm^{-1} respectively. This shows the coordination of the ligand to the metal ion in the enolate form [20, 21]. A new band has appeared at 1358 cm^{-1} and this is assigned to ν(C-O). An intense band observed at 975 cm^{-1} is due to V=O stretching. 2-Thiophenecarba benzhydrazone showed a band at 1549 cm^{-1}due to ν(C=N) of azomethine group and this has shifted to 1518 cm^{-1} in the spectrum of complex revealing the coordination of azomethine nitrogen to the central metal atom. The band at 1597 cm^{-1} is due to new ν(C=N). The complex has exhibited a broad band at 3062 cm^{-1} and this is due to the presence of lattice water. In 2-thiophenecarba benzhydrazone, the thiophene ring shows an absorption band at 852 cm^{-1} and this has shifted to 824 cm^{-1} in complex, due to the participation of thiophene sulphur in complexation.

The electronic absorption spectrum of the complex is recorded in acetonitrile solution (10^{-4} M). The spectra exhibited bands around 280 and 400 nm and these are assigned to n-π^* and π-π^* transitions respectively for the complexed hydrazone derivatives. The absorption bands are characteristic of the square pyramidal environment around vanadium(IV). The solid-state diffuse reflectance spectrum of the complex displayed bands around 800 nm corresponding to d→d transitions [22–25].

The EPR spectrum of [VOL(OMe)]H_2O was recorded in DMF at LNT and the spectra are given in Figure 3.3. In the complexes, the metal ion has oxidation state IV and hence has d^1 electron configuration.

Vanadium-51 has nuclear spin (I) 7/2. The unpaired electron spin can couple with the nuclear spin and this leads to splitting of each electron spin energy levels (Ms = +1/2, −1/2) in to (2nI+1 = 2×1×7/2+1 = 8) eight electron nuclear spin interaction energy levels or hyperfine splitting energy levels corresponding to M_I ranging from +7/2 to −7/2

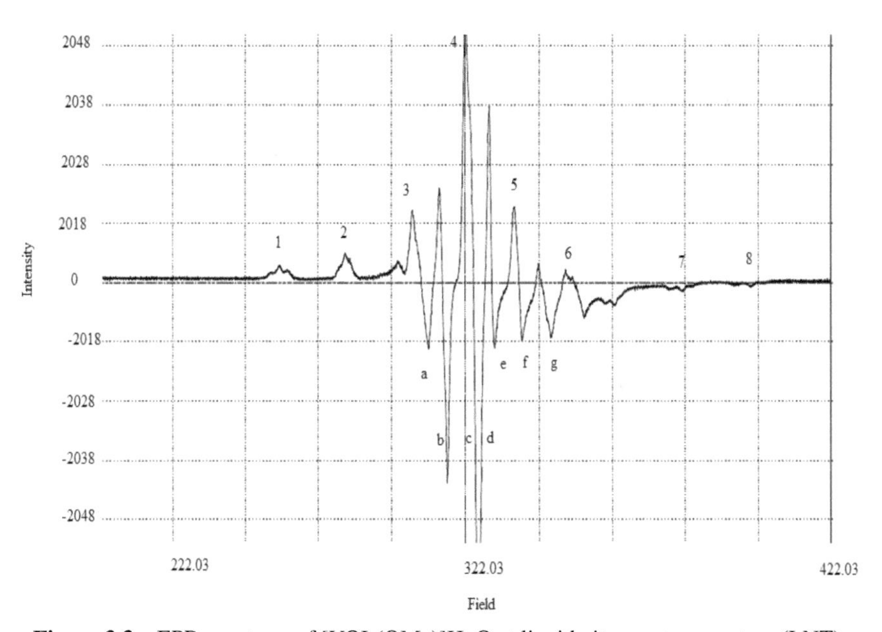

Figure 3.3 EPR spectrum of [VOL(OMe)]H_2O at liquid nitrogen temperature (LNT).

Table 3.3 EPR spectral data of oxovanadium complex

Parameters	[VOL(OMe)]H_2O
g_{II}	1.94
$g_{\perp}$	2.001
g_{av}	1.981
A_{II}	190×10^{-4} cm^{-1}
$A_{\perp}$	70×10^{-4} cm^{-1}
A_{av}	110×10^{-4} cm^{-1}

(ie,7/2,5/2,3/2,1/2,−1/2,−3/2,−5/2,−7/2). There will be eight for Ms = 1/2 (higher) and eight for Ms = −1/2 (lower) and totally sixteen energy levels. Based on the selection rules, Ms = ± 1 and $\Delta M_I = 0$, there will be 8 allowed transition corresponding to

$$-7/2 \longrightarrow -7/2, -5/2 \longrightarrow -5/2, -3/2, -1/2 \longrightarrow -1/2, 1/2 \longrightarrow 1/2,$$
$$3/2 \longrightarrow 3/2, 5/2 \longrightarrow 5/2, \text{ and } 7/2 \longrightarrow 7/2.$$

These will be obtained as first derivative signals, the adjacent ones separated by hyperfine splitting constant A, and center of the spectrum corresponding to g. The EPR spectra of the complex have shown two sets of signals suggesting that the complex has a square pyramidal geometry. As there are two sets of signals, one set (a to h) parallel to the applied magnetic field direction (g_{II}) and the other (1 to 8) perpendicular to the magnetic field direction ($g_{\perp}$), the latter set of signals being more intense than the former set of signals. In the same way, there are two hyperfine splitting constants A_{II} and $A_{\perp}$. The EPR spectral data of the complex is listed in Table 3.3. For square pyramidal complexes $g_{II}<g_{\perp}, A_{II}>A_{\perp}$. It is the characteristic of an axially compressed system with unpaired electron in dxy orbital. The values of A_{iso} and g_{iso} are measured by 1/3 (A_{II} + 2$A_{\perp}$) and (g_{II} + 2$g_{\perp}$).

Electrochemical studies of oxovanadium(IV) complex of 2-thiophene carba benzhydrazone was performed using DMF as the solvent, TBAP (tetra butyl ammonium perchlorate) as the supporting electrolyte at scan speeds of 50, 100 and 200 mV/s to understand the redox behavior. The electrode system consisted of glassy carbon as working electrode, platinum wire as counter or auxiliary electrode, and SCE as reference electrode. All the experiments were carried out at room temperature. The solutions were freshly prepared prior to use and were purged with nitrogen gas for 10 min to remove dissolved oxygen. The electrochemical nature of the complex was studied with three scan rates, i.e., 50, 100 and 200 mV/s. The redox nature of the complex was reversible and the cathodic peak current Ipc increased and the peaks shifted to more negative electrode potentials on increasing the scan rates from 50 to 200 mV/s (Figure 3.4).

The redox properties of oxovanadium(IV) complex of 2-thiophenecarba benzhydrazone was examined in a DMF solution (in nitrogen atmosphere) (Table 3.4). The cyclic voltammograms of the complex showed a reversible reduction-oxidation peak. The electrode process can be shown as

$$[V^{IV}OL(OMe)]H_2O \leftrightharpoons [V^{III}OL(OMe)]H_2O$$

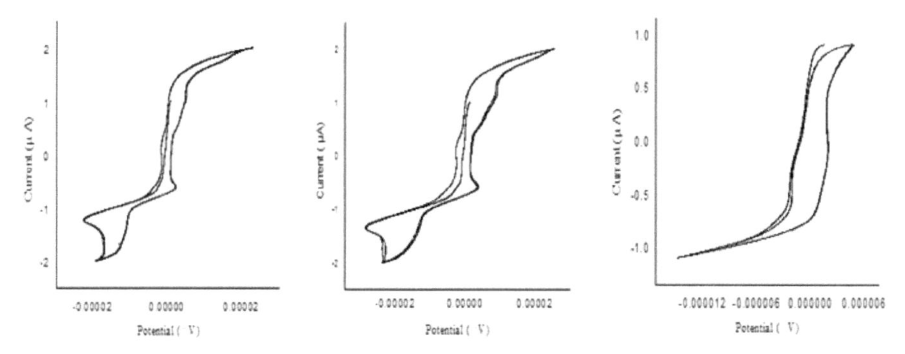

Figure 3.4 Cyclic voltammogram of [VOL(OMe)]H_2O (50, 100 and 200 mV/s).

Table 3.4 Cyclic voltammogram data of the complex (50 mV/s)

Compound	Epa (V)	Epc (V)	$E_{1/2}$ (V)	Ep (V)	ipc/ipa
[VOL(OMe)]H_2O	1.25	−0.65	0.3	−1.90	0.714

The obtained data from the electrochemical prospects of the metal complex is given in Table 3.5. The complex is reversible in nature. [VOL (OMe)]H_2O shows a reductive response at −0.65 V versus SCE assigned to VO(IV) to VO(III) and an oxidative response at 1.25 V versus SCE which is assigned to the VO(III) to VO(IV) change and the process is thus reversible.

The first decomposition step within the temperature range 115°C for [VOL (OMe)]H_2O may be attributed to the loss of hydrated water molecule. The second decomposition step with the range 267°C to 286°C for [VOL (OMe)]H_2O are reasonably accounted for by the partial removal of organic moiety from the complex. This corresponds with the decomposition of the ligand moiety. A horizontal zone beyond 500°C for the two complexes suggests the formation of ultimate pyrolysis product V_2O_5 (Figure 3.5).

Maldi spectrum of oxovanadium(IV) complex of 2-thiophenecarba benzhydrazone as α-cyano-4-hydroxy-cinnamic acid mix showed peaks at

Table 3.5 TG-DTG data of oxovanadium complex

Compound	First Weight Loss % Found (calc.)	Temp. °C	Fragment Lost	Nature of H_2O Lost	Second Weight Loss% Found (calc.)	Temp. °C	Fragment Lost
[VOL(OMe)] H_2O	5.16 (5.19)	115	H_2O	Lattice Water	66.73 (66.4)	286	$C_{12}H_{10}N_2OS$

m/z 346, 316 and 230. The molecular ion peak 346 was assigned to [VO $C_{12}H_{10}N_2OS$ (OCH_3)] H_2O, a peak at 316 corresponded to [VO $C_{12}H_{10}N_2OS$] H_2O and that at 298 was due to [VO $C_{12}H_{10}N_2OS$]. The peak at 231 corresponded to dissociated species $C_{12}H_{10}N_2OS$ (m/z 230) which was due to the degradation of the ligand (Figure 3.6).

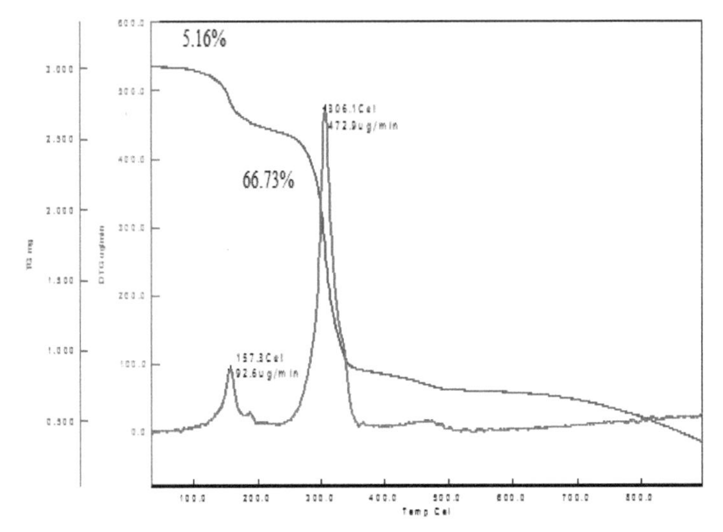

Figure 3.5 TG-DTG plot of [VOL(OMe)]H_2O.

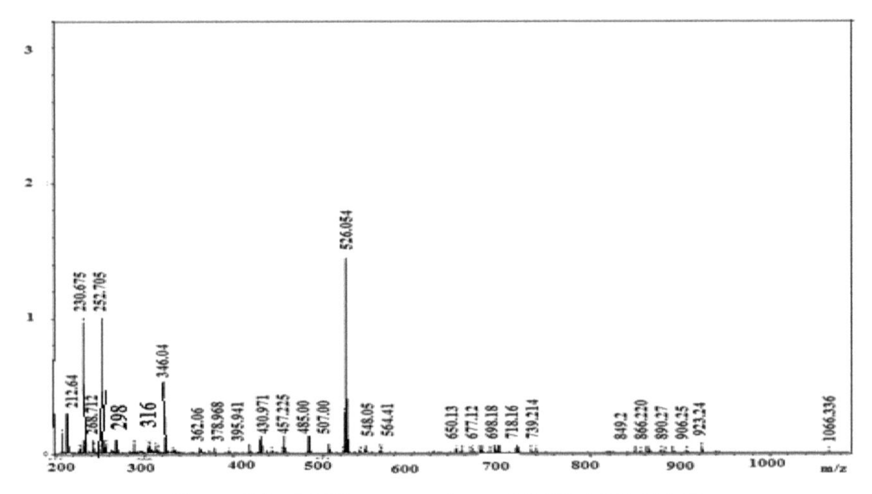

Figure 3.6 Maldi mass spectrum of [VOL (OMe)]H_2O.

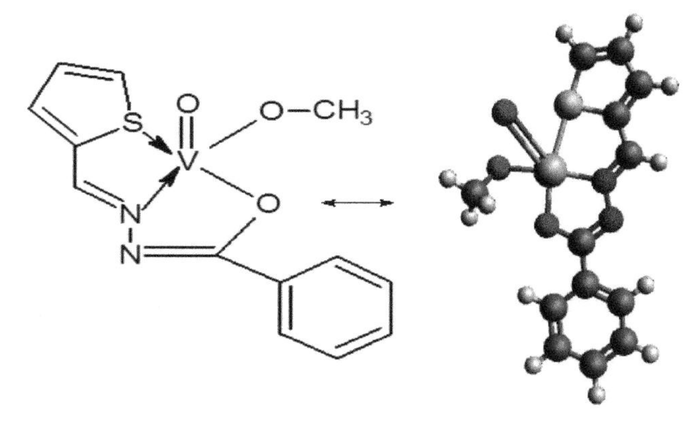

Figure 3.7 Proposed structure of [VOL (OMe)]H_2O (Water is not showed).

3.3.3 Proposed Structure of the Complex

The oxovanadium complex [VOL (OMe)]H_2O is colored and soluble in DMF, DMSO and acetonitrile. The proposed structure for the complex [VOL(OMe)]H_2O is square pyramidal (Figure 3.7).

3.4 Conclusion

The acid hydrazone was characterized by CHNS, IR, UV-Vis, 1D and 2D NMR. The oxovanadium(IV) complex was characterized by elemental analysis, IR, UV-Vis, molar conductance measurements, magnetic susceptibility measurements, atomic absorption spectra, diffuse reflectance spectra, EPR spectra, cyclic voltammetry, thermal studies and MALDI-TOF mass spectra.

Molar conductivity data of oxovanadium(IV) complex of 2-thiophene carba benzhydrazone revealed their non-electrolytic behavior in DMF and DMSO. Vanadium complex was found to be paramagnetic to the extent of one unpaired electron. IR spectral data showed the presence of lattice water and the coordination of hydrazone ligand to metal ion in enolic form. 1D and 2D NMR spectra displayed ^{1}H and ^{13}C resonances for the hydrazone derivatives. The complex was not used for NMR spectral studies, as they were paramagnetic. EPR spectra of 2-thiophenecarba benzhydrazonato oxovanadium(IV) was recorded in DMF at LNT, and g and A values were calculated. The complex was proposed to be square pyramidal in geometry. The g values calculated for the complex showed that the unpaired electron was in dxy orbital. Diffuse reflectance, UV-Vis spectra of oxovanadium complex studied with respect to n-π^*, π-π^* and d-d transitions.

Atomic absorption spectrum of the complex was carried out to determine the percentage of metal ion in the complexes. There was a close agreement between calculated and found values of oxovanadium complex. Cyclic voltammograms of the complex in DMF was studied by changing the scan rates to 50, 100, and 200 mV/s. ΔE values of the complex showed the reversible criterion and ipc/ipa values which were close to 1, indicating the redox couple as reversible. Thermograms of the complex were recorded to find the weight loss at different temperature ranges. Matrix-assisted laser desorption ionization time-of-flight (MALDI-TOF) mass spectrum showed mass number of the molecular ions. The molecular ion peaks were less intense suggesting that oxovanadium(IV) complex of 2-thiophenecarba benzhydrazone possesses lower stability in solution.

The prepared complex polymerized and studied for oxidation reactions. It was proved to be highly efficient. The main characteristic of oxovanadium(IV) complex is that it can be polymerized and used for the oxidation of olefins. Polymer anchored oxovanadium(IV) complexes were used for the catalytic oxidation of thioanisole, styrene, benzyl alcohol, cyclohexanol, ethyl benzene etc. It can act as a reusable catalyst in oxidation and bromination reactions.

Acknowledgements

I am grateful to Christ University, Bangalore for department research support. I acknowledge Sophisticated Analytical Instrumentation Facility, IISc Bangalore for NMR studies, magnetic studies and mass spectra, SAIF CUSAT for CHNS, IR and UV-Vis studies, IIT Indore for Cyclic voltammetry, IIT Bombay for EPR, Pondicherry University for DRS and fluorescence spectra, and Bangalore Test House for AAS.

References

[1] Katsuki, T. (2004). Unique asymmetric catalysis of cis-β metal complexes of salen and its related Schiff-base ligands. *Chem. Soc. Rev*. 33, 437–444.

[2] Gupta, K. C., Sutar, A. K., and Lin, C.C. (2009). Polymer-supported Schiff base complexes in oxidation reactions. *Coord. Chem*. 253, 1926–1946.

[3] Serbutoviez, C., Bosshard, C., Knopfle, G., and Wyss, P. (1995). Hydrazone derivatives, an efficient class of crystalline materials for nonlinear optics. *Chem. Matter*. 7, 1198–1206.
[4] Sanjay, E. R., Rao, A. K., Yadava, and Yadav. H. S. (2011). Synthesis and characterization of oxovanadium (IV) complexes with tetradentate schiff-base ligands having thenil as precursor molecule. *Curr. Res. Chem*. 3, 106–114.
[5] Collison, D., Gahan, B., and Garner, C. D. (1980). Electronic absorption spectra of some oxovanadium (IV) compounds. *J. Chem. Soc. Dalton Trans*. 1980, 667–674.
[6] Dickson, F. E., Kunesh, C. J., and McGinnis, E. L. (1972). Use of electron spin resonance to characterize the vanadium (IV)-sulfur species in petroleum. *Anal. Chem*. 44, 978–981.
[7] Clich, P. R., Daniher, A. T., and Challen, P. R. (1996). Vanadium (IV) complexes with mixed O,S Donor Ligands. Syntheses, Structures, and Properties of the Anions Tris(2-mercapto-4-methylphenolato)vanadate(IV) and Bis(2-mercaptophenolato)oxovana date(IV). *Inorg. Chem*. 35, 347–356.
[8] Davidson, A., Edelstein, N., and Holm, R. H. (1964). Differences between the four halide ligands, and discussion remarks on trigonal-bipyramidal complexes, on oxidation states, and on diagonal elements of one-electron energy. *J. Am. Chem. Soc*. 86, 2799–2813.
[9] Dianu, M. L., Kriza, A., and Musuc, A. M. (2013). Synthesis, spectral characterization, and thermal behavior of mononuclear Cu(II), Co(II), Ni(II), Mn(II), and Zn(II) complexes with 5-bromosalycilaldehyde isonicotinoylhydrazone. *J. Therm. Anal. Colorim*. 112, 585–593.
[10] Yaul, A. R., Rahangdale, M. K., and Aswar, A. S. (2012). Synthesis, characterization and thermal behaviour of VO (IV), MoO_2 (VI) and UO_2 (VI) complexes of hexadentate tetraanionic ligand. *J. Therm. Anal. Colorim*. 7, 109–124.
[11] Rao, T. R., Sahay, M., and Aggarwal, R. C. (1985). Synthesis and structural studies of some first row transition metal complexes of N-benzoylglycine hydrazide. *Synth. React. Inorg. Met-Org. Chem*. 15, 209–222.
[12] Bottari, B., Maccari, R., Montforte, F., Ottana, R., Rotondo, E., and Vigorita, M. (2000). Nickel (II) 2,6-diacetylpyridine bis (isonicotinoylhydrazonate) and bis (benzoylhydrazonate) complexes: structure and antimycobacterial evaluation. *Bioorg. Med. Chem. Lett*. 10, 657–660.
[13] Ainscough, E., Brodie, A., Ranford, J., and Waters, M. (1995). Copper (II) complexes of the antitumour-related ligand salicylaldehyde acetylhydrazone (H_2L) and the single-crystal X-ray structures of

[{Cu(HL)H_2O}$_2$] 2(NO_3) and [{Cu(HL) (pyridine) (NO_3)}$_2$]. *Inorg. Chim.* 83, 236–240.

[14] Ragnarsson, U., et al. (2001). Synthetic methodology for alkyl substituted hydrazines. *Chem. Soc. Rev.* 30, 205–213.

[15] Terzioglu, N., and Gursoy, A. (2003). Synthesis and anticancer evaluation of some new hydrazone derivatives of 2,6-dimethylimidazo [2,1-*b*] [1,3,4] thiadiazole-5-carbohydrazide. *Eur. J. Med. Chem.* 38, 781–786.

[16] Agarwal, R. K., Singh, L., Sharma, D. K., Singh, R. (2005). Synthesis, spectral and thermal investigations of some oxovanadium(IV) complexes of hydrazones of isonicotinic acid hydrazide. *Turkish Journal of Chemistry* 29, 309–316.

[17] Chaston, T. B., and Richardson, D. R. (2003). Iron chelators for the treatment of iron overload disease: relationship between structure, redox activity, and toxicity. *Am. J. Hematol.* 73, 200–210.

[18] Beenhardt, P. V., Chin, P., Sharpe, P. C., Wang, J. Y. C., and Richardson, D. R. (1985). TCM active ingredient oxoglaucine metal complexes: crystal structure, cytotoxicity, and interaction with DNA. *J. Bioinorg. Chem.* 51, 1998–2009.

[19] Xavier, A. V., (Ed.). (1985). *Frontiers in Bioinorganic Chemistry.* Weinheim: VCH.

[20] Fabian, M., and Palmer, G. (2001). *Biochemistry* 40, 1867–1874.

[21] Pozdnyakov, I., and Stafshede, P. W. (2001). *Biochemistry* 40:13728.

[22] Solomon, E. I., Szilagyi, R. K., George, S. D., and Basumallick, L. (2004). Electronic structures of metal sites in proteins and models: contributions to function in blue copper proteins. *Chem. Rev.* 104, 419–458.

[23] Mort, J., and Fister, P. (1982). Hole mobilities that decrease with increasing electric fields in a molecularly doped polymer. *G. Electro. Propert. Poly.* 153, 422–424.

[24] Dobler, M., and Weder, C. (1998). Synthesis and Properties of Poly (p-phenylene octylene). *Macro. Mol.* 31, 6184–6189.

[25] Rollas, S., and Kucukguzel, G. (2007). Biological activities of hydrazone derivatives. *Molecular* 12, 1910–1939.

4

Sorption and Desorption Analyses of Sorbents for Oil-spill Control

Neha Bhardwaj and Ashok N. Bhaskarwar

Department of Chemical Engineering, Indian Institute of Technology Delhi, New Delhi, India

Abstract

Oil-spills are known for the catastrophic effects and various countermeasures are employed to tackle them. One such method deals with the use of various kinds of sorbents, available in different structural forms. The sorption kinetics of sorbents is widely investigated to assess their performance as potential candidates for oil-spill control. However, few studies are devoted to desorption kinetics of sorbents. Understanding the behavioral characteristics of saturated sorbents such as oil retention and their tendency to release oil along with mechanical strength at saturation plays an important role in selection of the best sorbent for a specific hydrocarbon spill. This chapter considers the sorption-desorption behavior of various sorbents and different models dealing with it.

We also present here sorption and desorption analyses of different polyurethane sorbent samples. The analyses were conducted as per the ASTM procedures, and the factors crucial to the overall performance of the sorbent were investigated. The sorbents were tested for maximum sorption ability, various containment efficiencies, drainage and evaporation which led to release profiles, buoyancy at saturation, behavior under load, and possible release to other media.

To ascertain the applicability of these sorbents to a range of hydrocarbons, the sorbents were tested with four different test fluids. Most sorbents reached the maximum sorption capacity in a short duration. The desorption analysis of the samples revealed remarkably different release profiles for various test fluids. Various polyurethane samples also showed different release profiles for the same test fluid, due to differences in porosity. The saturated sorbents showed minimal test fluid release tendencies to the materials which came in immediate contact with them and had sufficient mechanical strength to be handled in the saturated state.

4.1 Introduction

Polymers are now pervasive in their applications, even in the fields which were once traditionally out of their scope. The molecular chains that Staudinger hypothesized have entered every aspect of modern life, and expanded their boundaries from commodity to engineering applications [1]. Recently, they have been designed for more sophisticated applications. The polymers cover a wide spectrum of applicability, ranging from clothes, coatings, paints, drug delivery, automobiles, polymer-based composites, self-healing materials, stimuli responsive materials, semiconductors, packaging, 3D printing, tissue engineering, and separation processes [2]. Advanced polymers, with improved material characteristics owing to their flexible construct, are being developed which can be tailor-made to suit a specific application. More than a hundred billion pounds of plastics were produced in the year 2000. This is an indicator of their usage, and of the dependence of modern life on synthetic macromolecules. The huge extents of production and consumption of polymers in various arenas have raised a series of serious concerns too. The humongous scale of production poses a looming threat of raw-material depletion (petrochemicals), which has propelled research on alternate raw materials. There are, however, several crucial concerns related to the toxicity of intermediates, products, and by-products of current syntheses and production techniques, which have resulted in a gradual shift toward the green-syntheses routes. The enormous scales of production and consumption inevitably lead to the waste generation and to the realization of inadequacies of current treatment techniques. One specific example reported that 86% of the plastic packaging are used only once [3]. This emphasizes the issues pertaining to reusability, recyclability, and the ultimate fate of a given polymer. Recent times have seen a gradual shift in terms of the origin of raw materials toward the agricultural sources, leading to cleaner synthesis techniques, and detailed toxicity analyses of the developed polymers. In this chapter, the pollution-prevention applications of polymers are highlighted with a specific focus on the role of polymers in marine oil-spill control.

Industrialization and the oil trade with their benefits have brought impairments to the environment. In the wake of anthropogenic environmental problems, the need of the hour is to bring in innovations in the field of materials research and their applications to curb the damages.

4.1.1 Pollution-prevention Application of Polymers

A few fascinating examples of the newly emerging classes of polymers oriented toward pollution-prevention applications have been mentioned in the following. The first example is of stimuli-responsive polymer-bound smart catalysts. These catalysts are particularly useful in waste minimization, catalyst recovery, and catalyst reuse [4]. The other examples include utilization of multifunctional polymeric smart coatings engineered for both detection and removal of hazardous nuclear contaminants [5]. Application of polymer nano-fibers in oil–water separation for cleaning up oil spills is discussed at length and suggested as an alternative to existing sorbents, in a very detailed review by Sarbatly et al. [6]. The important properties of these polymer nanofibers are specific surface morphology, porosity (voids among the fibers), diameter, and oleophilicity. One of the most widely known eco-friendly applications of polymers is their use as plastic mulching. Their long-term effects on the environment are however still under scrutiny [7]. In another fascinating investigation, functionalized polyurethane foam was used for oil-spill control. The functionalization was carried out using colloidal superparamagnetic iron oxide particles and particles of polytetrafluoroethylene(PTFE). This dual functionalization of foam resulted in improved sorption characteristics and rendered the foam magnetically active. A quicker rate of sorption was reported as a direct result of functionalization. In a short span of 6 s the foam was able to gain oil approximately up to its capacity, (13.35 times its weight with mineral oil as a test fluid) while being driven by an external magnet [8]. In another example of a similar kind, the polyurethane foam was synthesized using cellulose nano-whiskers, prepared by acidic hydrolysis of pine–leaf-derived cellulose and reacting the same with diphenylmethane diisocyanate. The foam was used in water effluent treatment using methylene blue dye as a model pollutant in its aqueous solutions. Complete dye removal was reported within 20 min, with a high maximum retention capacity of 554.8 mg/g and reusability through numerous cycles [9]. Further discussion is devoted to a brief introduction to the problem of oil spill, use of polymers in oil-spill control, and various factors affecting performance of the polymeric sorbents as a spill counter-measure.

4.1.2 Problem of Oil Spill

Oil spills are one of the leading environmental disasters. Over the past few decades, there have been many oil spill accidents during exploration, production, and transportation activities. Marine oil spills are always a potential threat as nearly half of the oil transportation is carried out through the sea routes [10]. Each year, an average of about 5 million tons of petroleum is transported across the seas around the world. These spills are notorious for the damage inflicted on the marine ecosystem, and they necessitate very expensive clean-up measures.

Oil spills may occur because of the following factors: structural failures, operational errors, weather-related events, earthquakes, human error, and negligence. Spills occur worldwide, but a vast majority of spills are relatively small in terms of volume. Table 4.1 shows the annual worldwide marine oil-spillage data. The table shows various sources of oil spills, and it also shows that the total volume of oil released into sea has reduced from the 70s to the 90s due to stringent policies and societal pressures. Around 72% of spills are 0.003–0.03 ton or lesser, and the total amount of these small spills makes only about 0.4% of the total spillage. The largest spills make 0.1% of incidents but involve nearly 60% of the total amount spilled [11]. Clean-up of oil spills is still a major challenge due to the limitations and high costs of the current clean-up practices. This environmental crisis has driven the research in multiple directions. In the section below, we have discussed a few major oil spills.

The biggest marine oil spill in the history of the mankind occurred in the Gulf of Mexico in the year 2010, where approximately 4.9 million barrels of oil got spilled. The spill killed 11 workers and injured 17 workers. The BP spill lasted for over 3 months. It soiled about 320 miles of beaches

Table 4.1 Annual worldwide marine oil-spillage data (estimated average annual tons spilled) ERC data [10, 12]

Source Type	1970s	1980s	1990s
Tank vessels	428,646	190,180	126,743
Non-tank vessels	2,735	23,811	10,248
Pipelines	59,087	36,744	85,664
Facilities	66,067	58,047	35,655
Offshore exploration/ production	69,111	68,099	38,351
Unknown	9,241	1,775	3,905
Total	634,887	378,656	300,546

and shorelines, and a year and a half later, a total of 491 miles (790 km) of shorelines was found to be affected [13]. This superseded the Bay of Campeche spill, where 3.3 million barrels of oil were spilled into the sea by the Mexican rig Ixtoc I in the year 1979 [14]. The event was hazardous to marine species and seagrasses. Out of 322 species, 53 species were threatened and 29 were nearly threatened, which included 16 species of sharks and eight corals. Another prominent example is the of sinking of the Torrey Canyon off the English Channel, which released 120,000 tons of crude oil and contaminated 100 miles of coast lines. Approximately 25,000 birds died as an aftermath of the disastrous oil-spill event. The following treatment methods were employed: natural weathering, use of dispersants, in-situ burning; straws and gorse were also used on sandy beaches to soak oil [15]. In the year 1978, Amoco Cadiz released 230,000 tons of light crude oil, off the coast of Brittany in France, and this contaminated 300 km long coast line. Expenditure involved in the entire operation was of the order of $282 million, out of which an amount $85 million was paid as fine. The spill killed over 3500 sea birds, and it severely damaged fisheries, oysters, and seaweed beds. Treatment techniques included microbial degradation, and manual removal of oil from contaminated beaches. Rubber powder and chalk sinking agents were also employed, but they did not prove very successful [16]. In the year 1989, Exxon Valdez released 10.9 million gallons of oil into Alaskan waters, which contaminated roughly 1900 km long coast line. The cost incurred was of the order of $7 billion for fines, penalties, and claims, of which over $2.1 billion was used for clean-up operations. The casualties include 250,000 sea birds, 2800 sea otters, 250 baldeagles, and 22 killer whales. The treatment methods included use of booms and dispersants. The dispersants were later on proved to be unsuccessful, due to emulsion formation, which had remarkably different properties compared to those of the initial oil. In-situ burning was successful, but could not continue because of the change in the state of oil as a result of the storm. Sorbents were used where other mechanical means were less practical. However, use of sorbents was labor-intensive and sometimes led to the generation of secondary waste in this particular case. Bioremediation agents were helpful in effectively cleaning over 70 miles of shoreline [17]. In another example, the collision between a tanker and another vessel led to release 3000 barrels of oil into Sundarbans Nature Reserve in Bangladesh, in the month of December 2014. The above-mentioned ruinous incidences are well documented and their impacts are carefully accessed, whereas in cases of war-related spills, the damages are not well evaluated and clean-up operations are scanty. One such prominent example is that of the Gulf War of 1990, where 650 oil wells in Kuwait were set ablaze, and roughly a million tons of oil was lost, and approximately 20,000 sea birds killed [16].

The ten biggest oil spills in the history of mankind have released approximately 40 million barrels of crude oil into the marine environment. The problem is relevant to India as cases of tar-ball deposition along the west coast of India have been steadily reported since the 1970s. Approximately 200 million tons of oil is being transported to the Indian coast every year. The Indian coast has experienced about 80 medium-sized oil spills, over the last 30 years [18]. Oil spills in the sea are more difficult and dangerous to tackle than the oil spills on land, as the spread in case of the former is greater due to the action of waves and wind. Figure 4.1 shows the burning rig which resulted in BP spill, along with the aerial view of the spill trajectory. The image also shows the severely affected fauna.

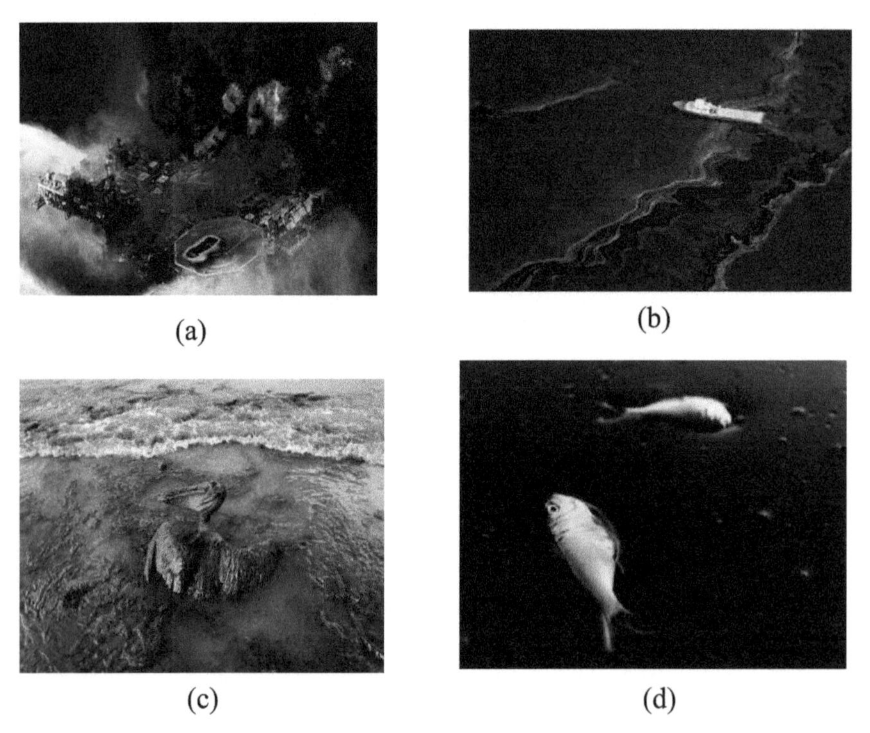

Figure 4.1 (a) Burning rig which resulted in BP spill, (b) aerial view of BP oil spill (c) seagull severely affected by the oil spill, and (d) dead fish floating on the surface of oil layer (AP Photo/Charlie Riedel) (2015, HNN) [19–20].

The broader research areas dealing with the marine oil spills are the detection of spill, modeling the fate and trajectory of the spill, investigating the prolonged behavior of oil in the marine environment, preventive measures of both passive and active types, and the overall impact on the marine ecosystem. Two types of clean-up measures currently in practice are active and passive techniques. Passive techniques and equipment are utilized to contain the oil spills on a water surface, for example booms and barriers. Active techniques remove oil from a water surface, for example the use of skimmers and sorbents. Usually a complete oil-spill operation implements hybrid (passive/active) methods [21]. Several counter-measures are commercially available, and broadly can be classified into the following categories: containment on water with the help of booms and barriers, use of skimmers, manual recovery, addition of spill-treating agents (biodegradation enhancers, detergents, surface-washing agents and emulsion breakers), in-situ burning (allowed in few countries), and sorbents. The efficiency of the cleaning operations depends heavily on the prevailing environmental conditions, temperature, wind speed, and tidal currents. These factors also affect the properties of the spilled oil. The chemical and physical changes undergone by the oil are termed as "weathering of the oil." This is an umbrella term encompassing various prominent mass transfer processes such as evaporation, emulsification, natural dispersion, dissolution, photo-oxidation, sedimentation and adhesion to materials, biodegradation, and formation of tar-balls.

The clean-up measures should be carefully examined and selected, as in a few cases alarming damage was reported due to a faulty clean-up method. The national oceanic and atmospheric administration investigations have revealed that most of the damage from the oil spill was caused by the cleaning operation following the disaster in the case of the Exxon Valdez spill. It was reported that pressure-washing was responsible for killing most of the marine life. On the stretches of beach that were un-cleaned, life seemed to recover after 18 months, whereas on the cleaned parts of the beach it did not recover for the next 3–4 years. This oil spill clean-up technique is still in practice because in the public opinion it is a way to save most animals. Hence the selection of clean-up operation is very crucial as further damages can pose very serious problems. The use of sinking agents, which contaminate sea beds, is another such problematic clean-up measure. One of the cleaner and more efficient ways of treating oil spills is the use of inert sorbent materials, which can easily be retrieved after sorption. The oil can also be extracted from such units by compression and/or treatment with solvents. Further discussion is devoted to the application of sorbents in oil-spill control.

Sorbents are one of the essential parts of the entire spill-cleaning operation and are utilized throughout the entire duration of the operation. They are especially helpful in cleaning the last traces of oil. Sorbents are the class of materials which preferentially adsorb or absorb the oil from water surface. Multifarious sorbent materials have been developed to counter the oil spills. Sorbents are products or materials that are oleophilic and hydrophobic in nature. The efficiency of a sorbent depends on its sorption capacity, sorption rate, wettability, density, geometry, and recyclability. These properties also determine the time required to spread and harvest the sorbents. The advantage of sorbents is their relative insensitivity to sea conditions. Sorbents have been recorded to be one of the most effective and cheapest methods of cleaning oil-spills on the shorelines, whose contamination has always had the highest economic and environmental impact because of the difficulty in cleaning the oil spilled on them. Figure 4.2 shows the deployment of such sorbent articles to protect shorelines.

The oil sorbent materials can be classified into three main categories, explained in the section below. First category comprises of inorganic sorbents namely, various mineral products, exfoliated graphite, silica aerogels, oleophilic clays, and zeolites. Second category is of natural organic sorbents such as vegetable fibers, peat, cellulose, treated rice husk, straw, corn cob, wood fiber, cotton fiber, cellulosic kapok fiber, kenaf, milkweed floss, and peat moss. Third category includes synthetic organic sorbents which are most widely commercialized due to their excellent sorption properties, reusability feature, and resistance to rot and mildew. However, a major disadvantage of these materials is that they degrade very slowly in comparison to the mineral or vegetable products and are of synthetic origin. A prominent example of synthetic organic sorbents is polypropylene (PP)

Figure 4.2 (a and b) Sorbent articles as oil spill counter-measures [22].

fibers both in woven and non-woven forms, and polyurethane (PU) sponge [23]. A few limitations of vegetable and mineral sorbents are their poor buoyancy characteristics, low sorption capacity, and hydrophilicity. Mineral sorbents are generally not preferred as they have numerous shortcomings, such as the contamination of sea beds and harmful effects on aquatic habitats. They also tend to release some of the sorbed oil while sinking because of the low oil-retention capacity. Mineral sorbents are expensive, highly dense and their transportation to the spill-site requires much effort [24]. However, in a few cases their relative abundance makes their usage cost-effective. The most widely used sorbents are synthetic sorbents made from high molecular weight polymers, such as PU and PP. They are available under various trade names, have good hydrophobic and oleophilic properties, and possess high sorption capacity. Examples of synthetic organic sorbents are PU, shredded foam with open pores, urea formaldehyde foam, polyethylene (PE) in fibrous, granular, and powdered form, PP in various fibrous forms, sheet and white powder, polystyrene free-flowing particles, synthetic fiber mixtures like nylon, rayon, polyester, polyethylene terephthalate in shredded form, and shredded tire [25–32]. In order to assess the available sorbent materials and devise a standard laboratory procedure to test sorbent materials, the US coast guard tested 49 different sorbent materials belonging to the following categories: inorganic, natural organic, and synthetic organic. The properties under consideration were oil and water sorption capacity, oil retention, buoyancy retention with and without absorbed oil, the effect of petroleum product, sorbent/oil adhesion, and reusability. The sorption experiments revealed that the polymeric foam and the other polymeric hydrocarbon products were most efficient in the removal of oil from the water surface. Among those two groups, the resilient polyurethane foams and the polypropylene fibers were the best materials. Figure 4.3(a) shows the containment of an oil-spill with the help of booms, Figure 4.3(b) shows sorbent material housed in a polymeric netting, and Figure 4.3(c) is an image of manual oil-spill clean-up operation.

4.2 Factors Affecting the Performance of Sorbents

While all the properties mentioned in the section above are desirable in a sorbent, the ultimate selection among the available materials will involve a compromise between a set of several desirable properties and other factors such as material availability, available equipment for distribution and harvesting, spill location, and cost considerations. Effective use of sorbents involves ease of transportation to the scene of the spill, dispersal onto the

(a) (b) (c)

Figure 4.3 (a) Containment with boom, (b) sorbents applied to a site of oil-spill, and (c) manual oil-spill clean-up operation (NOAA) [33].

spill, recovery, and disposal [34]. One of the most desirable attributes of the sorbents is oil sorption capacity, which is a measure of oil uptake by a sorbent. Under field conditions, the amount of oil absorbed depends not only on the properties of the sorbent, but also on the available mixing energy from waves and wind, oil-film thickness, nature of the petroleum product, and water quality. Another desirable feature of the sorbent is low water sorption capacity. This property tells us about the hydrophobic nature of a sorbent. If potential oil sorbents have a tendency to sorb significant amounts of water, their capacity for oil sorption reduces. Oil-sorbent adhesion and retention is a measure of the interaction between the sorbent matrix and sorbed oil. It is also an indicator of the ease or difficulty encountered while recovering the oil-sorbent mixture from the water. Quantitative measurement of oil-sorbent adhesion is expressed in terms of oil retention by the matrix.

One of the most significant attributes taken into account is sorbent reusability. Unless sorbents can be reused many times during the clean-up operation, the entire measure does not usually become cost effective. As large numbers of sorbents have to be stock-piled, transported, and disposed after being soaked with oil, efficient removal of oil from the sorbent after sorption is a significant aspect. One of the most facile and economical way to remove oil from sorbents is by application of mechanical force. Given the sizes of the spill, in the marine environment usually a large number of sorbent units are deployed in a clean-up operation. Hence, the most desired property of a sorbent apart from high sorption capacity is easy on-site separation of the oil and reuse (recycling) of the sorbent. The other property of prime interest is buoyancy in saturated state, in the absence of which sorbents themselves can become a serious source of sea bed contamination.

The above-stated properties are largely dependent on the structural and morphological properties of a sorbent material. The composition, along with the properties and state of weathering of spilled hydrocarbons strongly influence their interaction with the sorbent and final volume of hydrocarbon accommodated within the sorbent matrix. Sorption mainly occurs because of the capillary action, adsorption, and absorption. The rate of penetration of oil into a capillary is inversely proportional to the viscosity of oil, and directly proportional to the capillary radius. The open-ended porous sorbent should have optimum pore sizes, which allow easy access and penetration of oil, but at the same time should be small enough to retain oil within the matrix. High surface-to-solid volume ratio will result in high capacity for oil.

The sorbent should have the ability to retard the spreading of the oil. The rate of oil sorption should be sufficiently high so that sorption is completed in a shorter duration and residence time of such units can be minimized. Easy separation of oil from the sorbent permits reuse and is helpful in selection of the ultimate disposal means for the sorbent.

4.3 Sorption and Desorption Kinetics

The sorption kinetics of sorbents are widely researched as their sorption behavioris valuable in the determination of the residence time of sorbents in oceanic conditions and in understanding the nature of interaction between sorbent and test fluid. However, only a few studies are devoted to the desorption kinetics of sorbents. An understanding of the behavioral characteristics of saturated sorbents such as oil retention and the tendency to release oil along with mechanical strength at saturation plays an important role in selection of the best sorbent for a specific hydrocarbon spill. In view of this, we have discussed the detailed sorption and desorption analysis of various natural and synthetic polymeric sorbents.

4.3.1 Sorption Kinetics

The detailed understanding of sorption kinetics in the oil-spill treatment is significant as it provides useful insight into the mechanisms of various sorption processes. In addition, the kinetic analysis determines the oil uptake rate, which in turn controls the desired time of contact between sorbent and oil.

Three models were largely used to understand the nature of oil sorption. These models were first given by Lagergren, and later on adopted and modified by various research groups in order to understand the sorption kinetics of sorbents [35]. The models are discussed one by one in the section given below:

Pseudo first order model.

$$\frac{dq_t}{dt} = k_1(q_{e1} - q_t) \tag{4.1}$$

Where, q_t is the sorption capacity at any time t, $q_{e1}$is the equilibrium sorption capacity, and k_1 is the rate constant of the pseudo-first order model. With the initial condition of $q_t = 0$, at $t = 0$ and on solving, the above equation will take the following form:

$$\ln(q_{e1} - q_t) = \ln(q_{e1}) - k_{1t} \tag{4.2}$$

The pseudo-second-order model can also be used to understand the rate of sorption and can be written in the following form.

$$\frac{dq_t}{dt} = k_2(q_{e2} - q_t)^2 \tag{4.3}$$

With the initial condition of $q_t = 0$ at $t = 0$ and by integrating the above equation takes the following form:

$$\frac{1}{q_{e2} - q_t} = \frac{1}{q_{e2}} + k_{2t} \tag{4.4}$$

On rearrangement, this becomes

$$\frac{t}{q_t} = \frac{1}{k_2\, q_2 e_2} + \frac{t}{q_{e2}} \tag{4.5}$$

Where, q_t is the sorption capacity [g/g] at time t, q_{e2} is the saturated sorption capacity, t is the time of contact, and k_2 is the sorption rate constant. This constant depends on the properties of the liquid such as viscosity and surface tension and is also a function of the pore structure of the sorbent.

The third model is the intra-particle diffusion model which is defined as follows.

$$q_t = k_i t^{0.5} + C \tag{4.6}$$

Here, q_t is the sorption capacity at time t and k_i is the rate constant of sorption kinetics in the intra-particle diffusion model.

These models were applied to the experimental data for expanded perlite as an oil sorbent by Bastani et al. [36]. They have reported that the correlation coefficient obtained from the least square fit of the kinetic models to the experimental data shows better correlation of the pseudo-second-order model with the experimental sorption data as compared to the other two models.

In another investigation, treated rice husk was employed as a sorbent and investigated in detail for its sorption-desorption behavior by Bazargan et al. [37]. They have also reported the pseudo-second-order model to be in better agreement with the experimental results as compared to pseudo-first-order model or intra-particle diffusion model.

Ho and Mckay [38] have illustrated that in order to understand the sorption kinetics, a detailed knowledge of the rate law describing the sorption system is required. One can reach the rate law of any process using prior knowledge of the following factors. The first is the reaction details at the molecular level along with the effects of stereochemistry, interatomic distances and angles throughout the course of the reaction, and most importantly the individual molecular steps involved in the mechanism. Their pseudo-second-order rate expression has been applied extensively to the sorption of herbicides, metal ions, dyes, oils, and organic substances from aqueous systems. The model is based on the sorption capacity of solids but contrary to the previous models, it describes chemisorption as a mode of sorption over the whole adsorption time. The advantage of this model is that the equilibrium capacity can be calculated from the model without need of any elaborate experimentation.

In another study, Gui et al. [39] have developed highly porous CNT sponges as sorbents for oil absorption. Ferrocene and dichlorobenzene were used as the catalyst precursor, and the CNT sponges were prepared by chemical vapor deposition (CVD). The sorption analyses were carried out with the following test fluids: mineral oil, vegetable oil, and diesel oil. The analyses revealed that the sponges had high sorption capacities along with high sorption rates. The sorption capacities are reported to be greater than 100 g/g. In their case too, the sorption kinetic process has been described by a second-order kinetic model. Thompson et al. [25] have studied sorption behavior of crude oil on acetylated rice husks. It was reported that the mechanism for oil sorption by lignocellulose fibers is adsorption, prevalent on the surface, and capillary action through its lumen. Initially, at the time of contact, the oil is sorbed by oleophilic interaction and van der Waals forces between the oil and the natural sorbent in the fiber surface. Further sorption of oil within the fiber occurs by diffusion through internal capillary movement within the sorbent lumens. Another important property is the role of equilibrium sorption especially in case of biodegradable materials, in reference to their storage conditions. The sorption kinetics which describes the solute uptake rate is vital in evaluating the efficiency of sorption. The adsorption isotherm will also give the equilibrium temperature curve. They studied the kinetics of the sorption and the pseudo-second-order rate model equation was found to fit the sorption process. The two isotherm models used to fit the data were the Langmuir and the Freundlich model. The Langmuir model was selected for the estimation of maximum sorption capacity. The Freundlich model was chosen to estimate the adsorption intensity of the sorbent. An important assumption was made, to consider the sorption of a chemical on a solid from a water-based solution as a reversible reaction (sorption-desorption) which reaches a final equilibrium condition between the concentration of the chemical in the two phases. It was reported that the experimental data better fitted the Langmuir model than the Freundlich model as the correlation values of the Langmuir model tend to be much closer to 1 than those obtained from the Freundlich isotherm. Hence they have reported that the adsorption can be described as monolayer kinetic studies, and also the mode of sorption of oil was chemisorption.

Toyoda and Inagaki [40], have studied the sorption behaviors of four kinds of heavy oils using exfoliated graphite. The sorption capacity depended on the bulk density and pore volume of the exfoliated graphite. Although the time to attain maximum sorption, as well as sorption capacity, depended strongly on the grade of heavy oil, heavy oils sorbed by exfoliated graphite were recovered either by a simple compression or suction filtration with a recovery ratio of 60–80%.

4.3.2 Desorption Models

The desorption behavior is crucial during the selection of sorbent for a given oil-spill, as it is a measure of the following properties; sorbent-oil adhesion, buoyancy at saturation, material strength in a fully swollen condition, and reusability. A few desorption models formulated in order to understand the nature of the process are discussed next:

The unsteady state loss of liquids such as oils from sorbents in gravimetric experiments have been modeled by Bazargan et al. [41]. They have fitted an experimental data set for 62 sorbents-sorbate systems. It was reported that even with the wide variation among the retention capacities of sorbents, they all followed a similar trend for the loss of test fluid with respect to the time. This trend was divided into four different stages. At the beginning, the rate of liquid loss was highest and in this stage the retention capacity drops quickly, closely followed by a transition zone where the rate of release declines. The final zone is defined by the gradual approach of the curve to a horizontal asymptote. The desorption profile ends by reaching the steady state condition under which no additional liquid is lost due to dripping, with the passage of time. The following assumptions were made during the model formulation: the flow of the sorbate within the sorbent was approximated as flow within a porous medium. Also, Darcy's law was applied due to smaller pore size and high fluid velocity. The mean flow was assumed to be one-dimensional, and in the vertical direction. The liquid properties of test fluids such as viscosity and density were considered constant. The properties of the sorbent were also assumed to be constant, and this included the void fraction, pore size, intrinsic permeability, and capillary pressure. An important assumption was made, to neglect the losses because of evaporation/drying. The two-phase interactions of air and liquid within the sorbent through surface tension are considered to be, on average, invariant for all interface heights $z(t)$, and are represented through a constant capillary pressure.

The desorption model is represented by the equation given below.

$$U_t = U_L e^{-kt} + U_e \tag{4.7}$$

Here, U_t(kg/kg) is the uptake capacity of the sorbent at time t (s), U_L(kg/kg) is the loss in uptake capacity due to dripping, and U_e(kg/kg) is the equilibrium uptake capacity. K(1/s) was defined as the Kamaan coefficient and controls the curvature of the retention profile. The Kamaan coefficient is a function of sorbent and sorbate properties, such as intrinsic permeability, density, and viscosity. This model highlighted the role of macro-pores in the uptake capacity. The capillary and adsorption phenomena both were found to affect the shape of the desorption curve. The role of higher porosity in saturation uptake capacity and in the poor retention of sorbates was mentioned. For porous samples, the rate of release could be reduced if the solid had fine pores. An important observation was reported in the reduction of the rate of release, due to the fibrous sorbent structure. It was also deduced that the saturation uptake capacity is a function of the liquid density as well as the sorbent's physical properties such as size, density, and void fraction. Based on the derived model, it was concluded that the viscosity of the sorbate and the affinity between the sorbent and sorbate do not influence the initial saturated uptake capacity. However, the final amount of liquid retained by the sorbent (equilibrium uptake capacity) depends on the sorbent structure and capillary forces. Hence the affinity between the sorbent and the sorbate can strongly influence the final practical uptake capacity. The derived model can be used in determination of the practical applicability of sorbents. One of the important parameters that are determined by the model is the lost uptake capacity, U_L. This value is of prime importance in real-world applications as it is a measure of liquid lost from the used sorbent. In case of oil-spill clean-up, this value shows how much liquid will seep out of the used sorbent while it is being transported back to the depot. Ultimately, although many other factors are of importance, the most effective sorbent would be that with the highest equilibrium uptake capacity (U_e).

In a detailed investigation carried out to understand the sorption-desorption behavior of alkali treated rice husks, Bazargan et al. [37] have pointed out that the sorbent retains the liquid by capillary forces. These strong capillary forces prevent the loss of oil due to dripping. Their experiments showed the significant loss of the oil during the initial drainage period. This is due to the loss of the adhered oil and the presence of the larger macropores in the sorbent. After the initial loss, the remaining oil which was retained within the sorbent matrix due to stronger forces resulted in the appearance of a plateau in the desorption profile. Certain amounts of sorbate would always be retained within the matrix due to the action of forces stronger than that of gravitational pull, and an external force must be applied to take this fraction of oil out from the sorbent matrix. Following three models were applied to desorption profiles: Lewis model, page model, and the two-term model. These models were initially developed to model the retention of moisture by solids in drying experiments [42]. The data fitting revealed that the two-term model had a better fit to the sorbent-sorbate experimental data.

In another investigation, the desorption results for one of the most commercialized sorbents for oil-spill control were presented by Wei et al. [43]. Crude oil was selected as a sorption medium. An important aspect of their work was to consider the effect of weathering of oil on the sorption and desorption behavior of the sorbent. Oil was weathered by heating on a hot plate, at a low heating rate to make the evaporated oils with 25 and 50% weight loss respectively. Commercial nonwoven PP sorbents were used to conduct the experiments. Wide variation among sorbents in terms of retention capacities was observed, although the retention behavior of all sorbents follows almost a similar trend. Wei et al. have talked about three distinct zones in each retention profile as shown in Figure 4.4. The initial stage of release occurred over the first minute, signified by the higher rate of release. The second, or transition zone, occurred from 1 to 5 min, and the profile started flattening in this zone. The last zone represented the steady-state period, characterized by no significant release of oil. The nature of the release profiles was heavily governed by the properties of test fluid. For example, rate of release for the lighter oil is higher as compared to that of the heavy oil. The sorbent properties along with internal structure play a significant role in the

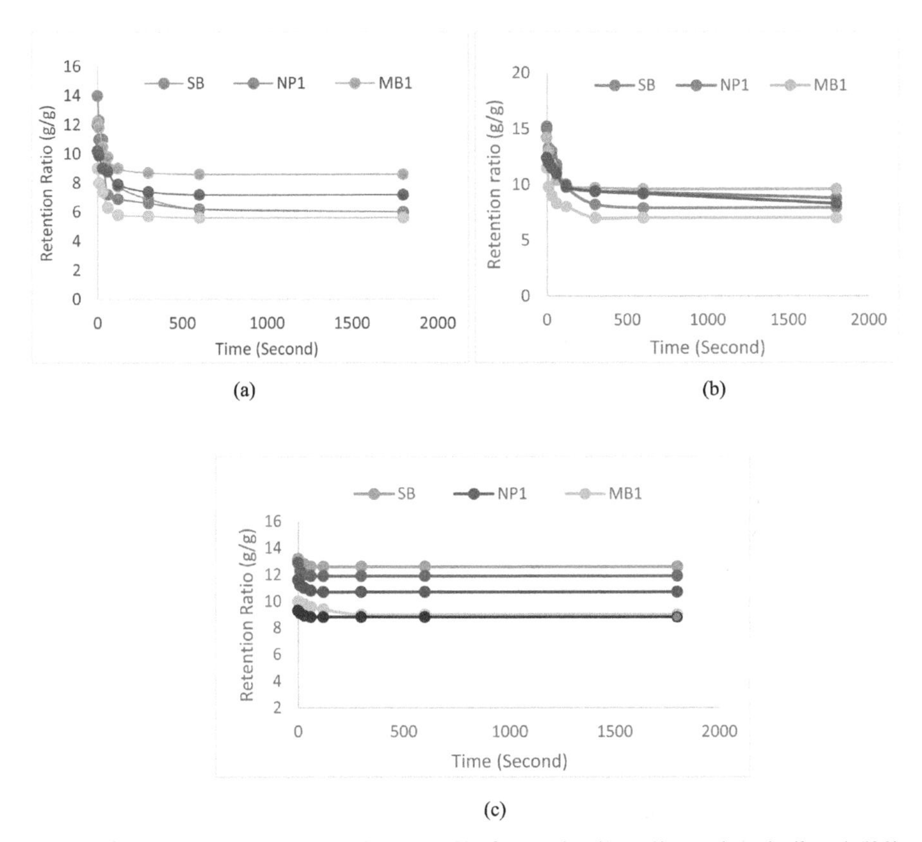

Figure 4.4 Plots (a–c) are the release profile for crude oil, 25% weathered oil and 50% weathered oil respectively [43].

retention behavior of a sorbent. The sorbent with greater porosity had high initial oil pickup, but poor retention capacity. Fine fibrous structure led to a reduction in rate of release.

4.3.3 Sorption-desorption Analysis of Polyurethane Foam

The retention of liquids by sorbents is a topic concerning a wide variety of research areas ranging from hydrogels and ceramics to cosmetics and biological tissues [41]. If a sorbent has poor liquid retention and loses much of the sorbed liquid, it is usually not suitable for use in conditions which require a long duration of handling and transport of the saturated sorbent.

Till now, we have discussed the properties, advantages and disadvantages of various kinds of sorbents. A tabulated view of some of these sorbents is given below in Table 4.2. The table shows that PU is among the best materials for oil-spill control, and ultralight open-cell polyurethane foams are capable of sorbing oil 100 times their weight from oil–water mixtures. Keeping this in view, further sorption and desorption analyses on PU samples were conducted as per ASTM protocols, to understand their long-term behavior. All the factors crucial to the overall performance of the sorbent were analyzed.

Sorbents for analyses were prepared by using polyoxypropylene glycol and polyurethane diol as the polyol monomers and toluene diisocyanate as an isocyanate monomer. Dimethylaminopyridine, *N,N*-dimethyl benzyl amine, and dibutyltin dilaurate were used as catalysts. All chemicals used were from Sigma–Aldrich. The sorbents were tested for maximum sorption

Table 4.2 Comparison of properties of various sorbents

Material	Structure	Type of Oil	Sorption Capacity Gram of Crude Per Gram of Sorbent, g/g.	Reusability
Clay	Granular	Diesel Hydraulic Engine	5.2–7.2 2.2–3.6 2.1–3.6	No
Polyurethane foam (PU)	Foam	Crude	100	Yes
Polypropylene (PP)	Fiber	Light crude	10	Yes
Expanded perlite	Granular	Light crude Heavy crude	3.5 3.25	Two cycles
Exfoliated graphite	Device	Heavy crude	86	Yes
Cotton fiber	Fiber	Crude	30–40	Three cycles
Shredded tire	Powder	Motor	10–20	Up to 100 times
Cellulose fiber	Chips	Heavy crude	5	Several cycles
Rice husk	Particulate	Heavy Crude Diesel	4.6–6.7 2.98–6.22 2.78–5.02	5 cycles
Recycled wool-based non-woven material	Non-woven material	Diesel Crude Vegetable Motor	9.62 11.06 13.16 15.8	5 cycles

capacity, various containment efficiencies, drainage and evaporation which led to release profiles, buoyancy characteristics at saturation, behavior under load, and possible release to adjacent media.

To ascertain the applicability of these sorbents to a range of hydrocarbons, the sorbents were tested with four different test fluids. Most sorbents reached maximum sorption capacity in a duration of 7 min. The desorption analyses of the samples revealed remarkably different release profiles for various test fluids. Various PU samples also showed different release profiles for the same test fluid, due to differences in porosities. The saturated sorbents showed minimal test-fluid release tendencies to the materials which came in immediate contact with them, and had sufficient mechanical strengths to be handled in their saturated states.

While a large number of studies have dealt with the sorption kinetics of sorbents, the importance of desorption kinetics and of the other behavioral features of oil-saturated sorbents still needs to be emphasized. Oil-sorption ability should not be the sole criterion in determining the potential of a material as a sorbent. Other factors such as the release of oil through evaporation, ease of handling the material post-sorption, change in buoyancy characteristics upon sorption, retention of oil, and flammability hazard all equally influence the selection and overall performance of a sorbent [13]. Based on the properties of the test fluid and its state of weathering, the sorption and desorption characteristics of a specific sorbent will vary. The test-fluid release profiles for different hydrocarbons, measured for different sorbents, would help in selecting a suitable sorbent for a given spill. We have attempted and established these release properties for a set of sorbents recently synthesized in our laboratory.

Another aspect of the current work deals with the determination of the ability of the sorbent bed to inhibit the flow of fluid through it. Some materials have a tendency to release the oil to an adjacent material due to a low adhesion between the material and oil. This factor is of significance in minimizing the possible contamination of beaches and back-release of oil into the waters. In the present contribution, we have chosen in-house synthesized absorbents with PU functionality. The sorption and desorption characteristics of different test fluids were measured based on the ASTM F716-09.

Four different tests were conducted to investigate the sorption and desorption behaviors of the PU sorbents. The different test fluids used were xylene, butyl acetate, crude oil, and diesel. In the first test, the absorption efficiencies of sorbents were measured by placing a volume of 2 ml of the sorbent in

100 ml of a test fluid for 2 h. The weight of the sorbent was measured after the period of 2 h. The absorption efficiency was calculated as follows.

$$A = \left[\frac{w_1 - w_2}{w_1}\right] \times 100 \quad (4.8)$$

Where, w_1= weight of material, after sorption of the test fluid.

w_2= weight of the material taken initially, and A is the absorption efficiency.

In second test, the rate of test-fluid release from the sorbent was measured. Under this test, the sample was saturated in a test fluid for 30 min. The saturated sample had then been taken out of the test fluid and hung from one corner, to wait until the cessation of dripping. Post dripping, the sample was weighed after every 10 min for 2 h. This weight vs. time data showed the release profile of the test fluid from a sorbent. In the penetration test, a 15-cm high bed of a sorbent was taken in a glass tube filled with 10 cm bed of sand followed by a cork. On top of it, 7cm of bed of sand was placed, and a 15-cm head of test fluid was maintained throughout the experiment; the experimental set-up is shown in Figure 4.5. This test was conducted to check the ability of the sorbent bed to inhibit the penetration of the test fluid through it. The fourth test comprises of placing a saturated sorbent, with dyed test fluid, on a 3-mm thick stack of blotting paper. The sorbent was moved to different places until the stain of the dyed fluid did not appear on the other side of the paper. The weight of the sorbent was noted at that point, and was used in calculating the practical containment efficiency. A similar procedure was repeated for the absorbent with 2 g/cm^2 uniform load on top of it; the weight

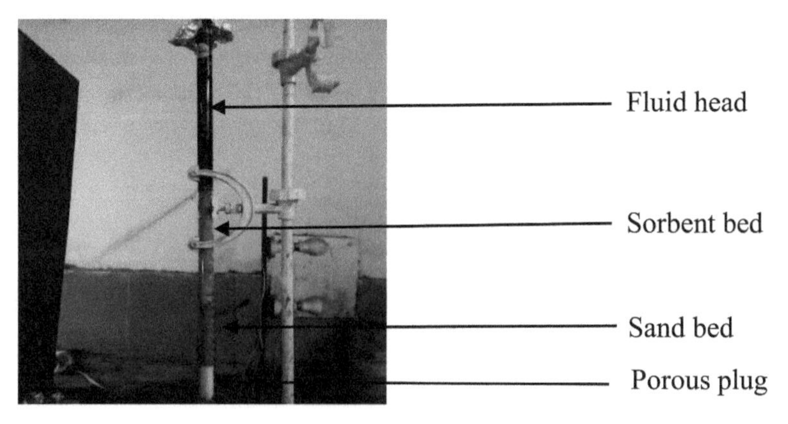

Figure 4.5 Experimental set up.

of the sorbent was noted at that time and utilized in calculating the maximum containment efficiency. This test helped in determining the tendency of the sorbent to release the fluid to adjacent materials. Each test was repeated thrice to ensure the reproducibility of the measurements.

Initial sorption analysis revealed that the sorbents belong to the class of absorbents, as the weight gain was more than 50% in each case. Four different samples were prepared by varying the types of polyol and their binary combinations. Figure 4.6 shows images of such samples. The oil-absorption efficiency for each of them is shown in the Table 4.3 below. The values of absorption efficiencies showed that the sample PU1 had better sorption capacities as compared to the other samples. Also, the sorption values were not significantly different for polar and non-polar solvents. A possible explanation for this observation could be a relatively uniform porosity distribution throughout the sorbent matrix, ascribable to the specific technique [44].

The results of second test which resulted in release profiles can be plotted as release profiles for four different test fluids as shown in Figure 4.7.

The profiles show that the retention of diesel and crude oil was fairly high for all the samples as compared to that of xylene and butyl acetate due to the low viscosities of the latter two, which resulted in lesser adhesion

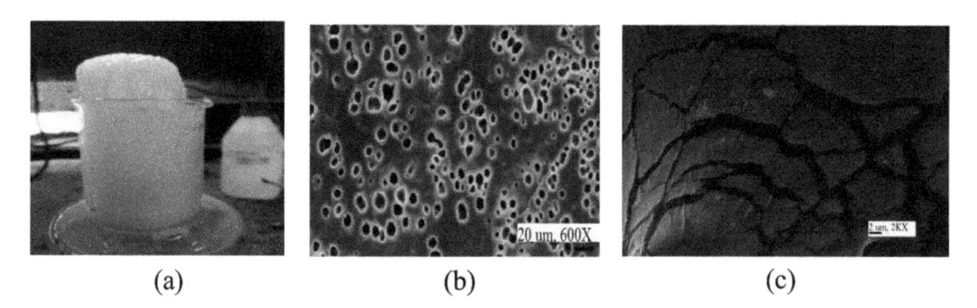

(a) (b) (c)

Figure 4.6 (a) PU sample, and (b and c) are the SEM micrographs of the same sample.

Table 4.3 Absorption efficiencies of different sorbents

Sorbents	Absorption Efficiencies%			
S. N.	Xylene	Butyl Acetate	Diesel	Crude Oil
PU-1	85.8	93.2	87.8	80.3
PU-2	60.5	51.4	56	52
PU-3	77.7	68.4	76.6	75.2
PU-4	71.4	79.3	53.4	59

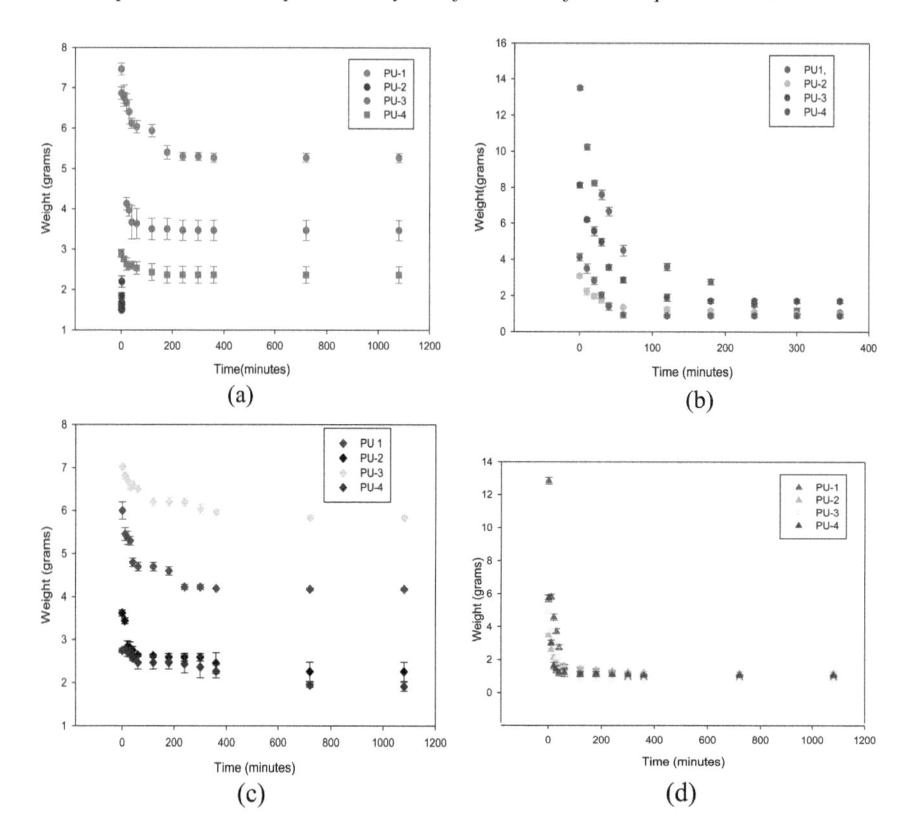

Figure 4.7 Plot (a) release profiles for crude oil as the test fluid, (b) release profile of xylene as test fluid, (c) release profile of diesel as a test fluid, and (d) release profile of butyl acetate as a test fluid.

among lighter oils and the sorbent matrix. This would ultimately be reflected in terms of oil retention by the porous sorbent matrix. The penetration tests were conducted on all the four samples. All these sorbent beds were effective in inhibiting the flow of the test fluids through them. The last tests were conducted to ascertain the sorption efficiencies of the four samples, and the corresponding values are shown below in Table 4.4.

The differences between the absorption efficiencies and the practical containment efficiencies were not significant for the same sorbent and test fluid, as these values show that the adhesion of various test fluids to the sorbent matrix is good. The differences between the practical and maximum containment efficiencies were also small and, hence the materials have a minimal tendency to release oil to the adjacent materials even under load.

Table 4.4 Practical and maximum containment efficiencies of different sorbents

Sorbents	Practical Containment Efficiencies				Maximum Containment Efficiencies			
	Xylene	Butyl Acetate	Diesel	Crude Oil	Xylene	Butyl Acetate	Diesel	Crude Oil
PU-1	85%	93%	87.2%	80.1%	85%	92%	84%	80%
PU-2	60%	50%	54%	50%	60%	49.7%	52%	48%
PU-3	76%	67%	74%	75%	75.3%	64%	72.5%	74%
PU-4	70%	78%	52%	59%	67.8%	76%	52%	58.4%

4.4 Conclusion

The literature on polymeric sorbents, for the application of oil-spill control is discussed in detail along with various models for explaining the sorption-desorption behavior of sorbents. The sorbents with urethane functionality were experimentally tested for their sorption and desorption characteristics. Sorption abilities and release tendencies were quantified in terms of various absorption efficiencies and test-fluid release profiles. A contact-time period of 2 h was sufficient to reach maximum sorption, as no further change in weight was observed after this time. Also the saturated material had sufficient mechanical strength to be handled and was buoyant even after sorption. Significant changes in sorption values for various test fluids were attributed to the differences in porosities. Release profiles for different test fluids revealed that the lighter crude was retained less and significant volumes could be released into the environment. Increased vapor content in the environment could possibly pose a fire and health hazard. The maximum and practical containment efficiency values were not much smaller than the absorption efficiencies, thus proving that the chosen materials had a minimal tendency to release oil to the adjacent materials such as beach sands. This reduces the possibility of secondary pollution generation and makes the material suitable for oil-spill applications. The sorption-desorption analyses revealed that the polyurethane sorbents had good practical promise of a performance based on several pertinent criteria considered.

References

[1] Mulhaupt, R. Hermann Staudinger and the origin of macromolecular chemistry. *Angewandte. Chemie Int. Ed. Engl* Wiley-VCH Verlag. 2004. 43, 1054–1063.

[2] Pourjavadi, A., Doulabi, M., and Hosseini, S. H. Novel polyelectrolyte gels as absorbent polymers for nonpolar organic solvents based on polymerizable ionic liquids. *Polymer*. 2012. 53, 5737–5742.

[3] Peplow M. The plastics revolution: how chemists are pushing polymers to new limits. Nature news Ed. 2016. 536, 266–267.

[4] Behrendt, J., and Sutherland, A. Polymer-supported reagents and catalysts. *Porous Polym.* John Wiley & Sons, Inc. London. 2011. Vol. 1, pp. 387–434.

[5] Gray, H. N., and Bergbreiter, D. E. Applications of polymeric smart materials to environmental problems. *Environ. Health Perspect.* 1997. 105, 55–63.

[6] Sarbatly, R., Krishnaiah, D., and Kamin, Z. A review of polymer nanofibres by electrospinning and their application in oil – water separation for cleaning up marine oil spills. *Mar. Pollut. Bull.* 2016. 106, 8–16.

[7] Steinmetz, Z., Wollmann, C., Schaefer, M., Buchmann, C., David, J., Tröger, J., Muñoz, K., Frör, O., and Ellen, G. Trading short-term agronomic benefits for long-term soil degradation. *Sci. Total Environ.* 2016. 550, 690–705.

[8] Calcagnile, P., Fragouli, D., Bayer, I. S., Anyfantis, G. C., Martiradonna, L., Cozzoli, P. D., Cingolani, R., and Athanassiou, A. Magnetically driven floating foams for the removal of oil contaminants from water. *ACS Nano*. 2012. 6, 5413–5419.

[9] Kumari, S., Chauhan, G. S., and Hyeon, J. Novel cellulose nanowhiskers-based polyurethane foam for rapid and persistent removal of methylene blue from its aqueous solutions. *Chem. Eng. J.* 2016. 304, 728–736.

[10] Fingas, M. Introduction. *Oil Spill Science and Technology*. Boston, MA: Gulf professional publishing, Boston. 2011. Vol.1. pp. 3–5.

[11] Fingas, M. Physical spill countermeasures. *Oil spill science and technology*. Gulf professional publishing, Boston. 2011. Vol.1. pp. 303–337.

[12] Etkin, D. S. Analysis of oil spill trends in the united states and worldwide. *Int. Oil Spill Conf.* 2001. 1291–1300.

[13] Al-Majed, A. A., Adebayo, A. R., and Hossain, M. E. A sustainable approach to controlling oil spills. *J. Environ. Manage*. 2012. 113, 213–227.

[14] Gupta, S., and Tai, N. Synthesis and performance of polymerisable room-temperature ionic liquids as gas sepration membranes. *J. Mater. Chem. A*. 2016. 4, 1550–1565.

[15] Bourne, W. R. P. The impact of *Torrey Canyon* and *Amoco Cadiz* oil on north French seabirds. *Mar. Pollut. Bull.*1979. 10, 120–124.
[16] Enzler, S.M. Top 10 of Anthropogenic and Natural Environmental Disasters. 2010. Lenntech, Delft, the Netherlands. http://www.lenntech.com/environmental- disasters.htm.
[17] Cutler, J., Cleveland, C.J., Saundry, P. Exxon Valdez oil spill. *Encyclopedia of Earth.* 2010. http://www.eoearth.org/article/Exxon_Valdez_oil_spill.
[18] Vethamony, P., Sudheesh, K., Babu, M. T., Jayakumar, S., and Manimurali, R. Trajectory of an oil spill off Goa , eastern Arabian Sea: field observations and simulations. J. *Environ. Pollut*. 2007. 148, 438–444.
[19] Mcnamee, W. Gulf Oil Spill Pictures: Birds, Fish, Crabs Coated. *Nat. Geo*. 2010. https://news.nationalgeographic.com/news/2010/06/photo galleries/100608-gulf-oil-spill-environment-birds-animals-pictures/
[20] Reuters. BP oil spill: 30 pictures of the Deepwater Horizon Gulf of Mexico disaster one year ago. The Telegraph. 2011. http://www.telegraph.co.uk/news/earth/earthpicturegalleries/8453474/BP-oil-spill-30-pictures-of-the-Deepwater-Horizon-Gulf-of-Mexico-disaster-one-year-ago.
[21] Oebius, H. U., Berlin, and Wum, T. U. Physical properties and processes that influence the clean up of oil spills in the marine environment. *Spill Sci. Technol. Bull*. 1997. 5, 177–289.
[22] ITOPF. Containment & Recovery. 2015. The international tanker owners pollution federation. http://www.itopf.com/knowledge-resources/documents-guides/response techniques/containment-recovery/.
[23] Adebajo, M. O., and Frost, R. L. Porous materials for oil spill cleanup: a review of synthesis and absorbing properties. *J. Porous Mater*. 2003. 10, 159–170.
[24] Ornitz, B. E., and Champ, M. A. Oil Spills First Principles: Prevention and Best Response, *Elsevier Science Ltd., Oxford.* 2002. Vol.1., pp. 345–367.
[25] Thompson, N. E., Emmanuel, G. C., and Adagadzu, K. J. Sorption studies of crude oil on acetylated rice husks. *Schol. Research. Libr*. 2010. 2, 142–151.
[26] Jang, J., and Kim, B.E. Studies of crosslinked styrene–alkyl acrylate copolymers for oil absorbency application. I. effects of polymerization conditions on oil absorbency. *J. Appl. Polym. Sci*. 2000. 77, 903–913.
[27] Li, H., Liu, L., and Yang, F. Hydrophobic modification of polyurethane foam for oil spill clean-up. *Mar. Pollut. Bull.* 2012. 64, 1648–1653.

[28] Lin, J., F., Shang, Y., Wang, F., and, Ding, B. Co-axial electrospun polystyrene/polyurethane fibres for oil collection from water surface. *Nanoscale*. 2013. 5, 2745–2755.
[29] Somani, K., Patel, K., Kansara, S., and Rakshit, A. Effect of chain length of polyethylene glycol and crosslink density (NCO/OH) on properties of castor oil based polyurethane elastomers. *J. Macromol. Sci., Part A: Pure and Appl. Chem.* 2006. 43, 797–811.
[30] Atta, M., Brostow, W., Datashvili, T., El-Ghazawy, R., Lobland, H.E.H., Hasan, A.M., and Perez, J.M. Porous polyurethane foams based on recycled poly (ethylene terephthalate) for oil sorption. *Polym. Int.* 2013. 62, 116–126.
[31] Yang, J. S., Cho, S. M., Kim, B.K., and Narkis, M. Structured polyurethanes for oil uptake. *J Appl. Polym. Sci*. 1998. 98, 2080–2087.
[32] Ajith kumar, S., Patel, N.K., and Kansara, S. S. Sorption and diffusion of organic solvents through interpenetrating polymer networks (IPNs) based on polyurethane and unsaturated polyester. *Euro. Polym. J.* 2000. 36, 2387–2393.
[33] National oceanic and atmospheric administration [NOAA]. *NOAA Joins Response to Pipeline Oil Spill at Refugio State Beach Near Santa Barbara, California. National Ocean Service*. 2015. http://response.restoration.noaa.gov/about/media/noaa-joins-response-pipeline-oil-spill-refugio-state-beach-near-santa-barbara-california.html.
[34] Schatzberg, P., and Nagy, K. V. Sorbents for oil spill removal. *Oil Spill Conf. Proc.* (Washington, DC: American Petroleum Institute). 1997. 221–233.
[35] Azizian, S. Kinetic models of sorption: a theoretical analysis. *J. Colloid Interface Sci*. 2004. 276, 47–52.
[36] Bastani, D., Safekordi, A. A., Alihosseini, A., and Taghikhani, V. Study of oil sorption by expanded perlite at 298.15 K. *J. Sep. Purif. Technol.* 2006. 52, 295–300.
[37] Bazargan, A., Wai, C., and Gordon, H. Marine residual fuel sorption and desorption kinetics by alkali treated rice husks. *Cellulsose*. 2014. 21, 1997–2006.
[38] Ho, Y. S., and Mckay, G. Pseudo-second order model for sorption processes. *Process Biochem*. 1999. 34, 451–465.
[39] Gui, X., Zeng, Z., Lin, Z., Gan, Q., Xiang, R., Zhu, Y., et al.Cao, A., and Tang, Z. Magnetic and highly recyclable macroporous carbon nanotubes for spilled oil sorption and separation. *ACS Appl. Mater. Interfaces*. 2013. 5, 5845–5850.

[40] Toyoda, M., and Inagaki, M. Heavy oil sorption using exfoliated graphite new application of exfoliated graphite to protect heavy oil pollution. *Carbon*. 2000. 38, 199–210.

[41] Bazargan, A., Sadeghi, H., Garcia-mayoral, R., and Mckay, G. Journal of colloid and interface science an unsteady state retention model for fluid desorption from sorbents. *J. Colloid Interface Sci*. 2015. 450, 127–134.

[42] Gunhan, T., Demir, V., Hancioglu, E., and Hepbasli, A. Mathematical modelling of drying of bay leaves. *Energ. Convers. Manage*. 2005. 46, 1667–1679.

[43] Wei, Q. F., Mather, R. R., Fotheringham, A. F., and Yang, R. D. Evaluation of nonwoven polypropylene oil sorbents in marine oil-spill recovery. *Marine Pollut. Bull*. 2003. 46, 780–783.

[44] Bhardwaj, N., and Bhaskarwar, A.N. A novel polymeric foam and process for making it. Indian Patent Application No. 3691/DEL/2014, Filing Date: December 12, 2014.

5

Polyhexahydrotriazines: Synthesis and Thermal Studies

Nitish Paul Tharakan[1], J. Dhanalakshmi[2], and C. T. Vijayakumar[1]

[1]Department of Polymer Technology, Kamaraj College of Engineering and Technology, Virudhunagar, India
[2]Department of Chemistry, Kamaraj College of Engineering and Technology, Virudhunagar, India

Abstract

Hemiaminal (HA-MDA) from 4,4'-methylene dianiline (MDA) and formaldehyde was prepared and characterized using Fourier transform infrared (FTIR), differential scanning calorimetry, and thermogravimetric analysis. The hemiaminal showed a strong bimodal exotherm which starts around 211°C and ends around 297°C. The maximum was found to be at 268°C. The hemiaminal was thermally polymerized at 250°C. The polyhexahydrotriazine (PHT-MDA) formed from HA-MDA was characterized for its structural aspects by FTIR. The thermogravimetric curves recorded at different heating rates (β = 10°C/min, 20°C/min, and 30°C/min) were used to calculate the change in apparent activation energy for the thermal degradation (Ea-D) of the material at different extents of degradation (α) using model-free kinetics developed by Vyazovkin. The results are presented and discussed.

5.1 Introduction

5.1.1 Polymer

A polymer is a large molecule or macromolecule, composed of many repeated subunits. Because of their broad range of properties, both synthetic and natural polymers play an essential and ubiquitous role in everyday life. Polymers range from familiar synthetic plastics such as polystyrene to natural biopolymers such as deoxyribonucleic acid and proteins that are fundamental to biological structure and function. Polymers, both natural and synthetic, are created via polymerization of many small molecules, known as monomers. Their consequently large molecular mass relative to small molecule compounds produces unique physical properties, including toughness, viscoelasticity, and a tendency to form glasses and semicrystalline structures rather than crystals [1].

5.1.2 Classification based on Thermal Behavior

There are two general classes of polymers based on their behavior when exposed to heat.

- Thermoplastic polymers
- Thermoset polymers.

Thermoplastic polymers are normally produced in one step and then made into products in a subsequent process. They become soft and formable when heated. The polymer melt can be formed or shaped when they are in this softened state. When cooled significantly below their softening point, they again become rigid and usable as a formed article. This type of polymer can be readily recycled because each time it is reheated it can again be reshaped or formed into a new article.

On the other hand, thermosetting polymers are normally produced and formed in the same step. Upon heating, thermosetting polymers will become soft, but cannot be shaped or formed to any great extent and will definitely not flow. Though thermoset and thermoplastics sound similar, they have very different properties and applications. Understanding the performance differences can help you make better sourcing decisions and improve your product designs. The primary physical difference is that thermoplastics can be melted back into a liquid, whereas thermoset plastics always remain in a permanent solid state. Think of thermoplastics as butter – butter can be melted and cooled multiple times to form various shapes. Thermoset is similar to bread in that once the final state is achieved, any additional heat would lead to charring.

5.1.3 Thermosetting Polymer

A thermosetting polymer also known as a thermoset is a prepolymer material that cures irreversibly. The cure may be induced by heat, generally above

200°C, through a chemical reaction or suitable irradiation. Thermoset materials are usually liquid or malleable prior to curing and designed to be molded into their final form or used as adhesives. Others are solids like that of the molding compound used in semiconductors and integrated circuits. Once hardened, a thermoset resin cannot be reheated and melted to be shaped differently. Thermosetting resin may be contrasted with thermoplastic polymers which are commonly produced in pellets and shaped into their final product form by melting and pressing or injection molding.

The curing process transforms the resin into a plastic or rubber by a crosslinking process. Energy and/or catalysts are added that cause the molecular chains to react at chemically active sites (unsaturated or epoxy sites, for example), linking into a rigid, 3-D structure. The crosslinking process leads to a molecule with a larger molecular weight, resulting in a material with a higher melting point. During the reaction, the molecular weight has increased to a point so that the melting point is higher than the surrounding ambient temperature, and the material forms into a solid material.

Uncontrolled reheating of the material results in reaching the decomposition temperature before the melting point is obtained. Therefore, a thermoset material cannot be melted and reshaped after it is cured. This implies that thermosets cannot be recycled, except as filler material.

Thermoset materials are generally stronger than thermoplastic materials due to this three-dimensional network of bonds (crosslinking) and are also better suited to high-temperature applications up to the decomposition temperature. However, they are more brittle. Since their shape is permanent, they tend not to be recyclable as a source for newly made plastic [2].

Merits

- More resistant to high temperatures than thermoplastics
- Highly flexible design
- Thick to thin wall capabilities
- Excellent esthetic appearance
- High levels of dimensional stability
- Cost effective.

Demerits

- Cannot be recycled
- More difficult to surface finish
- Cannot be remolded or reshaped.

5.1.4 Thermoset Materials

5.1.4.1 Phenol formaldehyde

Phenol formaldehyde resins are formed by the polycondensation between phenol and formaldehyde [3]. The condensation reaction can be catalyzed either by acids or bases. Phenolics were among the earliest commercial synthetic materials and have been the workhorse of the thermoset molding compound family. In general, they provide low cost, good electrical properties, excellent heat resistance, and fair mechanical properties (they suffer from low impact strength), coupled with excellent moldability. They are generally suitable for use at temperatures up to 150 C. Phenolic molding compounds can be reinforced or filled at levels higher than 70 %. In order to increase the impact strength, paper, chopped fabric or cord, and glass fibers are used. Glass fiber grades also provide substantial improvement in strength and rigidity. Glass-containing grades can be combined with heat-resistant resin binders to provide a combination of impact and heat resistance. Their dimensional stability is also substantially improved by glass. Phenolics are processed by thermoset injection molding, compression molding, and transfer molding. Phenolics are amenable to pultrusion also. Electrical grades are generally mineral- or flock-filled materials designed for improved retention of electrical properties at high temperatures and high humidity.

5.1.4.1.1 *Novolacs*

Novolacs are phenol–formaldehyde resins with formaldehyde to phenol molar ratio of less than one. The polymerization is brought to completion using acid catalysis such as oxalic acid, hydrochloric acid, or sulfonate acids. The phenol units are mainly linked by methylene and/or ether groups. Novolacs are commonly used as photoresists. The molecular weights are in the low thousands, corresponding to about 10–20 phenol units. Hexamethylenetetramine or “hexamine” is a hardener added to crosslink novolac. At a temperature $>90^{\circ}$C, it forms methylene and dimethylene amino bridges.

5.1.4.1.2 *Resoles*

Base-catalyzed phenol–formaldehyde resins are made with formaldehyde to phenol ratio of greater than one (usually around 1.5). These resins are called resoles. Phenol, formaldehyde, water, and catalyst are mixed in the desired amount, depending on the resin to be formed, and are then heated. The first part of the reaction, at around 70°C, forms a thick reddish-brown tacky material, which is rich in hydroxymethyl and benzylic ether groups. The rate of the base-catalyzed reaction initially increases with pH and reaches a maximum at about pH = 10. The reactive species is the phenoxide anion ($C_6H_5O^-$) formed by deprotonation of phenol. The negative charge is delocalized over the aromatic ring, activating sites 2, 4, and 6, which then react with the formaldehyde. Being thermosets, hydroxymethyl phenols will crosslink on heating to around 120°C to form methylene and methyl ether bridges through the elimination of water molecules. At this point, the resin is a three-dimensional network, which is typical of polymerized phenolic resins. The high crosslinking gives this type of phenolic resin its hardness, good thermal stability, and chemical imperviousness. Resoles are referred to as "one-step" resins as they cure without a crosslinker unlike novolacs, a "two-step" resin. Resoles are major polymeric resin materials widely used for gluing and bonding building materials. Exterior plywood, oriented strand boards, engineered laminated composite lumber, etc., are typical applications.

5.1.4.2 Urea–formaldehyde resin

Urea–formaldehyde, also known as urea-methanal, so named for its common synthesis pathway and overall structure, is a non-transparent thermosetting resin or plastic, made from urea and formaldehyde heated in the presence of a base. These resins are used in adhesives, finishes, particle boards, and molded objects [4]. Urea–formaldehyde and related amino resins are considered a class of thermosetting resins of which urea–formaldehyde resins make up 80% produced globally.

Examples of uses of amino resins include automobile tires in order to improve the bonding of rubber to tire cord, paper for improving tear strength, molding electrical devices, molding jar caps, etc. Urea–formaldehyde resin's attributes include high tensile strength, flexural modulus and a high heat distortion temperature, low water absorption, mold shrinkage, high surface hardness, elongation at break, and volume resistance.

The chemical structure of urea–formaldehyde resins can be described as that of an aldehyde condensation polymer. This description leaves the details of the structure undetermined, which can vary linearly and branched. These are grouped by their average molar mass and the content of different functional groups.

Changing synthesis conditions of the resins give good designing possibilities for the structure and resin properties. Generally, the polymer is prepared by reacting the urea and formaldehyde in a 1:1 ratio.

The result is a polymer with -NCH_2N- repeat units, not -NCH_2OCH_2N- repeat units found in melamine formaldehyde.

5.1.4.3 Melamine formaldehyde resin

Mineral-filled grades provide improved electrical properties; fabric and glass fiber reinforcements provide higher shock resistance and strength. The chemical resistance of the material is relatively good, although it is attacked by strong acids and alkalis. Melamines are tasteless and odorless and are not stained by pharmaceuticals and many food stuffs. Like the other member of the amino family, urea, the melamines find extensive use outside of the molding area in the form of adhesives, coating resins, and laminating resins.

5.1.4.4 Unsaturated polyester resin

Polyester resins are unsaturated synthetic resins formed by the reaction of organic acids and polyhydric alcohols. Polyester resins are used in sheet molding compound, bulk molding compound, and the toner of laser printers. Wall panels fabricated from polyester resins reinforced with fiberglass – so-called fiberglass reinforced plastic – are typically used in restaurants, kitchens, restrooms, and other areas that require washable low maintenance walls.

Unsaturated polyesters are condensation polymers formed by the reaction of polyols (also known as polyhydric alcohols), organic compounds with multiple alcohol or hydroxy functional groups, with saturated or unsaturated dibasic acids. Typical polyols used are glycols such as ethylene glycol; acids used are phthalic acid and maleic acid. Water, a byproduct of esterification reactions, is continuously removed, driving the reaction to completion. The use of unsaturated polyesters and additives such as styrene lowers the viscosity of the resin. The initially liquid resin is converted to a solid by crosslinking chains.

This is done by creating free radicals at unsaturated bonds, which propagate in a chain reaction to other unsaturated bonds in adjacent molecules, linking them in the process. The initial free radicals are induced by adding a compound that easily decomposes into free radicals. This compound is usually and incorrectly known as the catalyst. Substances used are generally organic peroxides such as benzoyl peroxide or methyl ethyl ketone peroxide.

Polyester resins are thermosetting and, as with other resins, cure exothermically. The use of excessive catalyst can, therefore, cause charring or even ignition during the curing process. Excessive catalyst may also cause the product to fracture or form a rubbery material.

5.1.4.5 Epoxy resins

These resins offer excellent electrical properties and dimensional stability, coupled with high strength and low moisture absorption. The most important of the epoxy resins are based on bisphenol-A and epichlorohydrin; they can be produced in either liquid or solid form. Epoxy novolacs are made by reacting novolacs from phenol formaldehyde with epichlorohydrin; the resulting resins have high heat resistance and low levels of ionic impurities [5].

All epoxies have to be reacted with a hardening agent (e.g., an aliphatic amine an aromatic amine or a polyamide or dicarboxylic acid anhydride) to crosslink into thermoset structures. Epoxies are used by the plastics industry in several ways. One is in combination with glass fibers (i.e., impregnating fiber with epoxy resin) to produce high-strength composites that provide excellent strength, electrical and chemical properties, and heat resistance.

- Epoxy and anhydride forms ester linkages
- Good stability at elevated temperature
- Stable in organic acids
- Better electrical insulation than amine curing.

Typical uses for epoxy-glass are in aircraft components, filament wound rocket motor casings for missiles, pipes, tanks, pressure vessels, and tooling jigs and fixtures. Another major application area for epoxies is adhesives. The two-part systems cure with minimal shrinkage and without emitting volatiles. Formulations can be created to cure at different rates at room or elevated temperatures. The auto-industry has started to use epoxy adhesives in place of welding and for assembling plastic body parts. Epoxies can further be transfer or injection molded. Transfer presses, for example, are being used today to encapsulate electronic components in epoxy or to mold epoxy electrical insulators. Injection molding presses are also available for liquid epoxy molding. Epoxy molding compounds are available in a wide range of mineral-filled and glass, carbon, and aramid fiber reinforced versions.

5.1.4.6 Bismaleimides

The bismaleimides (BMIs; Figure 5.1) are monomers on thermal curing lead to very high temperature-resistant condensation polymers and are made from maleic anhydride and a diamine. Bismaleimide (BMI) resin is one of the most important classes of high-performance thermosetting resins, with its outstanding properties including high tensile strength and modulus, excellent heat, and chemical and corrosion resistance [6]. However, the BMI resins are

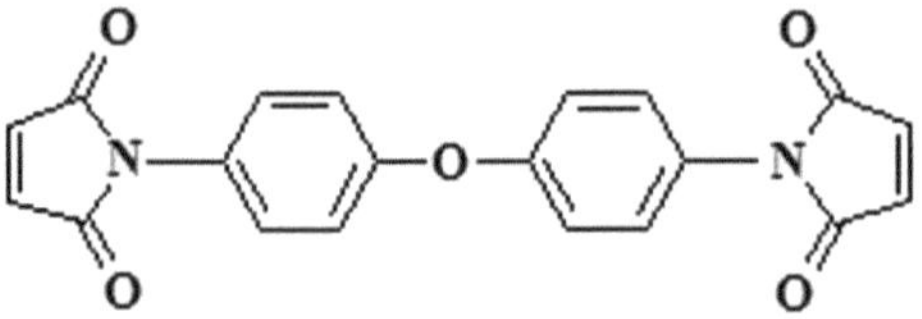

Figure 5.1 Structure of bismaleimide.

brittle as a result of the aromatic nature and the high crosslink density of the network. To overcome its brittleness (the major disadvantage) and improve its thermomechanical properties, the performance enhancement is needed. They possess excellent thermal and oxidative stability, flame retardation, and low propensity to moisture absorption. Thermosetting BMI resins are used as matrices for advanced composites in aerospace and electronics industries. Composite structures made from BMI resins are used in military aircraft and aerospace applications. These resins are also used in the manufacture of printed circuit boards and as heat-resistant coatings [7].

5.1.4.7 Bispropargyl ethers

Bispropargyl (Figure 5.2) terminated compounds shows promise for use in the preparation of matrix resins and adhesives for advanced aircraft and aerospace systems [6]. The compounds can be polymerized thermally without the evolution of volatile byproducts, thereby obviating the problems of void formation in composite structure and molded articles.

The propargyl ethers are prepared by the reaction between a polyhydric phenolic material and propargyl chloride in aqueous sodium hydroxide solution. The reactions occur at the interface between the aqueous basic solution of phenolic material and the propargyl chloride which is insoluble in water [8].

There have been growing interests in this kind of resin due to their excellent properties, such as good thermal stability, low moisture absorption,

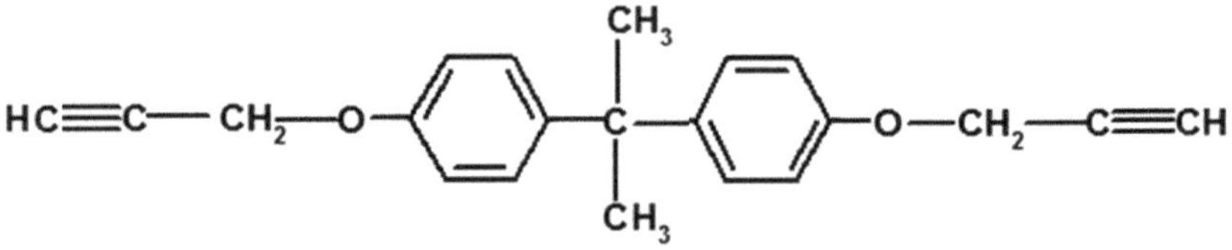

Figure 5.2 Structure of bispropargyl ether.

excellent adhesion, and low dielectric constant. In addition, they could be cured via addition reaction to avoid producing volatiles, which makes it possible to get void free materials without high pressure.

Bispropargyl ether resins are used as matrices for composites in aerospace and electronic industries. Such polymers with a properly designed backbone could serve as matrices in carbon composites and in various high-performance polymer composite structures for space applications. These resins are also used in the manufacture of prepregs [9].

5.1.4.8 Cyanate ester

Cyanate ester is a thermosetting polymer that can withstand very high temperatures, as well as maintain good thermal, mechanical, and electrical properties at these temperatures [10]. These properties make cyanate ester a very attractive choice for use in many applications as well as the manufacture of composite materials.

They are chemical substances generally based on a bisphenol or novolac derivative, in which the hydrogen atom of the phenolic OH group is substituted by a cyanide group. The resulting product with an –OCN group is named a cyanate ester. Cyanate esters can be cured and postcured by heating, either alone at elevated temperatures or at lower temperatures in the presence of a suitable catalyst. The most common catalysts are transition metal complexes of cobalt, copper, manganese, and zinc. The result is a thermoset material with a very high glass transition temperature (Tg) of up to 400°C and a very low dielectric constant, providing excellent long-term thermal stability at elevated end-use temperatures, very good fire, smoke, and toxicity performance and specific suitability for printed circuit boards installed in critical electrical devices.

This is also due to its low moisture uptake. This property, together with a higher toughness compared to epoxies, also makes it a valuable material in aerospace applications. Product properties can be fine-tuned by the choice of substituent in the bisphenolic compound. Bisphenol-A (2,2-bis(4-hydroxyphenyl) propane) and novolac-based cyanate esters are the major products; bisphenol-F (bis(4-hydroxyphenyl) methane) and bisphenol-E (1,1-bis(4-hydroxyphenyl) ethane) are also used. The aromatic ring of the bisphenol can be substituted with an allylic group for improved toughness of the material. Cyanate esters can also be mixed with BMIs to form BT-resins (BMI–triazine resin) or with epoxy resins to optimize the end-use properties.

This special type of polymerization, called cyclotrimerization, gives the resulting cured resin very interesting properties. The structure of the triazine rings makes them non-dipolar, giving cyanate ester its low dielectric loss. The three-dimensional polymerization of triazine rigs creates a very sound structure while still maintaining a low crosslinking density, which prevents brittleness and promotes a high glass transition temperature.

5.1.4.9 Triazines

Triazine is a class of nitrogen containing heterocycles. The parent molecules' molecular formula is $C_3H_3N_3$. They exist in three isomeric forms: 1,3,5-triazines, being common. The triazines have planar six-membered benzene-like ring but with three carbons replaced by nitrogen's.

The three isomers of triazine (Figure 5.3) are distinguished by the positions of their nitrogen atoms and are referred to as 1,2,3-triazine, 1,2,4-triazine, and 1,3,5-triazine. Other aromatic nitrogen heterocycles are pyridines with one ring nitrogen atom, diazines with two nitrogen atoms in the ring, and tetrazines with four ring nitrogen atoms.

The best known triazines are derivatives of the 1,3,5-triazine derivatives melamine and cyanuric chloride (2,4,6-trichloro-1,3,5-triazine). With three amino substituents, melamine is a precursor to commercial resins. Another triazine extensively used in resins is benzoguanamine. Chlorine-substituted triazines are components of reactive dyes. These compounds react through a chlorine group with hydroxyl groups present in cellulose fibers in nucleophilic substitution; the other triazine positions contain chromophores. Triazine compounds are often used as the basis for various herbicides.

Hexahydro-1,3,5-triazine (Figure 5.4) is a class of heterocyclic compounds with the formula $(CH_2NR)_3$. They are reduced derivatives of 1,3,5-triazine, which have the formula $(CHN)_3$, a family of aromatic heterocycles.

5.1.4.10 Polyhexahydrotriazine

Polyhexahydrotriazines (PHTs) are polymers of hexahydro-1,3,5-triazines, a class of heterocyclic compounds with the formula $(CH_2NR)_3$. They are among the strongest known thermosetting plastics and are stable to solvents

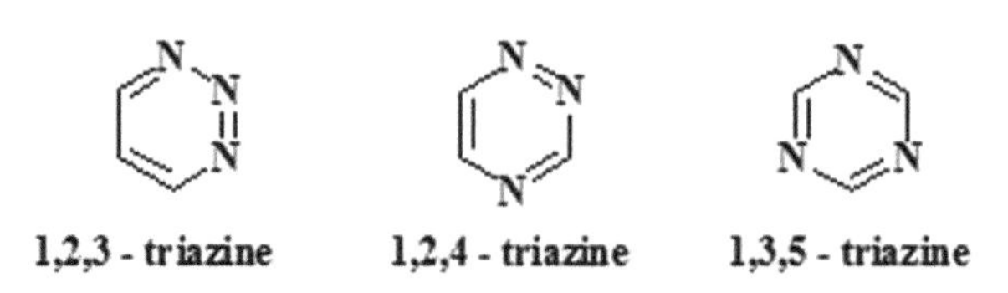

Figure 5.3 Structure of three isomers of triazines.

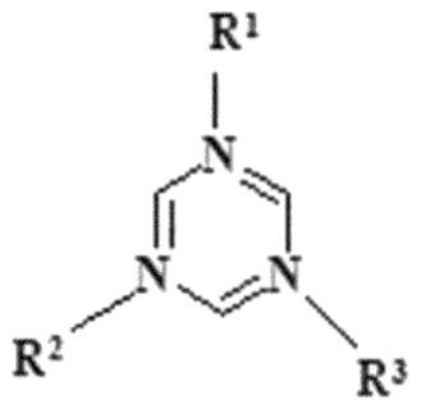

Figure 5.4 Structure of hexahydro-1,3,5-triazine.

at pH > 3, but decompose to the monomers in acidic solutions [11]. The PHT 1.6 has a yellowish color. It is resistant to solvents at pH > 3, but decomposes into monomers within a day in acidic solutions at pH < 2. This property is unusual for thermosetting plastics and allows easy recycling. This PHT has a Young's modulus exceeding 10 GPa, which is among the highest for a thermosetting plastic; it can be further increased by 50% by dispersing carbon nanotubes in the polymer. The PHT is brittle and cracks when strained to 1%. Upon heating, it softens at a glass transition temperature of 190°C and decomposes at 300°C.

The properties of PHTs were similar to those of the traditional, unrecyclable thermosets: rigid, heat-resistant, and chemically stable and displaying excellent resistance to solvents and environmental stress, especially when reinforced with carbon nanotubes [12]. Moreover, by varying the combinations and compositions of the monomers used in their reactions, the researchers were able to produce elastic gels with self-healing properties. These gels can also be broken down in strong acid. This work is an important advance as thermosets have long been considered impossible to recycle.

5.2 Experimental

5.2.1 Preparation of Hemiaminal Using 4,4'-Methylenedianiline (HA-MDA)

4,4'-Methylenedianiline (MDA, 0.20 g, 1.0 mmol) and paraformaldehyde (PF, 0.15 g, 5.0 mmol) were weighed out in a 100-mL round bottom flask. N-methylpyrrolidone (NMP, 6.2 g, 6.0 mL) was added. The reaction mixture was heated in an oil bath at 50°C for 24 h and hemiaminal is formed. After approximately 0.75 h, the hemiaminal begins to precipitate in NMP. After the stipulated time, the hemiaminal (HA-MDA) was precipitated in acetone, filtered, and dried in oven [3]. The material was collected as an off-white solid (yield = 0.22 g, >98%). The synthetic procedure is depicted in Figure 5.5.

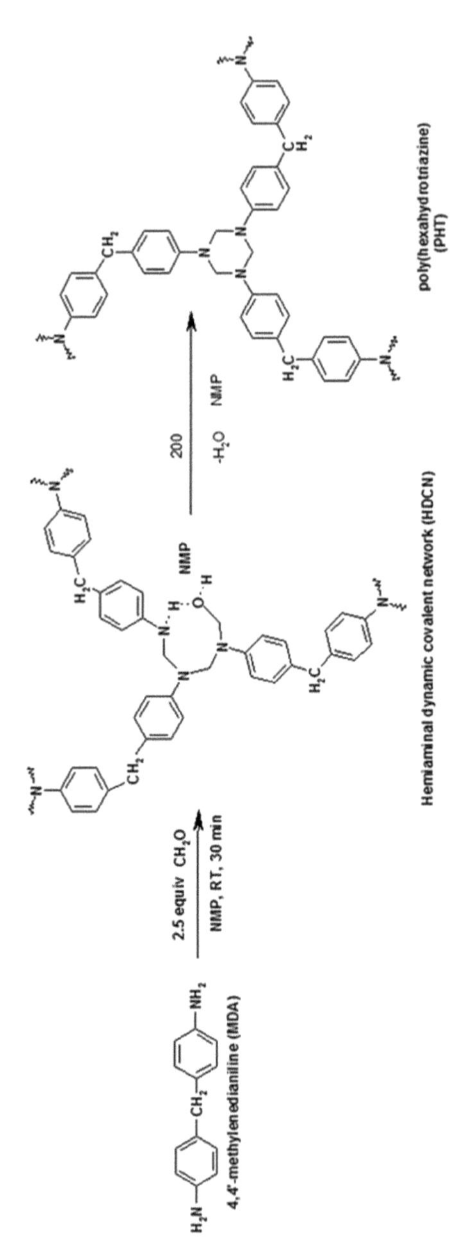

Figure 5.5 Preparation of polyhexahydrotriazine 4,4'-methylene dianiline.

5.2.2 Thermal Curing

The hemiaminal (HA-MDA) (0.22 g) is taken in a micro test tube and flushed with dry oxygen free nitrogen and then heated to 225°C and kept at this temperature for 3 h. After the thermal polymerization, the sample (PHT-MDA) was allowed to cool at room temperature and then removed from the micro test tube. The material was ground to coarse powder, packed, stored in a desiccator, and preserved for further analysis.

5.2.3 Methods

The Fourier transform infrared (FTIR) spectra of the materials (HA-MDA and PHT-MDA) were recorded in a Shimadzu-8400S infrared spectrophotometer using KBr pellet technique. The absorption bands in FTIR spectra were used to identify the functional groups present in the materials investigated.

The differential scanning calorimetry (DSC) curve for the compound HA-MDA was recorded in a TA Instruments DSC Q20. The material was hermetically sealed in an aluminum pan and heated at a rate of 10°C/min in dry nitrogen (flow rate = 50 mL/min) atmosphere.

The TG curves for the compounds (HA-MDA and PHT-MDA) were recorded in a TA Instruments TG Q50 system using 10°C/min heating rate in nitrogen atmosphere. Generally, 4–5 mg of the sample was used with a balance purge of 40 mL/min and a sample purge of 60 mL/min. The sample, PHT-MDA, was heated at different heating rates (β = 10°C/min, 20°C/min, and 30°C/min) from ambient to 800°C in nitrogen atmosphere.

By using the model-free kinetic method developed by Vyazovkin [13], the apparent activation energy for the thermal degradation (Ea-D) can be calculated at any particular degree of conversion (α) by finding the value of for which the objective function is minimized. The equation is given below:

$$\Omega = \sum_{i=1}^{n} \sum_{\substack{j=1 \\ j \neq i}}^{n} \frac{I(E_a, T_{\alpha i})\beta_j}{I(E_a, T_{\alpha j})\beta_i}.$$

5.3 Results and Discussion

5.3.1 FTIR Studies

The FTIR spectra of HA-MDA and thermally cured HA-MDA (PHT-MDA) are shown in Figure 5.6. The presence of an absorption band at 674 cm^{-1} in

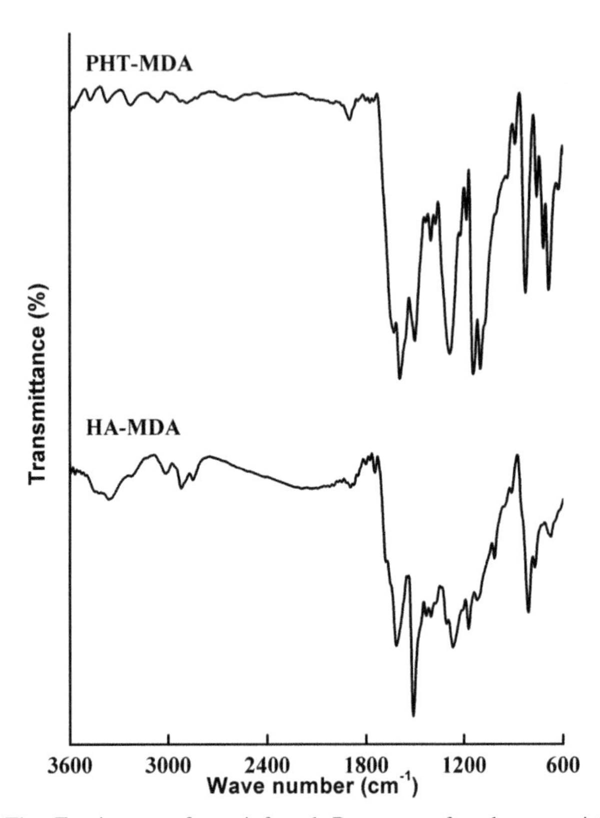

Figure 5.6 The Fourier transform infrared R spectra for the materials hemiaminal 4,4'-methylene dianiline and polyhexahydrotriazine 4,4'-methylene dianiline.

the FTIR spectrum of HA-MDA is attributed to the C–O–H bending vibration and the absorption bands at 1171 cm^{-1} and 1265 cm^{-1} are assigned to C-N stretching. A band at 1,615 cm^{-1} is due to the NMP present in the material. The presence of an absorption band at 2,924 cm^{-1} is associated to the $–CH_2–$ present in the compound (4).

The presence of an absorption band at 3,205 cm^{-1} is associated to the NH symmetric deformation. The appearance of a broad absorption band at around 3,300 cm^{-1} is due to the H-bonded -OH group stretching. The presence of absorption bands pertaining to NMP and the identification of -OH stretching confirms the proposed structure for HA-MDA.

The appearance of a new absorption band in the compound PHT-MDA is noted at 824 cm^{-1}. The absence of absorption bands at 3,205 cm^{-1} and

3,300 cm^{-1} in the FTIR spectrum of PHT-MDA indicates the occurrence of cyclization reaction in HA-MDA (Figure 5.6).

5.3.2 Thermal Studies

The DSC curve for HA-MDA recorded at a heating rate (β) of 10°C/min in an inert atmosphere is shown in Figure 5.7. The TG and DTG curves recorded for the material are presented in Figure 5.8. The parameters obtained from the

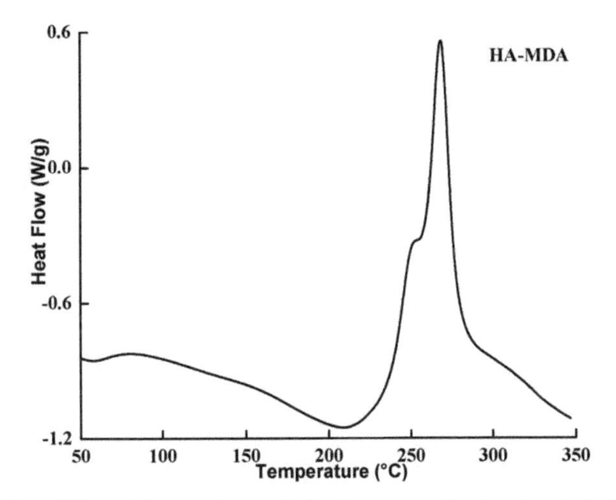

Figure 5.7 The differential scanning calorimetry curve for hemiaminal 4,4'-methylene dianiline.

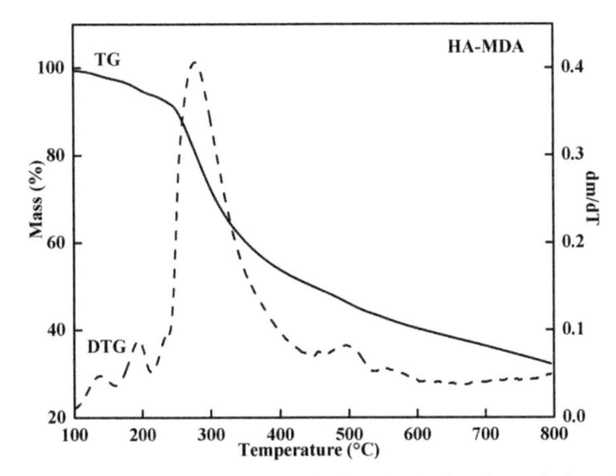

Figure 5.8 The TG and DTG curves for hemiaminal 4,4'-methylene dianiline.

DSC curves, namely, onset of exotherm (Ts), exotherm maximum (Tmax), endset of exotherm (Te), and the total enthalpy associated with the exotherm (ΔHc) are discussed below.

The DSC curve of HA-MDA shows a strong complex exotherm and it starts at around 211°C and ends at 297°C. Two maxima are identified in this exotherm, one at 253°C and the other at 269°C. The total enthalpy associated with the exotherm is 201 J/g.

The TG and DTG curves for HA-MDA are presented in Figure 5.9. From Figure 5.9, it is clear that the material starts to lose weight slowly at around 86°C and the weight loss proceeds up to 161°C. The amount of weight lost during this temperature region is around 3%. Again, the material slowly losses the weight from 161°C to 215°C and the amount of weight lost is 3%. The material shows the major weight loss from 215°C and proceeds till 445°C. The amount of weight lost in this region is 44%. After which the material undergoes very slow weight loss in the temperature region 443–535°C and the amount of weight lost is 7%. After 535°C, the thermal degradation proceeds very slowly up to 750°C and the amount of weight lost during this temperature region is 9%.

The conversion of HA-MDA to PHT-MDA is proceeding with the elimination of the included NMP and water from HA-MDA. Hence, the exotherm shown in the DSC curve (Figure 5.7) may be attributed to the elimination of NMP and water from HA-MDA and the conversion of hemiaminal structure to hexahydrotriazine structure.

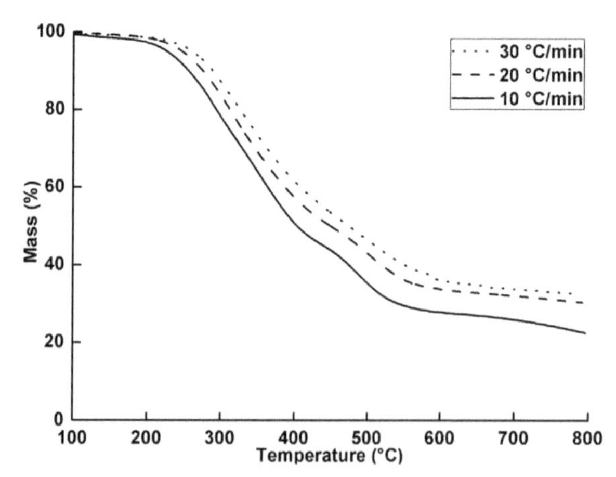

Figure 5.9 The TG curves for the sample polyhexahydrotriazine 4,4’-methylene dianiline at different heating rates in nitrogen atmosphere.

The TG and the DTG curves recorded for the material PHT-MDA at different heating rates (β = 10°C/min, 20°C/min, and 30°C/min) are shown in Figures 5.9 and 5.10, respectively. The parameters such as onset of degradation (Ts), degradation maximum (Tmax), end set of degradation (Te), and char residue (%) at 750°C obtained from the TG and DTG curves are tabulated in Table 5.1.

The DTG curves for PHT-MDA recorded at different heating rates show multiple degradation steps which overlap. As the heating rate increases, one can easily find the merging of the different degradation stages.

Using the TG and DTG curves recorded for the sample PHT-MDA at different heating rates (Figures 5.9 and 5.10), the apparent activation energy for the thermal degradation (Ea-D) at different extents of reaction (α) was calculated using the model free kinetics developed by Vyazovkin. The results are shown in Figure 5.11. Initially, the Ea-D values progressively increase from 100°C till 500°C (α = 0.1–0.5). The Ea-D value shows a slow decrease up to α = 0.65, and then the Ea-D value increases steeply till α = 0.8.

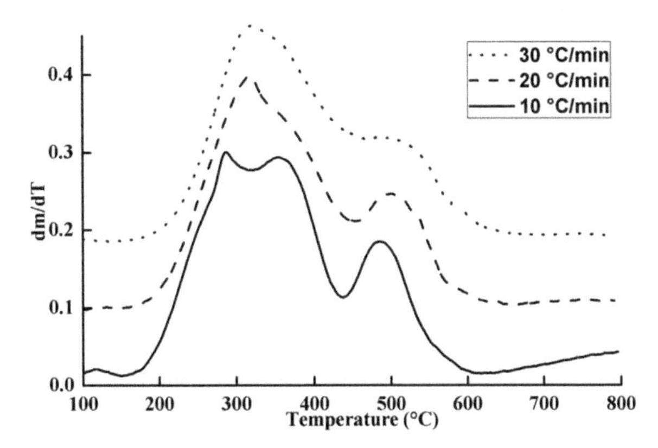

Figure 5.10 The DTG curves for the sample polyhexahydrotriazine 4,4'-methylene dianiline at different heating rates in nitrogen atmosphere.

Table 5.1 TG studies: The degradation parameters for the polymer PHT-MDA (β =10°C/min, 20°C/min, and 30°C/min)

Heating rate (β) (°C/min)	Ts (°C)	Tmax (°C)			Te (°C)	Char residue at 750°C (%)
10	154	287	354	387	608	24
20	155	318	356	498	612	31
30	156	320	358	503	639	33

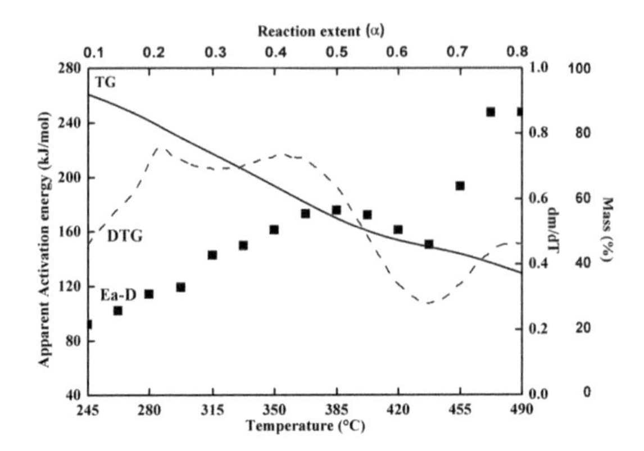

Figure 5.11 The TG and DTG curves for the sample polyhexahydrotriazine 4,4'-methylene dianiline. The apparent activation energy for the degradation (Ea-D) is plotted against the extent of degradation (α).

Owing to the degradation of the coked material, the Ea-D values increase after $\alpha = 0.65$.

Extensive studies by hyphenated techniques like TG-FTIR and TG-MS will give clear picture regarding the thermal degradation products and thermal degradation mechanism of PHT-MDA.

5.4 Conclusion

Hemiaminal is prepared from MDA, PF, and NMP. The hemiaminal (HA-MDA) is heated to 225°C and kept at that temperature for 3 h so as to convert it to the hexahydrotriazine form. The DSC and TG studies of HA-MDA indicated the loss of NMP and water during the formation of PHT-MDA. The variation in the apparent activation energies for the thermal degradation (Ea-D) of PHT-MDA for different extents of thermal degradation (α) shows a complex pattern, indicating the multistage overlapping degradation of PHT-MDA.

Acknowledgments

The authors express their sincere thanks to the Management and Principal of Kamaraj College of Engineering and Technology, S.P.G.C. Nagar, K. Vellakulam Post 625701, India, for providing all of the facilities to do the work.

References

[1] Flory, J. (1953). *Principles of Polymer Chemistry*. New York, NY: Cornell University Press.
[2] Billmeyer, W. (1994). *Textbook of Polymer Science*. Singapore: John Wiley & Sons.
[3] Pilato, L. (2010). *Phenolic Resins: A Century of Progress*, Dordrecht: Springer.
[4] Pizzi, A. (2003). *Handbook of Adhesive Technology,* 2nd Edn. New York, NY: Dekker.
[5] May, A. (1998). *Epoxy Resins Chemistry and Technology*, 2nd Edn. New York, NY: Marcel Dekker.
[6] Lin, S. C., and Pearce, E. M. (1993). *High Performance Thermosets – Chemistry, Properties, Applications.* New York, NY: Hanser Publishers.
[7] Vijayakumar, C. T., Surender, R., Pitchaimari, G., and Vinayagamoorthi, S. (2013). Bispropargyl ether and bismaleimide blends: part I – isothermal curing studies. *Sci. Lett.* 2:39.
[8] Dhanalakshmi, J., Pitchaimari, G., Alam, S., and Vijayakumar, C. T. (2014). High performance polymers. *High Perform. Polym.* doi: 10.1177/0954008314525973
[9] Vinayagamoorthi, S., Vijayakumar, C. T., Alam, S., and Nanjundan, S. (2009). Structural aspects of high temperature thermosets bismaleimide/propargyl terminated resin system-polymerization and degradation studies. *Eur. Polym. J.* 45, 1217–1231.
[10] Hamerton, I. (1994). *Chemistry and Technology of Cyanate Ester Resins*. London: Blackie Academic.
[11] Hedrick James, L., Horn Hans, W., Jones Gavin, O., O'Brien Jeanette, M., and Virwani Kumar, R. (2015). Methods of preparing polyhemiaminals and poly hexahydrotriazines. U.S. Patent 20150104579A1
[12] Jeannette, M., Gavin, O., Virwani, K., McCloskey, D., Boday, J., Gijs, M., et al. (2014). Recyclable, strong thermosets and organogels via paraformaldehyde condensation with diamines. *Science* 344, 732–735.
[13] Vyazovkin, S., and Dollimore, D. (1996). Linear and non-linear procedures in isoconversional computations of the activation energy of nonisothermal reactions in solids. *J. Chem. Inf. Comp. Sci.* 36, 42–45.

6

Influence of Cement Behavior with and without Polymer Nano Composites

Mainak Ghosal[1,2] and Arun Kumar Chakraborty[3]

[1]Adjunct Assistant Professor, JIS College of Engineering, Kalyani, India
[2]Research Scholar, Indian Institute of Engineering Science and Technology, Howrah, India
[3]Associate Professor, Indian Institute of Engineering Science and Technology, Howrah, India

Abstract

Cement is the only standard material in concrete and also accounts for its major costing. As the typical process of cement hydration (chemical reaction between cement and water) completes 90–95% in the first 28 days (typical setting time of concrete), the standard testing protocol for cement concrete suggests testing of various behavioral properties of cement and concrete at 28 days both in general and in particular. In this paper, an attempt has been made to study the effects of two different types of polymeric nano composites [(1) conventional polymers-melamine formaldehyde (MF) and (2) modern polymers-PolyCarboxylate ether (PCE)] for mechanical behavior of ordinary cement composites on the short term (1 day, 3 days, and 7 days), medium term (28 days), and long term (90 days, 180 days, and 365 days). Our test results shows that an optimized addition of carbon nanotubes (CNTs; dispersed in Superplasticizer PCE) gives a 37% gain in strength at 28 days and increasing the CNT % also provides for a latter age (1 year) strength from 38% to 69%, while the addition of CNTs (dispersed in Superplasticizer MF) failed to produce any satisfactory results at 28 days and showed an appreciable loss in strength at latter ages, except some flashy 44% strength gain at 90 days.

As CNTs were insoluble in aqueous medium, it was dispersed in PCE or MF using ultrasonication method for the latter to get more uniformity in dispersion.

6.1 Introduction

The keywords for good quality construction today are "Application of Polymers" as construction of durable structures is absolutely not possible without the right usage of construction chemicals. Good construction chemicals are not an additional cost but an investment into the long life of a structure, which should be incorporated from the planning stage itself. The earlier attitude of taking recourse to the use of polymers (admixtures) only after facing problems is changing fast, and now, in most of the large projects, the admixtures (polymers) are already included in the specifications. Construction industry has expanded by more than 5.3% in the last few years. Though extensive advancement in construction technologies has taken place in the last few decades, the construction industry is still found to lag behind in using nanotechnology when compared to others. As a result, the modern day structures are deficient in many aspects mainly – fresh properties, durability aspects, and early strength developments. Water – *"the universal solvent"* – on the one hand sustains life and nurtures the nature and on the other hand also acts as the most destructive element that has the power to disintegrate matter back to its elemental form. Construction materials, by nature, are water-loving. Whether it is soil or cement plaster or bricks or reinforced cement concrete, it has a tendency to absorb water. Such ingress of water inside concrete results in various chemical changes within a structure that eventually results in its degeneration and aging. Now, the question is how water enters inside concrete? Concrete by birth is porous in nature. And the concrete pore sizes are much larger than the size of a smallest water droplet (size of a water drop when in a cloud is 1,00,000 nm or 0.1 mm). What then results is that further breakdown of this water drop so as to make it penetrate inside most impervious concrete.

However, the above-stated phenomenon can be controlled (reduced) considerably, if not completely eliminated, by appropriate designing, selection of proper raw materials, following a methodical system of construction, and last but not the least addition of some polymeric compounds in insignificant small dosages with their performance increasing manifold with the applications of nanotechnology. Now, the question arises – What are polymers? "Poly" is a Greek word meaning "many" and "Mer" is also a Greek word meaning "unit."

On addition of polymers to the cement mix, the following mechanisms take place simultaneously:

1. *Reduction in surface tension* of water.
2. *Induced electrostatic repulsion* between particles of cement.
3. Lubricating film between cement particles.
4. *Dispersion of cement grains, releasing water trapped* within cement flocs.
5. Inhibition of surface hydration reaction of the cement particles, leaving more water to fluidify the mix.

There are various types of polymers, namely, thermoplastics, thermosets, long-chain polymers, natural polymers, homopolymer, copolymer, and inorganic polymers which are also called geopolymers (geopolymer concrete is called "concrete without cement"). As our chapter is dedicated to cement behavior, we would not go into details of the geopolymers. Polymers find application in a number of areas such as follows:

1. *Medicine*: Many biomaterials, especially heart valve replacements and blood vessels, are made of polymers like Darcon, Teflon, and Polyurethane.
2. *Consumer Science*: Plastic containers of all shapes and sizes are light weight and economically less expensive than the more traditional Containers. Clothing, floor coverings, garbage disposal bags, and packaging are other polymer applications.
3. *Industry*: Automobile parts, Windshields for fighter planes, pipes, tanks, packing materials, insulation, wood substitutes, adhesives, matrix for composites, and elastomers are all polymer applications in the industrial market.
4. *Sports*: Playground equipment, various balls, golf clubs, swimming pools, and protective helmets are often produced from polymers.
5. *Cement/Construction*: Apart from cement bags which replaced jute gunny bags, construction sector boasts of polymer-made doors/windows, overhead water tanks, roof tops, balustrades, sanitary ware, etc.

The application of polymers as shown in Figures 6.6–6.8 is increasing day by day but not as well as its production, so there exists a demand–supply gap. The dosages of polymers vary from 0% to 5% with respect to the weight of cement present in the mix. It is seen from Figure 6.1 that more the polymer addition, the more the water reduction, the low the w/c ratio,

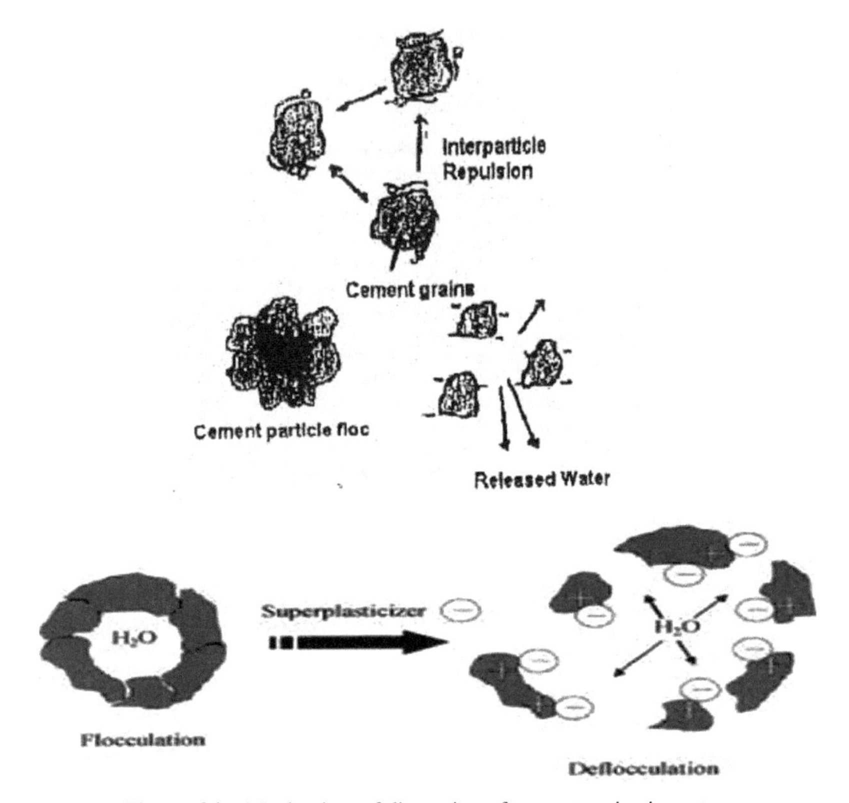

Figure 6.1 Mechanism of dispersion of cement grains in water.

and more better the fresh and long-term properties of cement concrete. Our test results confirm the fact that these issues can be addressed effectively by using polymer nano composites. As per the available literatures, normal conventional polymeric additions in cement concrete-melamine formaldehyde (MFs) give typically 16–25%$^+$ water reduction, while the modern 3rd generation polymer – polycarboxylate ether (PCE) – typically gives 20–35%$^+$water reduction giving ample scope for manipulation of the basic molecular structure of PCE at a micro/nano scale. This has currently lead to cement-concrete products targeted as precast, having very good water reduction, early strength developments, and higher compressive strength. These new generation polymers are far superior to the conventional ones with respect to the initial slump as well as the slump retention with time, which is a very important criterion for ready mix concrete (RMC) as RMC has a very high transportation time band for that transportation duration the cement should not set. The efficient working of these plasticizers/polymers is due to new molecular designs. PCEs have a main chain formed of very long polyethylenoxide-Pfropfpolymer condensate fused with carboxylate side chains, which are adsorbed on the cement particle surfaces. These side chains work on the principle "steric hindrance" (Figure 6.2). Further, the lower loss of slump with respect to time is also due to this phenomenon. In addition, since the adsorbed chains on cement grains are small, there is no retardation effect, and therefore this admixture poses no hurdles to developments to high early strengths. This would not be possible with conventional high-range water-reducing admixtures/polymers.

The mechanism of steric hindrance of PolyCarboxylate ether is given in Figure 6.3.

The side chains in PCEs act like physical spaces and produce the spatial repulsion of the cement particles, thereby providing flow to the binder paste. The disadvantages associated with longer setting times of conventional Superplasticizers are offset by PCE-based Superplasticizers, and therefore its use in cement concrete can also attain high early strengths. Moreover, the zeta potentials (the magnitude of the zeta potential indicates the degree

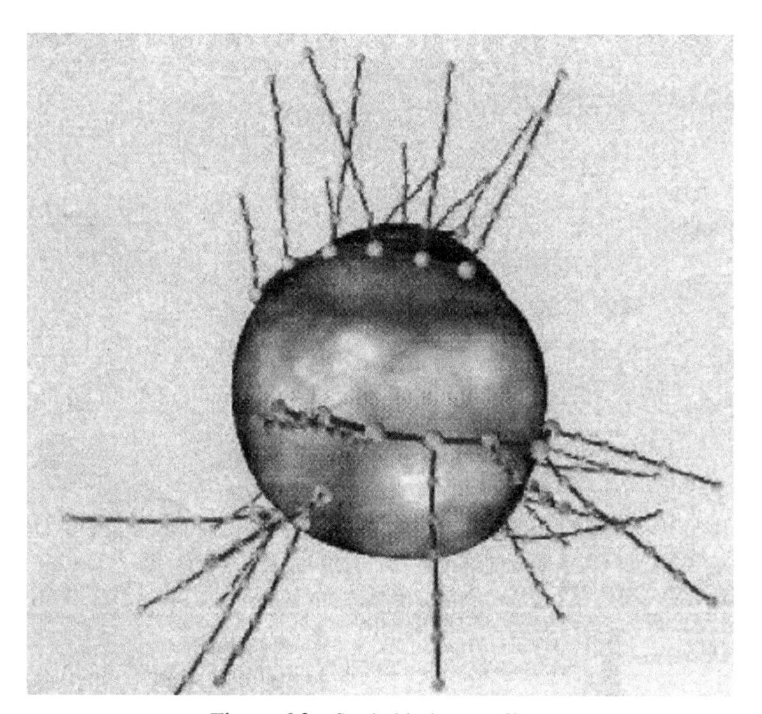

Figure 6.2 Steric hindrance effect.

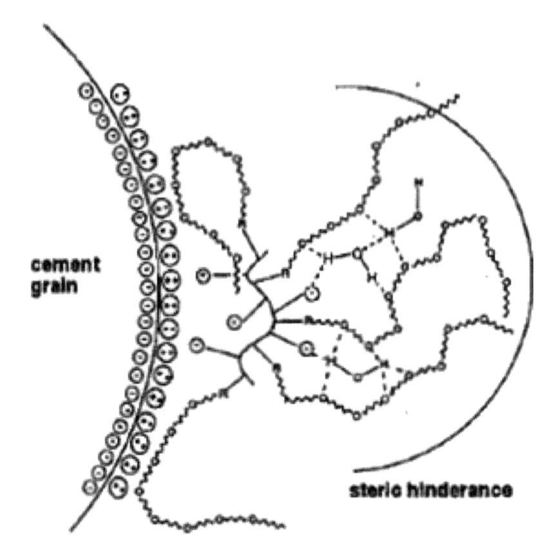

Figure 6.3 Mechanism of steric hindrance of polycarboxylate ether.

of electrostatic repulsion between adjacent, similarly charged particles in a dispersion; Figure 6.4) are negligible in case of PCE.

In studies carried out with acrylic polymer-based Superplasticizer, it seems that the mechanism of action is more related to adsorption and steric

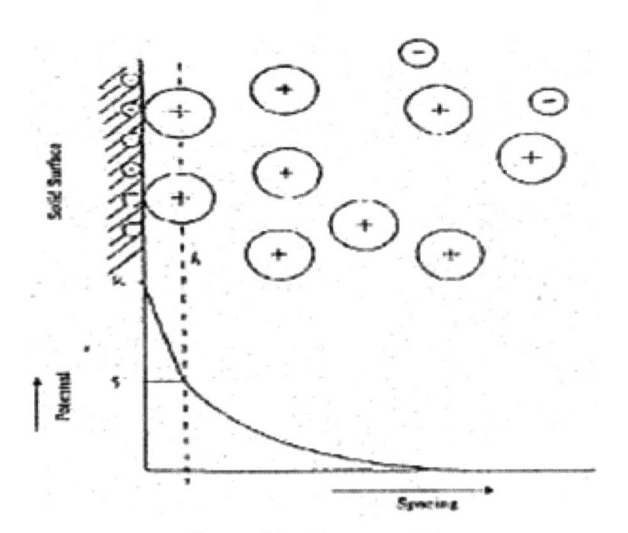

Figure 6.4 Zeta potential.

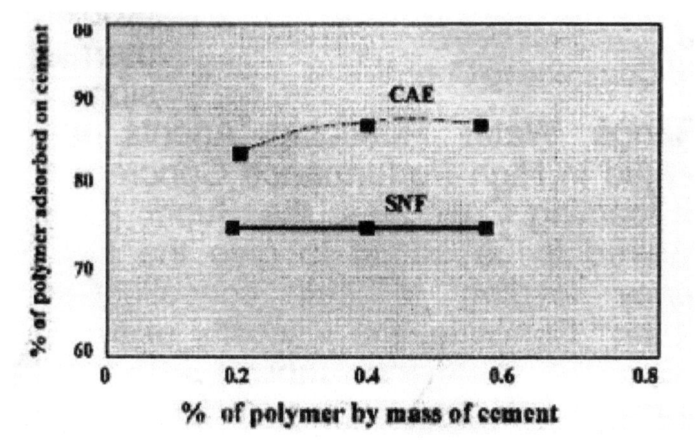

Figure 6.5 Mechanism of action in acrylic polymer-based polycarboxylate ether.

hindrance effect rather than electrostatic repulsion. Therefore, the mechanism of action in acrylic polymer (ACE)-based PCE is primarily due to higher adsorption of the polymer on cement surface when compared to the old generation-based polymer like melaminesulfonate as shown in Figure 6.5.

Concrete additives based on PCE offer advantages like

- Significant reduction of the water demand of the mix
- Little loss of consistency
- Short setting times
- High early strengths
- Low tendency to segregation
- Optimize cement contents, even in normal concretes
- Placing and compacting concrete much easier when compared to conventional polymers (applicable in self-compacting concrete)
- Can reduce water cement ratio (w/c) to less than 0.2 and thus can be applied to high-performance concrete.

The advantages of these new generation polymers are very clear, not only in terms of performance but also in terms of the dosages used for similar conditions, and this factor balances the disadvantages in economy, as new generation Superplasticizers are relatively expensive.

It is also a well-documented fact that PCE-based polymers/ Superplasticizers don't have the side effects of retardation often seen with normal retarding polymers/Superplasticizers. This is beneficial as workability time cement concrete can be controlled, but the hydration and setting time of

Figure 6.6 Images of polymers used as Superplasticizer in construction industry.

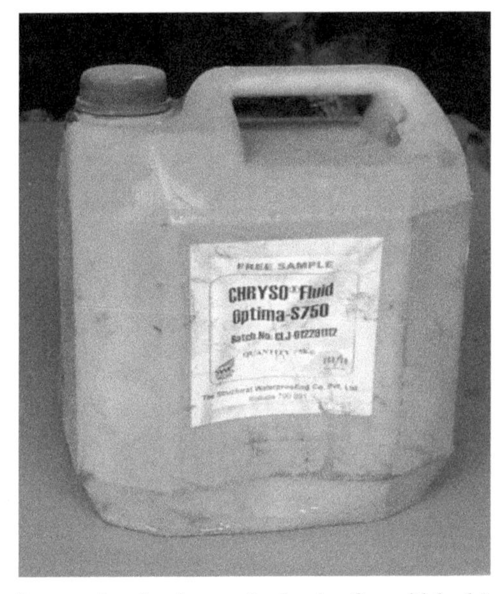

Figure 6.7 Image of conventional polymers (melamine formaldehyde) as used in laboratory.

cement will proceed unhindered. This ensures that any subsequent vibration to cement concrete after initial set will not open up capillaries, as is the case if concrete is retarded for a very long period of time, thereby rendering the concrete relatively waterproof. PolyCarboxylate ethers (PCEs) are the most recent Superplasticizer development (for the last 15 years) and continue to establish a dominant position by a growing market share in the polymer market despite its rising costs.

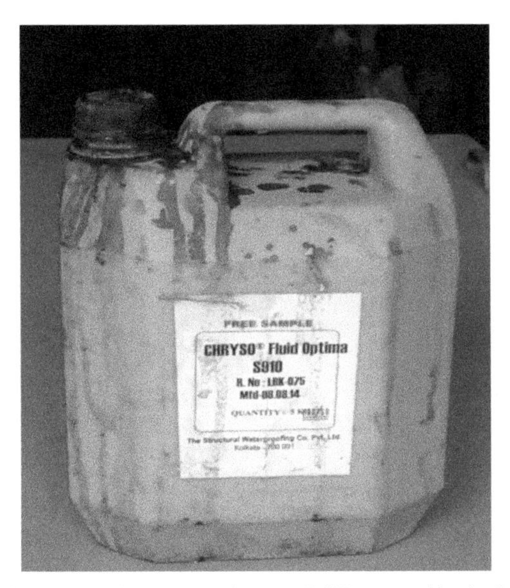

Figure 6.8 Image of modern polymers (PCE) as used in the laboratory.

Very limited research has been done on the use of nanomaterials in cement concrete involving a polymer which causes the cement matrix to densify by complex process. The term-strength of a cement-based materials is based on the hydration of its clinker components and we know that these clinker consists of various calcium silicates including Alite (tricalcium silicate, Ca_3SiO_5, sometimes formulated as $3CaO\ SiO_2$; C_3S in cement chemist notation) and Belite(dicalcium silicate, Ca_2SiO_4, sometimes formulated as $20CaO\ SiO_2$; C_2S in cement chemist notation). The final long-term strengths of cement-based materials are dependent on the amount of these cement hydrates, as well as on their crystal shapes, which may be platy, elongated, and rod-shaped. The C–S–H gel so formed along with interlocked hydration crystals forms the dense, impenetrable cement matrix.

As shown in Figure 6.9 incorporation of nanoparticles [10, 13, 17, 19] can boost up the quantity of these crystalline products and can produce remarkable improvements in the mechanical properties of cement-based materials on account of their nucleation effects thus acting like additional crystallization seeds (nucleus) facilitating further hydration at nanolevel and crack bridging effects at the nanoscale level for the formation of different types of cement hydrates. Various researchers have employed the incorporation of fibers [2, 3, 5, 6, 8, 11, 12, 18, 20] with different shapes and sizes to arrest the cracking

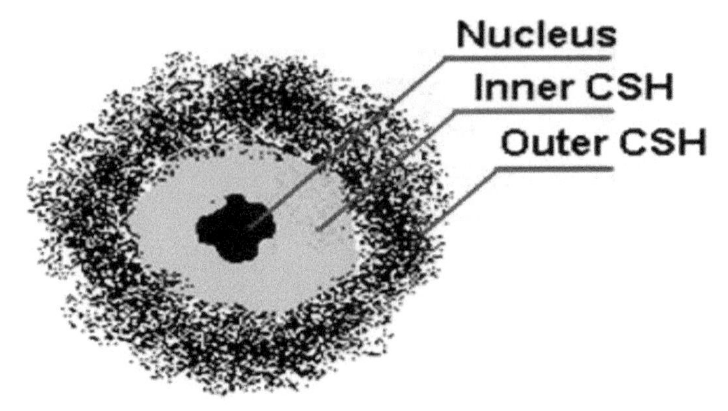

Figure 6.9 Nucleation seed for C–S–H hydrates.

phenomenon [9] but these reinforcing materials exhibit their reinforcement at a micro level and, therefore, cannot arrest the cracks originating at the nanoscale as shown in Figures 6.10 and 6.11.

The construction industry was the only industry to identify nanotechnology as a promising emerging technology in the UK Delphi Survey in the early 1990s ... ("Application of Nanotechnology in Construction", *Materials and Structures*, 37, 649 (2004), Springer). But now other industries like tyres, paints, medicines, etc., have paced ahead lagging construction industry behind. Till date, the usage of these new generation smart materials is not substantial in India mainly due to the cost factors and also due to its inherent problems in dispersion and distribution in a suitable medium as emphasized in Ministry of Commerce and Industry's (Government of India) 14th International Seminar on Cement and Building Materials, December,

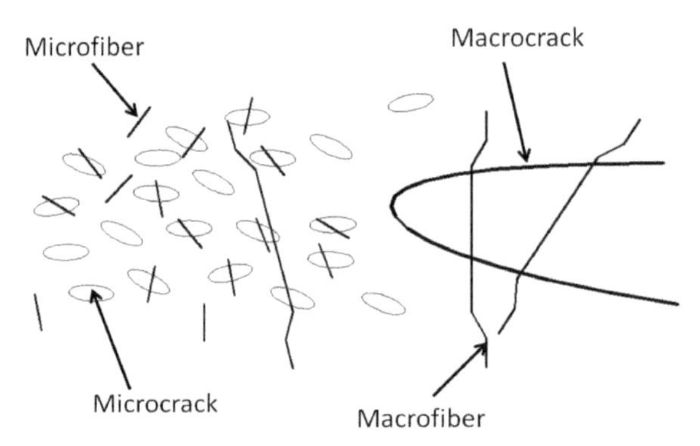

Figure 6.10 Difference between micro and macro cracks in cement concrete.

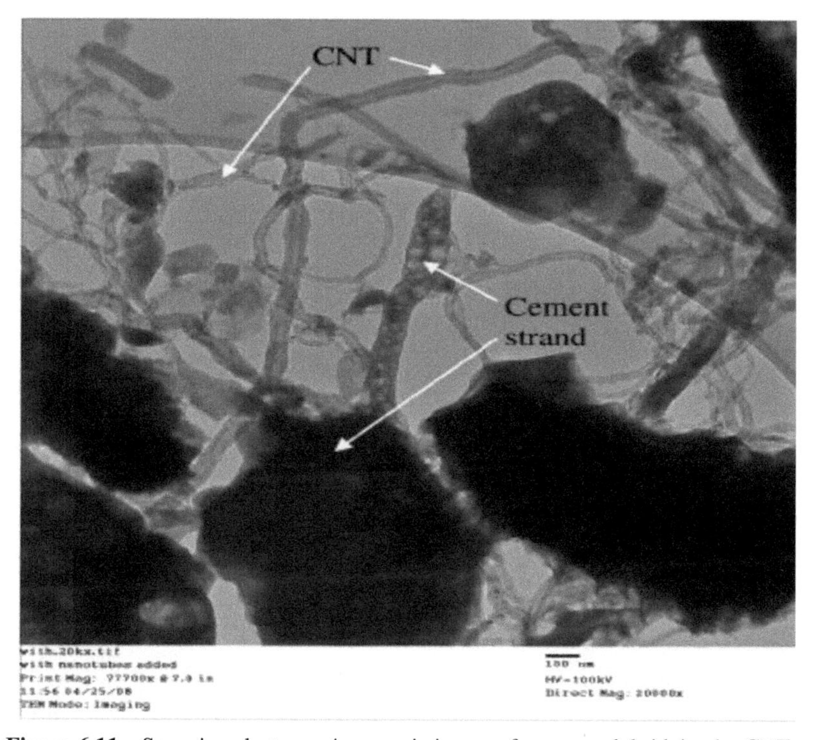

Figure 6.11 Scanning electron microscopic image of nano crack bridging by CNTs.

2015, New Delhi, that the biggest challenge faced by nanotechnology today is "distribution and dispersion." But soon the technology and the demand of infrastructure will pick up the speed with which projects are being executed in this country. Day by day, these infrastructural projects will be more specific and for that specific materials need to be added in specific concrete casting.

6.2 Experimental Program

The materials used were cement-ordinary Portland cement (OPC; 43 Grade), fine aggregate river sand conforming to Zone II of IS:383 – 1970, potable water, admixture (Superplasticizer) – MF and PCE – and nanomaterials (viz., CNTs and nano titanium oxide).

Tables 6.1 and 6.2 show the specific properties of CNTs and titanium di oxide used.

And Figures 6.12 and 6.13 show the X-ray diffraction images of nano silica, CNTs, and titanium di-oxide used.

Table 6.1 The specific properties of multiwalled carbon nanotubes (industrial grade) used

Item	Description
Diameter	20–40 nm
Length	25–45 nm
Purity	80–85% (a/c Raman spectrometer and scanning electron microscopic analysis)
Amorphous carbon	5–8%
Residue (calcination in air)	5–6% by wt.
Average interlayer distance	0.34 nm
Specific surface area	90–220 m^2/g
Bulk density	0.07–0.32 gm/cc
Real density	1–8 gm/cc
Volume resistivity	0.1–0.15 ohm.cm (measured at pressure in powder)

Table 6.2 The specific properties of nano titanium oxide (TiO_2) used

Item	Description
Nano titanium oxide %	97
Rutile content %	98
pH	7
Average particle size (transmission electron microscope)	30–40 nm
Treatment	Nil
Moisture %	1.75–2
Bulk density	0.31 gm/cc
Water solubility	Insoluble

6.2.1 Tests on Cement Mortar

Mortar cubes of 70.7 mm 70.7 mm 70.7 mm size were casted with one part of cement + three parts of sand with water added as per the normal consistency formula of Indian standards, i.e., according to the standard formula P' = (P/4 +3)(one part cement + three parts sand). Here P' = quantity of water and P = consistency of cement used, i.e., amount of water used to make 300 g cement paste to support a penetration of 5–7mm in a standard Vicat mold with a Vicat needle. CNTs added in proportions as per literature review, i.e., 0.02%, 0.05%, and 0.1% and nano titanium oxide added in proportions ranging from 1.0% and 2.5% with respect to cement weight after proper dissolutions in a suitable Superplasticizer (PCE) (for CNTs and TiO_2 as they were insoluble in water as shown in Figure 6.14) keeping the w/c ratio fixed at 0.4. The cubes

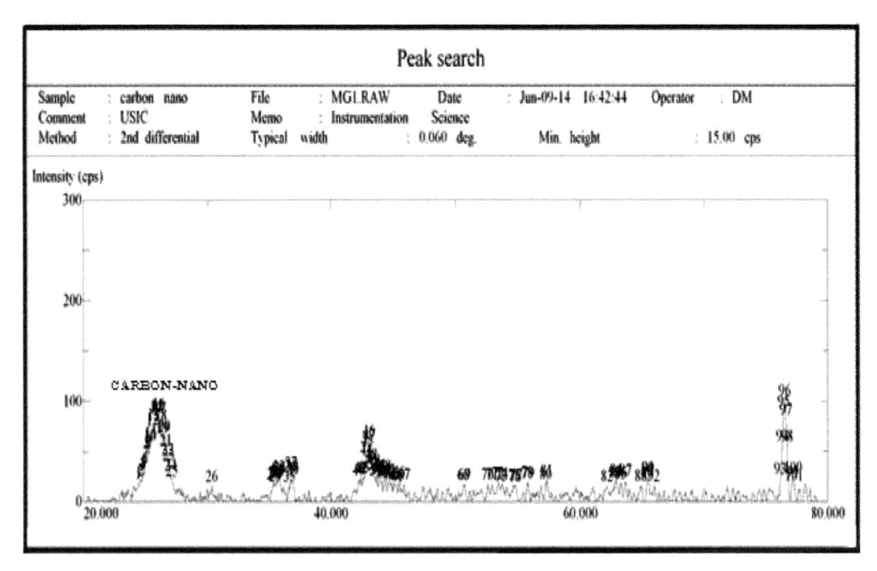

Figure 6.12 X-ray diffraction image of carbon nanotubes used.

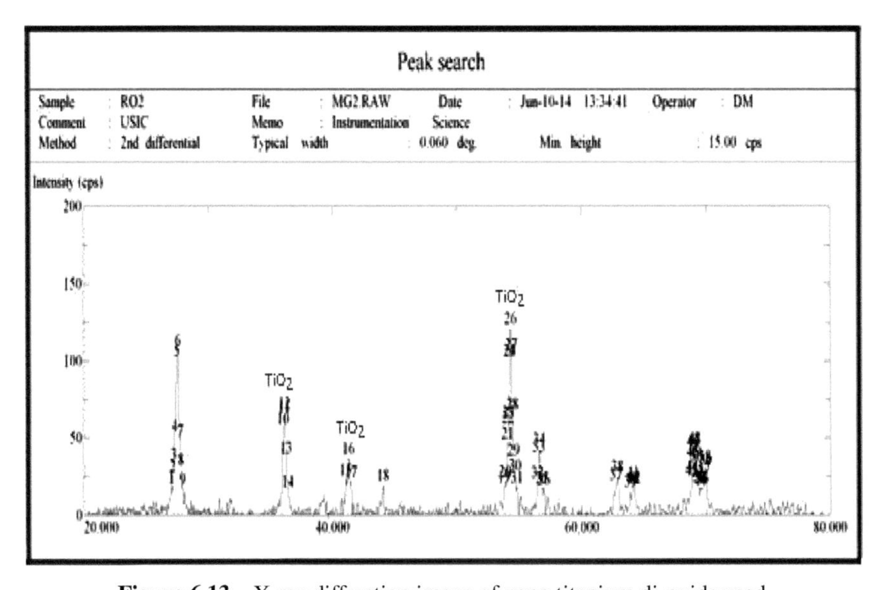

Figure 6.13 X-ray diffraction image of nano titanium di-oxide used.

were then ordinary cured under water and tested at 7 days, 28 days, 90 days, 180 days, and 365 days.

Figure 6.14 Some apparatus used in performing the tests for polymer nano composites on cement mortar.

Test Data

(A) Specific gravity of cement = 3.08 (as laboratory experiment suggests)
(B) Chemical admixture = Superplasticizer (MF /PCE) = solid content = 30%

6.3 Results

Table 6.3 shows the strength development at various ages in cement mortar.

6.4 Discussions of Test Results

The better performance of PCEs may be attributed to the fact that polymers such as PCEs cause dispersion of cement grains by steric hindrance [which is also increased multifold by CNTs addition] while electrostatic dispersion of cement grains takes place for MFs as shown in Figures 6.18 and 6.19.

Table 6.3 Strength (N/mm^2) for various proportions/ages of nano-added OPC mortar (% increase with respect to ordinary control cement cubes)

Sl. No.	% Nano Additions in Cement (OPC)	1 Day Strength	3 Day Strength (% Increase)	7 Day Strength (% Increase)	28 Day Strength (% Increase)	90 Day Strength (% Increase)	180 Day Strength (% Increase)	365 Day Strength (% Increase)
1.	OPC	19.59	23.72	21.08	31.89	31.2	30.01	30.01
2.	OPC+1%MF	Samples not demolded	13.68 (–42.33%)	28.59 (35.63%)	41.15 (29.04%)	33.00 (5.77%)	35.51 (15.49%)	29.06 (–3.15%)
3.	OPC+1%MF + 0.02% CNT	Samples not demolded	15.86 (–33.13%)	30.85 (46.35%)	31.84 (–0.15%)	45.16 (44.74%)	29.06 (–3.15%)	26.02 (–13.29%)
4.	OPC+1%PCE	Samples not demolded	Samples not demolded	18.88 (–10.43%)	28.73 (–9. 91%)	36.03 (15.48%)	33.23 (10.73%)	42.40 (41.28%)
5.	OPC+1%PCE+ 0.02% CNT (optimized at 28 days)	Samples not demolded	11.03 (–53.5%)	17.69 (–10.4%)	43.75 (38.7%)	35.59 (15.48%)	30.89 (10%)	28.53 (–4.93%)
6.	OPC+1%PCE+ 0.05% CNT	Samples not demolded	7.98 (–66.36%)	27.19 (–16.1%)	34.88 (37.2%)	31.85 (14.07%)	38.55 (3.0%)	41.69 (38.92%)
7.	OPC+1%PCE + 0.1% CNT	Samples not demolded	14.21 (–40.09%)	21.69 (28.9%)	24.83 (9.37%)	31.5 (2.08%)	30.16 (23.55%)	50.78 (69.21%)
8.	OPC+1%PCE + 1% TiO_2 (optimized at 28 days)	Samples not demolded	27.81 (17.24%)	25.24 (19.73%)	36.71 (12.59%)	35.92 (15.13%)	33.42 (11.36%)	41.16 (37.15%)
9.	OPC+1%PCE+ 2.5% TiO_2	Samples not demolded	21.67 (18.74%)	20.34 (–3.51%)	34.97 (9.58%)	37.80 (21.15%)	40.95 (36.45%)	28.16 (–6.16%)

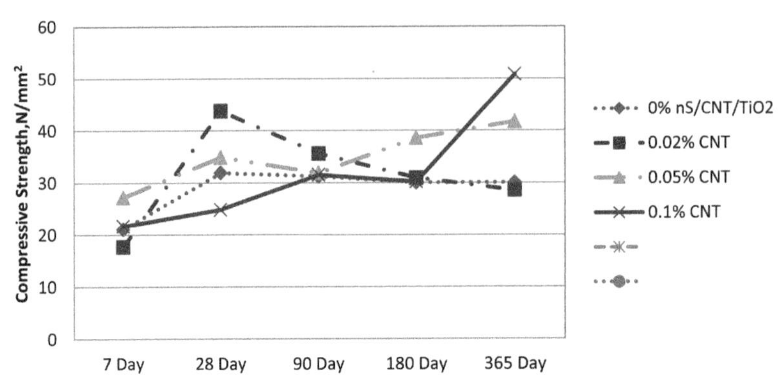

Figure 6.15 Chart showing strengths at various ages of different % of PolyCarboxylate ether (PCE)-carbon nanotube (CNT) addition in ordinary Portland cement (OPC) mortar.

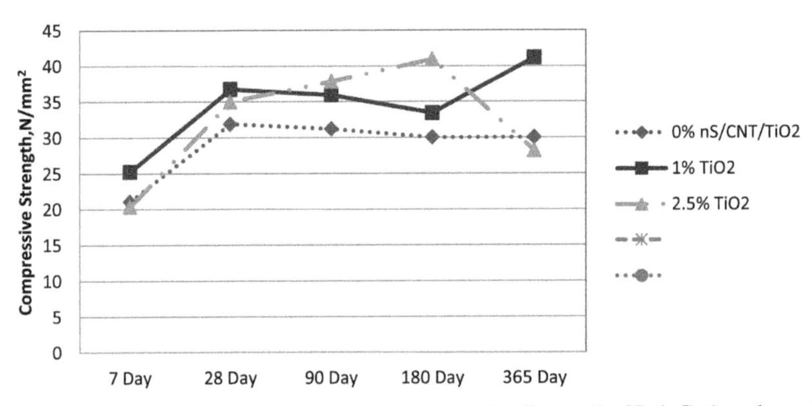

Figure 6.16 Chart showing strengths at various ages of different % of PolyCarboxylate ether (PCE)-nano TiO_2 addition in ordinary Portland cement (OPC) mortar.

(CNT: carbon nanotube)

6.4.1 Physical Characteristics

1. Polymeric additions [PCE] in cement resulted in a long setting time [more than 3 days]

2. Nano-additions considerably reduced the setting time [less than 1.5 days].

6.4.2 Dispersion Mechanism

Melamine formaldehyde (Figure 6.17a) has a single molecular structure where a melamine ring is predominant whereas PolyCarboxylate ether (Figure 6.17b) is a polymer-based structure with a polymer backbone and varying molecular weight. When water is added to cement, the grains are not uniformly distributed throughout the water but tends to form into small lumps or flocs. These flocs trap water within them causing the mix to be less mobile and fluid than would be the case if the cement were in the form of individual grains. Dispersants' MFs or PCEs adsorb into the cement surfaces and break up the flocs, leaving individual cement grains, which can pass each other easily, making the mix more fluid.

There is no change in volume of the mix, just the release of water trapped within the floc and the ability of individual cement grains to reorientate to

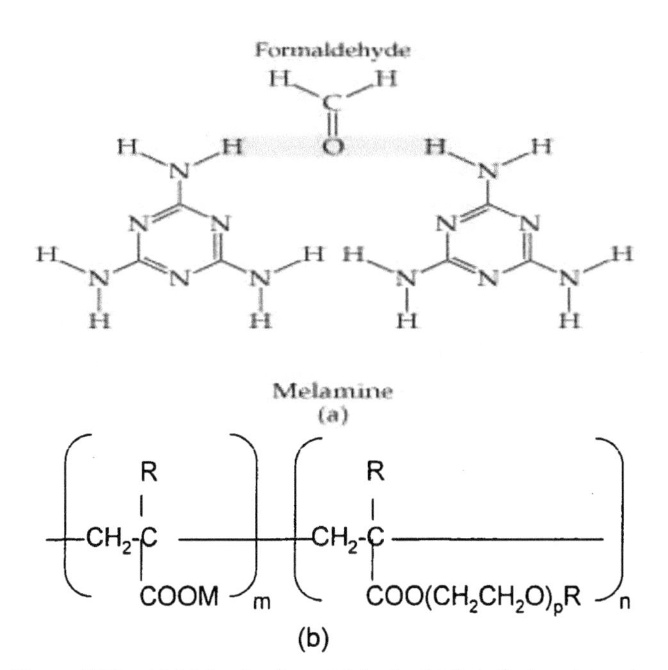

Figure 6.17 (a) Melamine formaldehyde. (b) PolyCarboxylate ether.

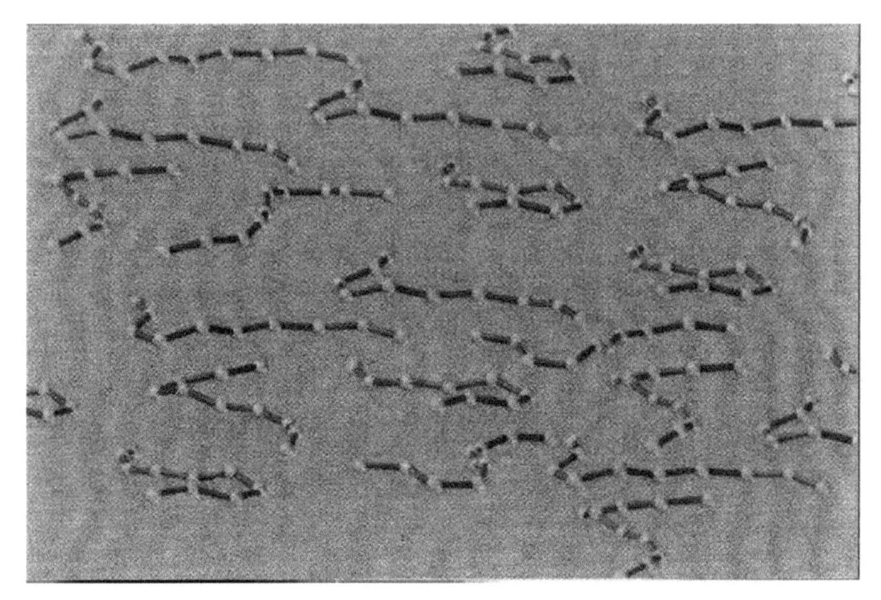

Figure 6.18 Electrostatic effect of melamine formaldehyde.

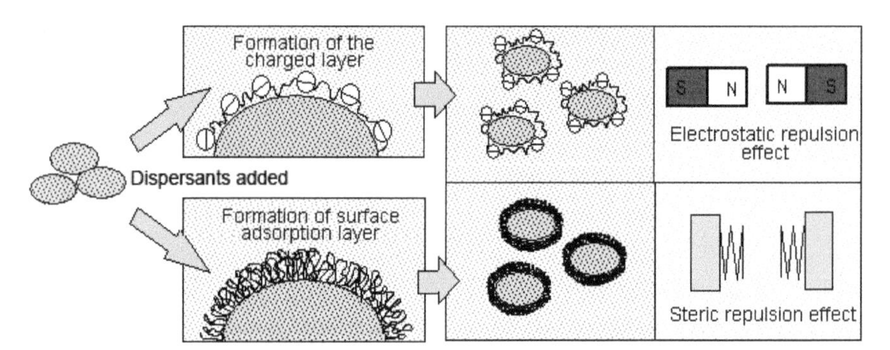

Figure 6.19 Difference between electrostatic repulsion and steric hindrance.

make more efficient use of all the available space within interparticle cement thus making the hydration process easier.

These dispersants especially PCEs are basically surface-active chemicals consisting of long-chain organic molecules, having a polar hydrophilic group (water attracting, such as –COO, $-SO_3$, and $-NH_4^+$) attached to a non-polar hydrophobic chain (water repelling) with some polar groups (–OH). The polar groups in the chain get adsorbed on the surface of cement grains, and the hydrophobic ends (with polar hydrophilic groups at the tip) project outward from the cement grain. The hydrophilic tip is able to reduce the surface tension of water and the adsorbed polymer keeps the cement particles apart by electrostatic repulsion [1]. This phenomenon results in separation of the PCE molecules from each other due to bulky side chains thus making steric hindrance more effective than electrostatic repulsion. The side chains of PCEs extend to the surface of cement particles, migrate in water, and the cement particles are dispersed by the steric hindrance of the side chains [4].

6.4.3 Compressive Strength

1. The mortar compressive strength determined as per IS:4031 shows that for CNTs, the gain in strength was 38.7% at 28 days but falling to 15.48% at 90 days and 10% at 180 days. Optimized TiO_2 indicated no such appreciable gain in strength as shown in Figure 6.16.
2. In the long-term strength (i.e., @ 365 days), it is noticed that with slight increased additions of nanomaterials in excess of the optimized, one produces more strength gain. For example, increasing 0.75% nS to 1% nS, a strength gain of 8% is seen instead of 4.96% whereas increasing the CNT to 0.05% and 0.1% gives a gain of 38.92% and 69.21% instead of a strength loss of 4.93% for the optimized 0.02% CNT as shown in Figure 6.15.

6.5 Conclusions

1. The results showed that the optimizations for nanomaterials in PCE are CNT = 0.02% and TiO_2= 1.0% for cement mortar up to 28 days. In the long-term strength, some contradictions were noticed for which the reasons are not clear. However, no conclusions can be drawn for nanomaterials in MF.
2. Polycarboxylate ethers (PCEs) are much superior to MF in performance, but the cost factor is still weigh in favor for MF and a balance of economy is the need for the hour.
3. Further research on microstructural studies is necessary for the characterization of nanomaterials in cement.

References

[1] Surlaker, S. (1989). Admixtures and curing for concrete durability. *Civ. Eng. Constr. Rev*. 1989, 24–29.

[2] Xu, G., and Hannant, D.J. (1991). Synergistic interaction between fibrillated polypropylene networks and glass fibres in a cement-based Composite. *Cem. Concr. Compos.* 13, 95–106.
[3] Marikunte, S., Aldea, C., and Shah, S.P. (1997). Durability of glass fiber reinforced cement composites:: Effect of silica fume and metakaolin. *Adv. Cem. Based Mater.* 5, 100–108.
[4] Newman, J., and Choo, S. B. (2003). *Advanced Concrete Technology.* Amsterdam: Elsevier, 4–12.
[5] Bezerra, E. M., Joaquim, A. P., Savastano, H. Jr., John, V. M., and Agopyan, V. 2006. The effect of different mineral additions and synthetic fiber contents on properties of cement based composites. *Cem. Concr. Compos.* 28, 555–563.
[6] Won, J. P., Hong, B. T., Choi, T. J., Lee, S. J., and Kang, J. W. 2012. Flexural behaviour of amorphous micro-steel fibre-reinforced cement composites. *Compos. Struct.* 94, 1443–1449.
[7] Kumar, S., Asce, F., Kolay, P., Asce, M., Malla, S., and Mishra, S. (2012). Effect of multiwalled carbon nanotubes on mechanical strength of cement paste. *J. Mater. Civ. Eng.* 24, 84–91.
[8] Uysal, M., and Tanyildizi, H. 2012. Estimation of compressive strength of self compacting concrete containing polypropylene fiber and mineral additives exposed to high temperature using artificial neural network. *Constr. Build Mater.* 27, 404–414.
[9] Yuvraj, S. (2012). Experimental research on improvement of concrete strength and enhancing the resisting property of corrosion and permeability by the use of nano silica flyashed concrete. *Int. J. Emerg. Technol. Adv. Eng.* 2, 105–110.
[10] Abyaneh, M. R. J., Mousavi, S. M., Mehran, A., Hoseini, S. M. M., Naderi, S., and Irandoost, F. M. (2013). Effects of nano-silica on permeability of concrete and steel bars reinforcement corrosion. Aust. J. Basic Appl. Sci. 7, 464–467.
[11] Abeysinghe, C., Thambiratnam, D. P., and Perera, N. J. (2013). Flexural performance of an innovative hybrid composite floor plate system comprising glass–fibre reinforced cement polyurethane and steel laminate. *Compos. Struct.* 95, 179–190.
[12] Yoo, D. Y., Lee, J. H., and Yoon, Y. S. (2013). Effect of fiber content on mechanical and fracture properties of ultra high performance fiber reinforced cementitious composites. *Compos. Struct.* 106, 742–753.

[13] Ghosh, A., Sairam, V., and Bhattacharjee, B. (2013). Effect of nano-silica on strength & microstructure of cement silica fume paste, mortar and concrete. *Indian Concr. J.* 2013, 11–25

[14] Madhavi, T., et al. (2013). Effect of multiwalled carbon nanotubes on mechanical properties of concrete. *Int. J. Sci. Res.* 2, 166–168.

[15] Rajmane, N. P., et al. (2013). "Application of nano-technology for cement concrete," in *Proceedings of National Seminar, Dec 15-18*, IE(I), Karnataka centre.

[16] Ghosal, M., and Chakraborty, A. K. (2014). "Effect of Nano-Materials in Building Materials," in *Proceedings of the Seminar Volume of National Seminar and 29th National Convention of Architectural Engineers-Innovative World of Building Materials, Jan. 31st & Feb 1st, 2014*, IE(I), West Bengal state centre.

[17] Ghosal, M and Chakraborty, A. K. (2014). "A technical comparison on different nano-incorporations on cement composites," in *Proceedings of the Seminar Volume of 2nd International Conference on Nano Structured Materials & Nano Composites (ICNM 2014), 19–21 Dec. 2014, International & Inter University Centre for Nanoscience & Nanotechnology (IIUCNN), Mahatma Gandhi University*, Kottayam.

[18] Tokgoz, S., Dundar, C., and Tanrikulu, A. K. (2014). Experimental behaviour of steel fiber high strength reinforced concrete and composite columns. *J. Constr. Steel Res.* 74, 98–107.

[19] Ghosal, M., and Chakraborty, A. K. (2015). A comparative study of nano embedments on different types of cements. Int. J. Adv. Eng. Technol. 8, 92–103.

[20] Shu, X., Graham, R. K., Huang, B., and Burdette, E. G. (2015). Hybrid effects of carbon fibers on mechanical properties of Portland cement mortar. *Mater. Des.* 65, 1222–1228.

7

Effect of Structure of Diphenol on Polymerization of Bis(isoimide)

V. Sarannya[1], R. Surender[2], S. Shamim Rishwana[3], R. Mahalakshmy[1], and C. T. Vijayakumar[2]

[1]Postgraduate and Research Department of Chemistry, Thiagarajar College, Madurai, India
[2]Department of Polymer Technology, Kamaraj College of Engineering and Technology, Virudhunagar, India
[3]Department of Chemistry, Kamaraj College of Engineering and Technology, Virudhunagar, India

Abstract

The compound 4,4'-bisisomaleimidodiphenyl methane (VS) was prepared. It was blended separately with different diphenols [resorcinol (VS 1), quinol (VS 2), 4,4'-dihydroxybiphenyl (VS 3), and bisphenol-A (VS 6)] in the mole ratio 1:1, and the materials were thermally polymerized. The thermal polymerization of these blends was studied using differential scanning calorimeter (DSC). The DSC studies show two discrete exotherms and also significant changes in the melting behavior. All the polymers from the blends show considerably lower char yield indicating lower thermal stability compared to PolyVS. The behavior may be attributed to the structure of the resulting copolymer from the blend.

7.1 Introduction

7.1.1 High-performance Thermosetting Resin

Thermosetting resins are reactive low-molecular-weight oligomers or high-molecular-weight polymers having reactive functional groups for crosslinking reactions. After thermosetting process, these resins become infusible and insoluble materials. Typical high-performance uncrosslinked polymers include polyimides, polybenzobisoxazoles, aromatic polyamides, polybenzimidazoles, liquid crystalline polyesters, polyether ether ketones, polyether sulfones, polyamide–imides, polyquinoxalines, etc. Most high-performance linear polymers suffer from their poor solubility in organic solvents, extremely high melting points, and high melt viscosities for fabrication. The processability of high-performance polymers can be achieved only by the reduction of their molecular weight. However, this approach results in a significant decrease in their final physical properties. The solution to this is to reduce the size of the polymer molecule for fabrication and to incorporate reactive functional moieties on the molecule for further reaction. The reactive groups eventually polymerize upon curing and give crosslinks to further improve material performance. The curing reaction should provide not only crosslinking but also additional rigid ring structures. This becomes the basic concept for high-performance thermosets.

The various high-performance oligomeric segments such as amide, amide-imide, ester, sulfone, ether ketone, imide, quinolone, and quinoxaline have been synthesized from a number of different monomers to meet performance requirements. Because of the molecular weight reduction, the properties of high-performance thermosets become significantly dependent on the types of reactive functional groups. The end-capping groups, such as epoxy, maleimide, nadiimide, cyanate ester (CE), acetylene, arylethynyl biphenylenebenzocyclobutene, phthalonitrile, styrenyl, etc., are used to prepare usual high-performance thermosets such as bismaleimides (BMIs), bis nadimides, CEs, arylethynyl resins, acetylene-terminated resins, and benzocyclobutene resins. Some of the most common thermosetting resins are discussed below.

7.1.2 Alkyd Resins

Alkyd resins are polyesters used primarily in organic paints, with some molding applications. The alkyd resin is based on acids (phthalic anhydride, maleic anhydride, isophthalic acid, and terephthalic acid) and polyhydric alcohols (glycols, glycerol, pentaerythritol, and sorbitol). These resins are modified by the addition of fatty acids derived from mineral and vegetable oils. Alkyds of the glycerol-phthalic acid type or glycerol-sebacic acid are used in lacquers along with natural resins such as shellac to impart flexibility. Drying alkyds contain, in addition to the glycerol phthalate components, some oils or acids of drying oils. They are used in varnish type coatings. Phenol-aldehyde or urea- or melamine-aldehyde resins may be added to improve hardness, and hard resins are made from maleic anhydride combined with glycerol and rosin. Their function is to improve film hardness and gloss in both lacquer and varnish coatings [1].

7.1.3 Amino Resins

Amino resins or aminoplasts are prepared from the condensation of either urea with formaldehyde (UF resins) or melamine with formaldehyde (MF resins). UF resins are prepared by the nucleophilic addition of urea to formaldehyde to give methylol derivatives. Subsequent condensation of these derivatives gives the final high-molecular-weight resin. As amino resins are clear and colorless, objects of light or pastel color can be produced. Amino resins are used as adhesives for plywood and furniture. The amino resins modify textiles such as cotton and rayon by imparting crease resistance, stiffness, shrinkage control, fire retardance, and water repellency. The urea-based enamels are used for refrigerator and kitchen appliances and the melamine formulations in automotive finishes.

7.1.4 Unsaturated Polyester Resins

Polyester resins are widely used in construction and marine applications. Unsaturated polyester resins, which are used as the matrix component of glass fiber composites, can be obtained by copolymerization of both saturated acids (phthalic anhydride) and unsaturated acids (maleic anhydride) with a diol such as propylene glycol or diethylene glycol. The low-molecular-weight product is soluble in styrene, which can then polymerize in the presence of peroxides at the double-bond sites of the prepolymer [2]. The major applications of glass reinforced polyester resins fall in the following categories: boat hulls, transportation including passenger car parts and bodies and truck cabs, and consumer products, including such diverse items as luggage, chairs, trays, pipes, and ducts.

7.1.5 Allyl Resins

Allylic monomers are not used directly in chain reaction or addition polymerization because of the stability and low reactivity of the allyl radical; diallyl esters can be crosslinked by polymerization through their double bonds to give thermosetting resins. Prepolymers of diallyl phthalate and diallyl isophthalate are used as molding compounds and in the production of glass fiber reinforced laminates. They are cured by peroxide catalysts to heat- and chemical-resistant products with good dimensional stability and electrical insulating properties.

7.1.6 Epoxy Resins

Epoxies are the most versatile class of polymers and have diverse applications; they are probably the best known thermoset polymers. The largest use of epoxies is in protective coatings; other applications include printed circuit boards, laminates, electronic materials, and structural composites. Cured epoxies provide excellent mechanical strength; toughness; outstanding resistance to chemicals; resistance to moisture and corrosion; good thermal, adhesive, and electrical properties; the absence of volatiles and low shrinkage on cure; and good dimensional stability. This unique combination of properties is, in general, not found in any other plastic material. The patent literature indicates that the synthesis of epoxy resins was discovered as early as the late 1890s. These materials are used as antenna reflectors, solar panel arrays, and optical support structures in satellites. Epoxy-polyaromatic composite motor cases have replaced the heavy metallic counterparts in modern launch vehicles. Satellites and other components are also made of epoxy/carbon and epoxy/polyaromatic composites [3].

7.1.7 Polyurethanes

Polyurethanes are carbonic acid derivatives and polymers containing the group –NH-COO– formed through the reaction of a diisocyanate and a dihydroxy compound. Polyurethanes are used for adhesives, coating, elastomer, fibers, and foams. Flexible foams are usually prepared from dihydroxy polyesters or polyethers and rigid foams from polyhydroxy prepolymers. The flexible urethane foams are used as cushions in furniture and automobiles. Rigid foams are used for thermal insulation in refrigerators, portable insulated chests, and for imparting buoyancy to boats.

7.1.8 Silicone Resins

Silicones are high-molecular-weight linear polymers, usually polydimethylsiloxane. They can be cured in several ways: i) by free radical crosslinking with, for example, benzoyl peroxide, through the formation of ethylenic bridges between chains, ii) by crosslinking of vinyl or allyl groups attached to silicon through reaction with silylhydride groups, and iii) by crosslinking linear or slightly branched siloxane chains having reactive end groups such as silanols which yields Si–O–Si cross links. Silicone elastomers are used as gaskets and seals, wire and cable insulation, and hot gas and liquid conduits. Also they are valuable in surgical and prosthetic devices. These resins are used as insulating varnishes, impregnating and encapsulating agents, and in industrial paints.

7.1.9 Cyanate Ester Resins

Cyanate ester (CE) resins have received considerable attention in the past few years due to their importance as thermosetting resins for use as encapsulants in electronic devices, high-temperature adhesives, and structural materials in aerospace applications because of their outstanding mechanical, thermal, and adhesive properties [4, 5]. This new generation of thermoset resins encompasses the processability of epoxy resins, thermal characteristics of BMI resins, and the heat resistance and fire resistance of phenolic resins [6]. The CE resins have their own unique properties such as good strength, low dielectric constants, radar transparency, low water absorption, and superior metal adhesion, which make them the resin of choice in high-performance structural applications in the electronics and aerospace industries [7, 8]. The CE resin is formed in excellent yields by the reaction of the corresponding phenols with cyanogen halides.

Like other thermoset resins, CE resin is amenable to processing by a large variety of conventional techniques. Their processing versatility has gained them wide spread acceptability in composites for various applications. The flexibility is further enhanced by blending with other resins such as epoxies, BMIs, additives, and toughening agents. The cure cycle is dependent upon the catalytic level. Partially polymerized thermosetting resins retain the ability to fuse and to form further crosslinks achieving good tackiness.

7.1.10 Phenolic Resins

Phenolics are the earliest synthetic polymers developed in the world. Since their introduction in 1910, the relative inexpensive phenolic resins have played an important role in construction, automotive, electrical, and appliance industries, and also in high technology applications in electronics and aerospace. Phenolic resins are almost always reinforced with fibers or fillers, such as wood flour, cotton flock or fibrous glass, and carbon fiber, mainly to improve properties and also to reduce cost [9–11].

Traditional phenolic materials are the crosslinked products of their low-molecular-weight precursors, either of novolac or of resole type. These approaches have been successful in bringing about the first class of synthetic plastic materials since the turn of century. The materials obtained exhibit good heat resistance, flame retardance, and dielectric properties. Thus, they have been widely used in construction, household, and electrical facilities. Another notable advantage is that the raw materials and the fabricating process are very inexpensive.

However, there are a number of shortcomings associated with these traditional phenolic resins, namely, the brittleness of the materials, release of water and ammonia during the curing processes due to condensation reactions, the use of strong acids as catalysts, and poor shelf life. The use of phenol and the process to synthesize novolac or resole precursors in aqueous medium create significant threats to environment as well as to human health. Thus, the revitalization of phenolic materials for high-performance composite and electronic applications depends, to a large extent, on our ability to create new chemistry and processes to overcome the aforementioned shortcomings. When compared with many known thermally stable polymers, their thermo-oxidative stability is low. The rigid aromatic units tightly held by short ethylene linkages make the matrix brittle.

7.1.10.1 Allyl functional phenols

Allyl phenol–formaldehyde novolac, synthesized by the allylation of novolac, can cure thermally at 180°C without the evolution of volatiles. The allyl derivatives of phenols have been used for the manufacture of glass fiber reinforced plastics and moldings, as well as casting or impregnating compositions with high heat resistance, mechanical strength, and chemical stability [12].

7.1.10.2 Bisoxazoline phenols

The unusual addition coreaction of novolac phenolic resins with phenylene bisoxazoline has been explored to derive a new class of non-conventional phenolic thermosetting resins by Culbertson et al. [13]. The polymerization involves a tertiary phosphine catalyzed reaction of bisoxazoline with a phenol free novolac resin, leading to an ether–amide copolymer. These materials have low cure shrinkage, high neat resin modulus, no volatiles during cure, low coefficient of thermal expansion, and excellent toughness. The fiber reinforced copolymers possess the low smoke and heat release requirements of materials for aircraft interior applications. Based on this chemistry, several compositions with many interesting properties have been patented. Physical, mechanical, and electrical properties of the neat resin suggest that these new thermosets could be useful in various electrical applications.

7.1.10.3 Phenolic resins epoxy systems

Curing of epoxy with novolac by making use of the OH–epoxy reaction appears to be the simplest way to design an addition cure phenol system [14, 15]. Although less preferred, polyphenols are used as curative for epoxies because the addition curing results in void-free products that are comparatively tougher due to the formation of flexible ether networks [16, 17]. Interest in these systems has been revived further by the need for void-free, low-moisture absorbing matrices with low dielectric properties for various electronic applications [18–20]. Their cure kinetics has been studied extensively [21].

7.1.11 Polyimide

Imide is a functional group with two acyl groups bound to nitrogen. They are less reactive, but on the other hand, commercially they are components of high-strength polymers. Many high-strength and electrically conductive polymers contain imide subunits. They are structurally related to acid anhydrides. An imide can be prepared by treating an acid anhydride with primary amine.

Polyimide is a polymer of imide monomers. Since 1955, its mass production has been started. The typical monomers used are pyromellitic anhydride and 4,4'-oxydianiline. Based on the types of interactions, they are classified as thermoplastic and thermosetting polyimides. Thermosetting polyimides are available as uncured resins, polyimide solutions, stock shapes, thin sheets, laminates, and machined parts. The exceptional chemical resistance and thermal stability of polyimide resins have found them in numerous applications. Polyimides are incredibly strong synthetic polymers that are also astoundingly heat and chemical resistant. Their properties are so great that these materials often replace glass and steel in many demanding industrial applications. They are used for the struts and chassis in some cars as well as some parts under-the-hood because they can withstand the intense heat and corrosive lubricants, fuels, and coolants cars require. Polyimides are even used in many everyday applications such as microwave cookware and food packaging because of their thermal stability and resistance to oils, greases, and fats. They can also be used in circuit boards, insulation, fibers for protective clothing, composites, and adhesives.

7.1.11.1 Classification of polyimides

Polyimides exist in two forms

(1) The first of these structures is a heterocyclic structure where the imide group is part of a cyclic unit in the polymer chain.

(2) The second is a linear structure where the atoms of the imide group are part of a linear chain.

Aromatic heterocyclic polyimides are typical of most commercial polyimides, such as Ultem from Sabic (originally invented by G.E. Plastics) and DuPont's Kapton resin. These polymers have such incredible mechanical and thermal properties that they have replaced some metals and glass in high-performance applications in the electronics, automotive, and even the aerospace industries. These properties come from strong intermolecular forces between the polymer chains. A polymer which contains a charge transfer complex consists of two different types of segments, a donor and an acceptor. The donor has a plenty of electrons to go around because of its nitrogen groups. The acceptor, carbonyl groups, takes away its electron density. The chains will stack together like strips of paper with donors and acceptors paired up and pulling the chains more tightly together. This charge transfer complex holds the chains so tightly that they can't move around very much. Hence, polyimides are so strong and they don't melt or soften except at really high temperatures.

The interactions are so strong, which makes the material difficult to process or do anything with. So it may become necessary to make the polymer a little softer so that it can be made into something useful. This is accomplished at the molecular level by using monomers that have more flexible segments such as the bisphenol-A derived linkage. This is another interesting property of polyimides which makes them excellent for use in construction and transportation industries.

7.1.11.2 Properties of polyimide

Polyimides are having the following properties.

1. Good mechanical properties
2. Good electrical properties
3. High wear resistance
4. Low creep at high temperatures
5. Good compression with glass or graphite fiber reinforcement

6. Good chemical resistance
7. Inherently flame-resistant character
8. Excellent resistance to most solvents and oils.

7.1.12 Bismaleimides (BMIs)

Condensation and addition imide resins are two different types of imides available for high-performance matrix applications. Condensation polyimides are typically based on high-molecular-weight poly amic acid precursors that release void forming volatile by products during the thermal imidization process. Addition polyimides are usually low-molecular-weight resins containing unsaturated moieties such as maleimido or nadimido or ethynyl end-capping groups for subsequent crosslinking reactions.

Among addition imides, BMIs are used for advanced material applications due to its high performance to cost ratio and provide fatigue resistance and prevention of fracture propagation. Owing to less mobility of atoms in cured BMIs, they have in general high glass transition temperature. In a small molecular unit, there are more double bonds. There lies the chance for all the double bonds to be crosslinked. Due to this, the crosslinking densities of BMIs are large. When the crosslinking density increases, the brittleness of the cured BMIs resin also increases. This reduces the toughness of the component and so the component will fail even when small loads are applied. Brittleness of cured BMIs resin is of major concern for structural applications.

The structural variations of BMIs resin depend only on the imagination of the researcher. The design of a structural material requires the consideration of several criteria such as toughness, tensile modulus, compression modulus, and dimensional stability in a predefined temperature range and long-term performance. Various approaches such as basic structure change of BMI resins, chain extensions, copolymerization, alkenyl phenyl compounds as coreactants, and rubber toughened BMI have been developed to improve the toughness of a cured BMI resin. Rubber toughening is the most economical way to improve the impact resistance of a cured epoxy resin. However, rubber toughening leads to a decrease in heat and thermo oxidative resistance due to the reduction of glass transition and the poor oxidative stability of the rubber tougheners.

BMI resins are prepared by reacting diamines with maleic anhydride in two reaction steps: 1) bismaleamic acid formation and 2) imidization (similar to synthesis of condensation polyimides). The formation of the bismaleamic acid is a fast and exothermic reaction, and the imidization is carried out thermally or chemically. The standard synthesis of an aromatic BMI involves the chemical imidization using acetic anhydride in the presence of a catalytic amount of anhydrous sodium acetate at temperature below 80°C. Some addition of amine to BMI double bonds (Michael addition) also takes place when the reaction is carried out at 140°C for a long duration.

BMI resins, low-molecular-weight compounds, and oligomers require significant molecular weight increase and structure modification to become useful. BMI resins are reactive toward various reactive species under mild conditions. A few curing reactions such as thermal polymerization, Michael additions (addition reactions), ene reaction, and Diels–Alder reactions have been developed to increase the application performance of BMI resins.

The most important curing reaction of BMI resins has been their homo polymerization at elevated temperatures. The reaction produces no volatile byproducts and provides void-free thermosets. BMI resins on curing give rigid, solvent resistant highly crosslinked thermosets. Aromatic BMI resins usually are high melting solids. They are usually dissolved in solvents to improve their processability.

BMI resins contain reactive unsaturated olefinic groups that are further activated by two strong electron withdrawing carbonyl groups and are known to undergo Michael additions with hydrogen active moieties such as phenols, thiols, and amines. The reactions have been used to prepare not only various high-molecular-weight linear polymers but also high-performance thermosets. During imidization, maleamic acid intermediates prepared from maleic anhydride and amines undergo reversible reactions to regenerate amines that eventually add to unsaturated olefinic bonds. Hence, the properties of a BMI resin strongly depend on the synthetic conditions.

One important class of high-performance thermosets is derived from BMI resins and allylphenyl oligomers. Upon thermal curing, the mixtures copolymerize to produce polymers with high glass transition temperature. For example, the mixture of 3,3'-diallylbisphenol-A and bis (4-maleimidophenyl)methane displays low- and high-temperature exotherms occurring at approximately at 130°C and 255°C, respectively. On the basis of the nuclear magnetic resonance (NMR) spectrum results, the low-temperature exothermic reaction was concluded to be an ene reaction between allyl and maleimide groups. The high-temperature exothermic reaction is related to a Diels–Alder reaction of the functional groups produced in the ene reaction. Diels–Alder polymerization requires a reacting diene monomer and a monomeric dienophile, which is normally activated by electron withdrawing substituents. During the polymerization, a stiff, heat-resistant ring structure is formed by every Diels–Alder reaction unit. Therefore, it is an excellent approach to increase the polymer performance, for example, the reaction of difuran end-capped monomers with BMI resins.

Rubber toughening is a well-known and relatively inexpensive technology to improve the durability of the highly crosslinked thermoset. Rubber, such as carboxyl-terminated copolymers of butadiene and acrylonitrile, has been used for the modification of some BMI resins. The well-dispersed rubber domain in a cured BMI resin absorbs fracture energy and improves impact strength by preventing crack propagation.

7.1.13 Isoimides

The increasing need for high service temperature adhesives and structural matrix resins has led to the development of many new polymeric systems in recent years. One of the most interesting and potentially useful of these new polymers is polyimides. Polyimides are noted for their excellent thermal limited due to problems with fabrication and processing of these polymers. Nevertheless, careful design of polyimides can lead to enhanced processability. In this respect, several approaches have been investigated and found to be useful. One design method which has improved the processability of linear aromatic polyimides is the introduction of meta-substituted aromatic diamines for para-substituted analogs. This procedure, while improving the processability, also has the possible detrimental effect of lowering the glass transition temperature. Another method which has been successfully utilized in improving polyimide processing and solubility characteristics is the incorporation of bulky side groups such as phenylated diamine monomers. Although these materials maintain a high glass transition temperature, their resistance to solvents may be sacrificed. Processability can also be improved by diluting the rigid imide functionality in the polymer chain through the use of block copolymerization with a flexible segment such as a siloxane. These approaches all rely on enhancing the thermoplasticity of the polyimides through incorporation of flexibilizing linkages. Finally, processability and fabrication aspects of polyimides have been improved through the use of low-molecular-weight imide oligomers terminated with acetylenic groups. The material (MC-600) is an example of a commercially available product (National Starch and Chemical Co.). These materials have improved solubility and processing characteristics while maintaining both a high glass transition temperature and good solvent resistance due to their highly crosslinked nature following thermal cure of the acetylene groups. This approach also has problems in terms of processing parameters. Preliminary reactions of the terminal acetylene groups during thermal cure lead to a restriction of flow and wetting properties before good contact is achieved. In addition, since the glass transition temperature of these imide oligomers is quite high (~200°C), the crosslinking reaction proceeds very rapidly resulting in an in fusible, rigid network. Once the glass transition temperature has been exceeded, enough mobility is available in the system for rapid crosslinking of the terminal acetylene groups. In this case, above 200°C (the Tg of the uncured polymer), this crosslinking reaction proceeds very rapidly. The gel time for MC-600 has been estimated at less than 3 min at a temperature of 250°C. This short gel time severely restricts the uses of the material in applications such as matrix resins and adhesives where good flow is necessary prior to the onset of gelation. In order to circumvent this problem of rapid gelation, an isomeric imide structure, termed isoimide, has been introduced into these systems materials with this functionality (i.e., IP-600) exhibit improved solubility as well as longer gel times and lower glass transition temperatures (~160°C versus ~200°C for the corresponding imide oligomer). Initially, it was thought that the presence of the isoimide structure, as an unfavorable side reaction product in polyimides, led to premature thermal decomposition of polyimides through loss of CO^2 from the iminolactone heterocyclic ring. However, later work showed that the isoimide functionality thermally isomerized to the imide functionality prior to any significant degradation of the polymer backbone. Since the utility of these materials is improved by the incorporation of these reactive functionalities without severely decreasing other favorable properties such as thermo-oxidative stability and solvent resistance, the chemistry of the isoimide isomerization and acetylene crosslinking reactions is of considerable interest. Research work was carried out to elucidate the relationship between the thermal isomerization and crosslinking reactions occurring in this acetylene-terminated polyisoimide oligomer, Thermid IP600. The techniques of Fourier transform infrared spectrometry and differential scanning calorimetry have been shown to be useful in determining the cure states of acetylene-terminated resins such as imides and sulfones. These techniques will be applied to the cure reactions occurring in the IP600 acetylene terminated isoimide system.

Isoimides are isomeric with imides. An isoimide is converted to an imide by Mumm rearrangement. Cyclic imides are important and have different chemical and biological applications such as antibacterial, antifungal, anticonvulsant, and antitumor. They are also applied in agriculture as herbicides, fungicides, or insecticides. But isoimides are crystalline with melting points much lower than the corresponding symmetrical imides; generally, cyclic isoimides are thermodynamically less stable than their corresponding imides, but high conjugation or steric factors contribute a great deal to the stability of some isoimides.

7.1.14 Polyisoimide

From early 1960s, polyisoimide synthesis was put into light. The novel copolymers of this invention were characterized by an excellent combination of properties, including hydrolytic stability, stiffness, toughness, shelf stability, electrical properties, thermochromicity, and convertibility to permanently creased or formed articles. After that, various properties of isoimides were studied and in near 1980s, a class of low-molecular-weight oligomers containing at least one isoimide group with terminal groups that are capable of undergoing an addition polymerization were synthesized. These oligomers were reported to have an excellent solubility in common solvents and a melting temperature considerably lower than their cure temperature, and so the cured polymers were formed without the evolution of deleterious gases. Further improved process for the synthesis of isoimide containing oligomers without any side reactions was also discovered. Aromatic polyimide and polyisoimide bisacetylene additives were found to be a good adhesive and sealant compositions including anaerobic curing compositions which exhibit improved strength properties at elevated temperatures as well as they showed improved resistance to thermal degradation.

7.1.15 Bis(isoimides)

Thermosetting polyimide resins based on bis(isoimides) have properties different from conventional polyimides. Resins based on bis(isoimide) have more advantages. Bis(isoimides) in combination with dihydric phenols form prepolymers which are readily soluble and stable in low boiling solvents. Prepregs can be manufactured from bis(isoimides). Advanced composites that are formed from isoimide precursors are high-strength, high-modulus materials which are finding increasing use as structural components in aircraft, automotive, and sporting goods applications. Typically, they comprise structural fibers such as carbon fibers in the form of woven cloth or continuous laments embedded in a thermosetting resin matrix. An improvement in the thermosetting bis(isoimide) resin compositions was further achieved in 1990s by replacing a small part of the starting material, i.e., unsaturated carboxylic acid anhydride. Polyisoimides were found to have more attractive applications compared to polyimides. Hence, the properties of isoimides were studied in detail. The thermal stability of bis(isoimides) was noted by John D. Harper in 1992 [22]. He suggested that thermal stability can be enhanced by the addition of small amounts of trihydroxyaromatic or polyhydroxyaromatic compounds such as novalacs and phloroglucinol which are termed as modifiers to the partially polymerized bis(isoimide) resin composition. The solution of prepolymer along with the modifiers was stable for weeks and can be used to impregnate fibrous materials like glass cloth.

7.1.16 Maleimide and Isomaleimide

Bismaleimides occupy a prominent and special position in the spectrum of thermosetting materials. The novel maleimide–isomaleimide compounds have a unique structure which makes it more important in different fields. The novel aromatic maleimide–isomaleimide was produced ordinarily in combination with the corresponding BMI or bis(isomaleimide),when the starting diamine is α,α'-bis(4-aminophenyl)-metadiisopropylbenzene. After the patented work based on a new class of thermosetting resin from bis(isoimides) by Wank and Harper, a detailed study on thermosetting resin based on bis(isomaleimides) and dihydric phenols was done in 1992 and it was reported that they had a high glass transition temperature compared to the previously known bis(isomaleimides) [23]. The formation of isoimides was mainly dependent on the dehydrating agents used. Different dehydrating agents like dicyclohexycarbodiimide (DCC), trifluoroacetic anhydride, and thionyl chloride were used to convert polyamic acid to isoimide. But they were found to give some byproducts which complicated the reaction. While using DCC, urea may be obtained as the byproduct and affects the reaction. So later on, different dehydrating agents which do not affect the formation of isoimide were selected. Finally, one such dehydrating agent which was less poisonous and could be easily handled was found in the late 1990s by Japanese scientists. A derivative of dihydroquinoline was reported by them to be more effective in producing isoimides [24]. The preparation and difference in steps between polyimide and polyisoimide were presented in a sequence by Stephen et al., which provided an easy and clear idea about the formation of isoimides. Isomaleimides and polyisomaleimides can be prepared by a method comprising the steps in sequence. The mixing of maleamic acid or polymaleamic acid with an acid halide is done followed by reacting the mixture with a tertiary amine at a temperature sufficiently low to suppress the formation of a maleimide, and finally the resulting isomaleimide or polyisomaleimide was isolated. It was also found that polymerization of a polyisomaleimide with a polynucleophilic monomer like polythiol, polyol, and polyamine gave rise to novel polymers [25]. Later on, complicated polymers based on isoimides were synthesized and their properties were studied. The formation of a bisisoimide is shown in Figure 7.1

Some other complex bisisoimide compounds synthesized successfully in late 1990s are shown in Figure 7.2.

A considerable amount of findings was started after 2005 based on isoimides. Till then, there were some problems encountered with BMI compositions. The major drawback regarding these compositions was that they were unable to satisfy both properties of being quick curing and easy to handle with low moisture uptake. So in 2006, thermosetting resins stable at elevated temperatures, highly flexible, and having low moisture pickup property were synthesized. An easy way of solving this problem is by adding reactive diluents. But reactive diluents have much hazardous effects. So maleimide compounds in liquid form with non-reactive diluents were successfully developed [26].

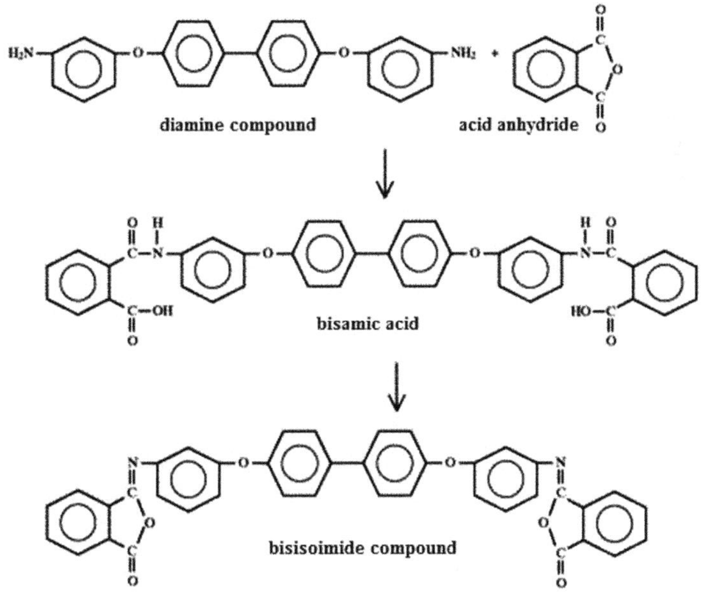

Figure 7.1 Formation of bisisoimide.

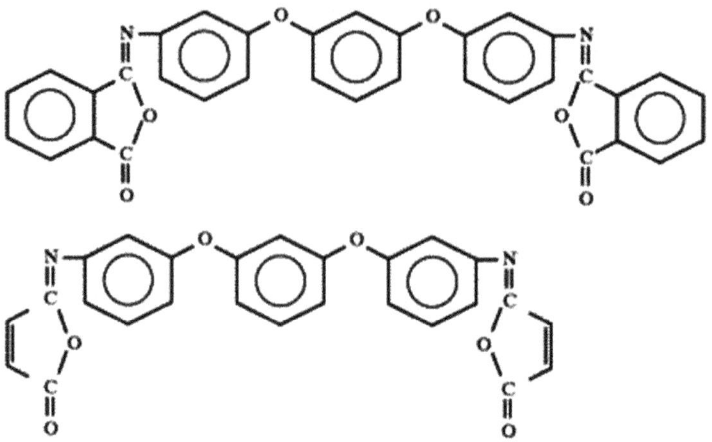

Figure 7.2 Complex structure of bisisoimide.

The dehydrated N-(4-hydroxy-1-napthyl)maleamic acid with trifluroacetic anhydride and obtained N-(4-hydroxy-1-napthyl)isomaleimide. This is the only substantiated example of an isomaleimide in the literature. Attempts to prepare other isomaleimides using the trifluroacetic anhydride reagent were unsuccessful. Mild dehydration of phthalanic acid with acetyl chloride was shown to give the hydrochloride of N-phenylisophthalimide which on careful treatment allowed the isolation of N- phenylisophthalimide. Roderick has shown that this method is inapplicable o the synthesis of N-substituted isomaleimides. Thus, no general method for the synthesis of N-substituted isomaleimides has been previously available. In contrast to the lack of syntheses for isomaleimides, maleimides have been prepared by a variety of methods, and their possible utility has been extensively investigated.

By analogy to the facile reactions of maleic anhydride, N,N'-biisomaleimide (BI) would be expected to undergo ring opening addition reactions at the carbonyl groups with nucleophiles. In fact, it was reported that BI reacts readily with amines to yield ring-opened adducts. However, little attention has been paid to the adaptation of this reaction to polymer formation. Previous investigations disclosed in patent literature have dealt almost exclusively with the formation of polymaleimides having a relatively low molecular weight.

The low molecular models, the structure of isoimide rings in the polymer has been determined and studied the influences of temperature on model compounds as well as on polymers shows that the conversion proceeds only in one direction, i.e., from isoimide into imide. These observations completely preclude the possibility of isoimide formation during thermal degradation of polypyromellitimides.

Polyisoimides are expected to possibly serve as novel and suitable precursors for polyimides with improved performance because of the transformation without producing harmful water. Polymers containing isoimide rings were first described in a patent, where the preparation of copolyimide-isoimide films was claimed from poly(amic acid) films and trifluroacetic anhydride. Although synthesis and some interesting properties of polyisoimides have been reported, the blending properties of these materials have not been reported. The solubility of bis(isoimide) prepolymers in low boiling solvents is an attractive feature for the advanced composites' manufacture.

Isoimide can polymerize either alone or in combination with other suitable monomers. The mechanical properties of bisphenol-A-based epoxy resin modified with phenolic novolac resins in various stoichiometric ratios (1:0.6, 1:0.7, 1:0.8, and 1:0.9) between phenol and formaldehyde were investigated. All the blended materials are seen to improve tensile strength, elongation, and energy absorption at break of the resin.

One of the most important areas within polymer science is that the synthesis of high thermally stable polymeric materials. Thermally stable polymers are placed under the category high-performance polymers because of their wide range of applications in demanding environments. Much importance has been given for the synthesis of new polymers to enhance properties of the polymers for new technology applications. One of the existing methods to improve polymer properties is the functionalization for effecting changes in polymer property.

Phenolic resins play a significant role in the polymer industry providing a diverse range of products with an equally diverse range of applications. Phenolic resins are used for a wide variety of applications from commodity construction materials to high technology applications in electronics and aerospace. In general, phenolic resins have high chemical resistance, good dimensional stability, and high rigidity. The high aromatic content of the phenolic resins is responsible for good strength retention at high temperatures and high char yield.

Isoimides belong to thermally stable polymers. Systematic studies on isoimides to fully explore the characteristic properties, however, have not been done primarily because of the instability and difficulty in preparation. Polyisoimides are expected to possibly serve as novel and suitable precursors for polyimides with improved performance because of the transformation without producing harmful water. Polymers containing isoimide rings were first described in a patent, where th preparation of copolyimide-isoimide films was claimed from poly(amic acid) films and trifluroacetic anhydride. Although synthesis and some interesting properties of polyisoimides have been reported, the blending properties of these materials have not been reported. The solubility of bis(isoimide) prepolymers in low boiling solvents is an attractive feature for the advanced composites manufacture. Our attention has been focused on the preparation of bis(isoimide) and blending this isoimide with different diphenols and it was thermally cured. The thermal properties of the materials were studied using differential scanning calorimeter (DSC) and thermogravimetric (TG) analysis technique.

7.2 Experimental

7.2.1 Materials

4,4'-Methylene dianiline was purchased from Himedia Pvt. Ltd., Mumbai. Resorcinol was obtained from Indepec Chemical Corporation (Pittsburgh, Pennsylvania, USA). Quinol was obtained from Qualigens Fine Chemicals (Mumbai, Maharashtra, India). 4,4'-Dihydroxy biphenyl and bis (4-hydroxy phenyl) sulfone were obtained from Alfa Aesar Johnson Metthey GmbH (Karlsruhe, Germany). 4, 4-Dihyroxy benzophenone was prepared in the laboratory. Bisphenol-A was purchased from SISCO Research Laboratory Pvt. Ltd., Mumbai-400099.

7.2.2 Preparation of Bis(isoimide) of 4,4'-Methylene Dianiline

A solution of 1.05 moles of 4,4'-methylene dianiline in methylene chloride was slowly added to a solution of 2.0 moles of maleic anhydride in the same solvent. After stirring at ambient temperature for 1 h, the precipitate of the bis(maleamic acid) was separated by filtration, washed (Figure 7.3) with methylene chloride to remove excess diamine, and oven dried. It was then subjected to dehydration with dicyclohexylcarbodiimide to form bis(isoimide) as yellow crystalline powder [27].

7.2.3 Blending of Bisphenols with Bis(isoimide) (VS)

The different bisphenols were separately blended with VS (1:1 mole ratio) in an agate mortar, and the mixture was ground repeatedly to have effective mixing. The mixture was preserved for polymerization.

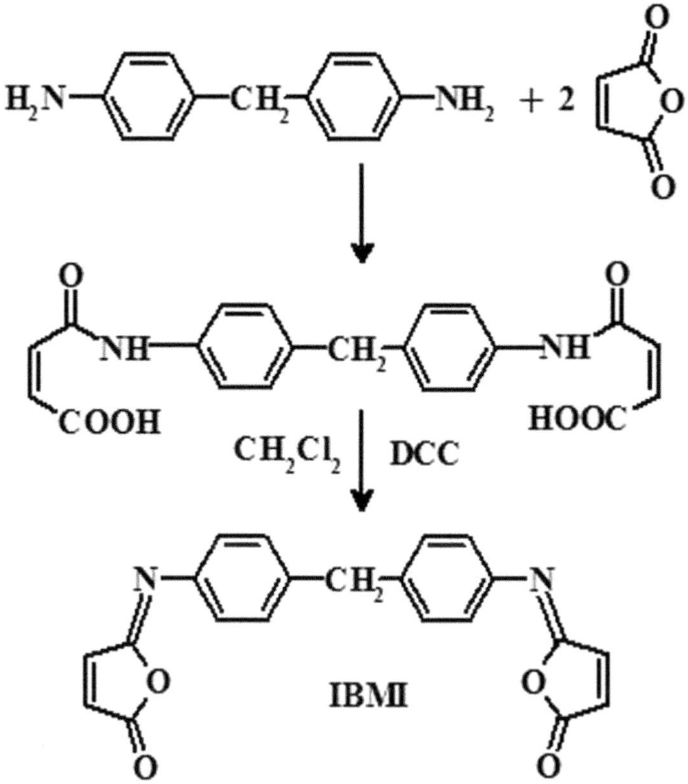

Figure 7.3 Preparation of IBMI.

7.2.4 Thermal Curing

The bis(isoimide) and the blends were taken in separate micro test tubes and flushed with dry oxygen-free nitrogen and polymerized for a period of 6 h. After the polymerization, materials were allowed to reach room temperature and the samples were removed from the micro test tubes, ground to coarse powder, packed, and stored for further analyses.

7.2.5 Fourier-transform Infrared (FTIR) Studies

The Fourier-transform infrared (FTIR) spectrum of the material was run on a Fourier transform infrared-8400S spectrophotometer, Shimadzu, Japan, using KBr disc technique. About 500 mg of potassium bromide was taken in a mortar. Approximately 10 mg of the sample was added and ground well with pestle. Approximately 100 mg of the above mixture was made into a transparent disc using pelletizer. The prepared disc was placed in the pellet holder and the IR spectrum was recorded using 16 scans with 4 cm^{-1} resolution.

7.2.6 Differential Scanning Calorimetric (DSC) Studies

Differential scanning calorimetric curves were recorded on TA Instruments DSC Q20. The sample (2–3 mg) was weighed, placed in the non-hermatic aluminum pan and sealed with an aluminum lid. The samples were heated from ambient to 350°C at a heating rate of 10°C min^{-1} in nitrogen (flow rate = 50 mL min^{-1}) atmosphere. The obtained DSC curves were analyzed using the Universal Analysis 2000 Software provided by TA instruments.

7.2.7 Thermogravimetric (TG) Studies

Thermogravimetric analyses were performed on a TA Instruments TG Q50 TG analyzer. To avoid the secondary reaction of evolved gases in TG analysis such as thermal cracking, recondensation, and repolymerization reactions, the nitrogen flow maintained at balance and samples were 50 mL min^{-1} and 60 mL min^{-1}, respectively. The sample (3–4 mg) was weighed into platinum crucible and was heated from ambient to 800°C at a heating rates 20°C min^{-1}. The obtained TG and DTG curves were analyzed using the Universal Analysis 2000 Software provided by TA instruments.

7.3 Results and Discussion

7.3.1 Fourier-transform Infrared Studies

The FTIR spectra of the monomer bis(isoimide) show lactone ring absorptions at 1795 $cm^{-1}$and 1680 cm^{-1}. The FTIR spectra of the polymerized compound are shown in Figure 7.4. The presence of intense band at 3361 cm^{-1} in blends indicates the presence of phenolic groups. The presence of phenolic groups indicates that the polymerization of the blends undergoes copolymerization, but the intensity of the OH peak in the blend varies. This may be due to the variation in the structure. There may be possibilities of the reaction between the OH group and isoimide group which may lead to the utilization of hydroxyl group in diphenols. The absence of band at 1635 cm^{-1} (C=C isoimide double bond) confirms the complete polymerization of the isoimide (VS).

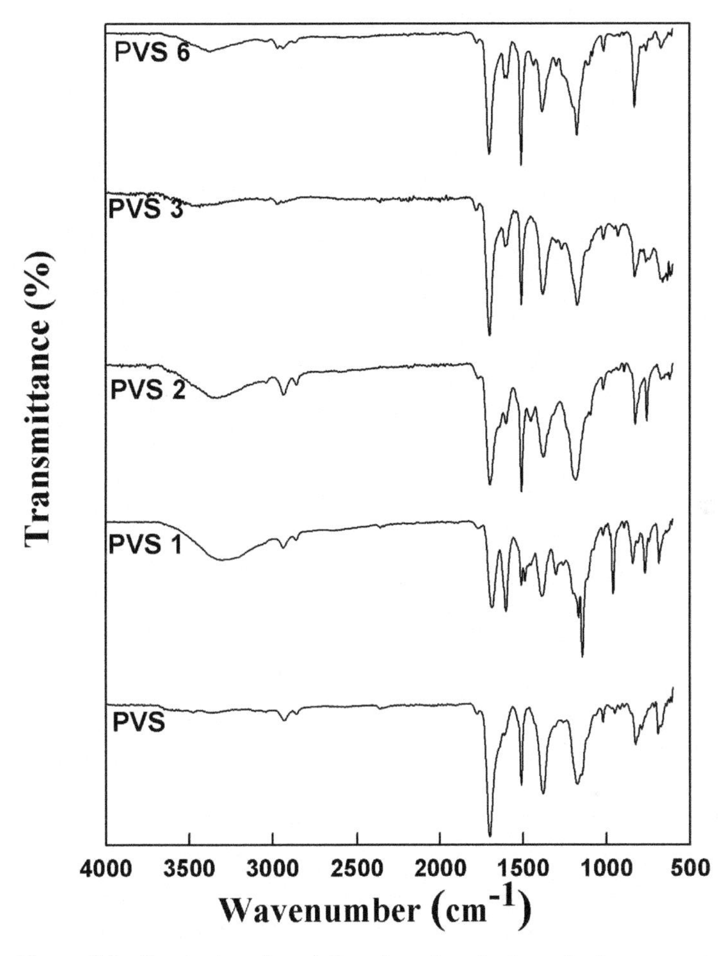

Figure 7.4 Fourier-transform infrared spectra of polymerized compounds.

7.3.2 Differential Scanning Calorimetric Studies

The DSC curves for the monomer and blends are shown in Figure 7.5. The values obtained for the parameters such as melting temperature (T_m), enthalpy of curing (ΔHc), onset temperature (T_i), temperature maxima (T_{max}), and final temperature (T_e) are tabulated in Table 7.1. The bis(isoimide) VS showed a sharp T_m at 144°C. The onset of the curing was around 161°C; curing attained a maximum at 232°C and ended around 295°C. The value of enthalpy of thermal curing (ΔHc) was 125 J/g and the

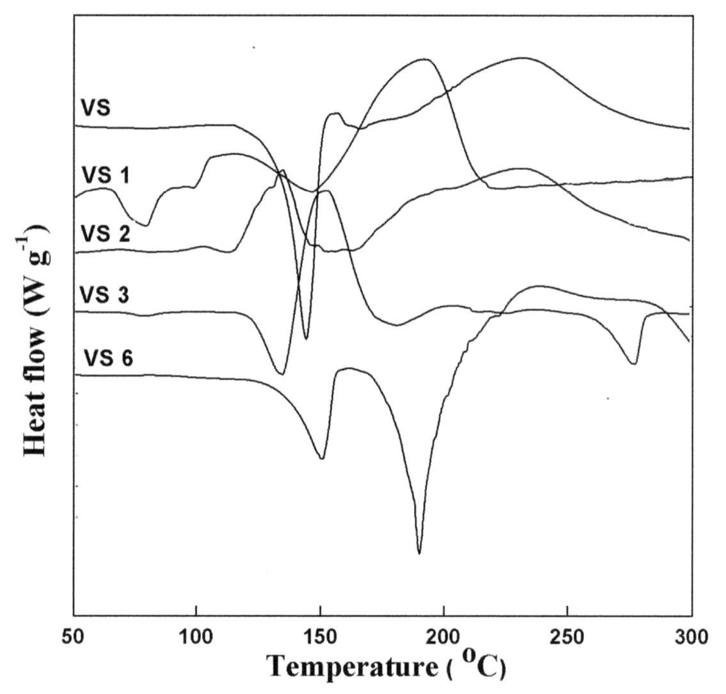

Figure 7.5 Differential scanning calorimeter curves of monomers and blend (heating rate (β) = 10°C/min).

temperature region of the curing window was 134°C. There was a characteristic change in the melting point of VS when different diphenols are added. There is a much variation in the melting point of VS 1 (79°C) and VS 2 (77°C), when comparing with VS 3(134°C). The compound VS 6 shows two melting points in the region 150°C and 191°C. The change that was observed in the melting region for pure VS and the diphenol blends provide very good evidence for the interactions existed between VS and diphenols at room temperature. The intensity of this interaction differed with the different types of diphenol systems.

From Table 7.1, it is evident that the introduction of different diphenols in pure VS affects the onset and curing maxima. For the compound VS 1, VS 2, and VS 3, the onset of curing and the curing maxima is shifted to the lower region and also shows two curing maxima. The compound bisphenol-A (VS 6) shows higher and two melting point region; similarly, the cure maxima is also higher compared to that of the other blends. This may be due to the influence of gem-dimethyl group present in the compound.

Table 7.1 Differential scanning calorimeter studies: Curing characteristics of isobismaleimide blends (β = 10°C/min)

Sample	MP (°C)	T_i (°C)	T_{max}(°C)	T_e(°C)	T_{max}–T_i(°C)	T_e–T_i(°C)	ΔH_C(J/g)	$\Delta H_C/T_e$–T_i (J/g °C)
VS	144	161	232	295	71	134	125	0.93
VS 1	79	98	111 192	218	13 94	120	248	2.06
VS 2	77	110	135 230	319	25 120	209	209	1.00
VS 3	134	134	152 201	220	18 67	86	132	1.53
VS 6	150 191	203	239 283	324	36 80	121	229	1.9

7.3.3 TG and DTG Studies

The TG and DTG curves of the polymers (heating rate = 20°C/min) are shown in Figures 7.6 and 7.7, respectively. The onset, maximum, end set temperatures for the degradation and the char residue obtained at 800°C for all the samples are tabulated in Table 7.2. The T_{max} values presented in Table 7.2 are obtained. The thermal stability of the blend was very less compared to the pure monomer. This may be due to the fact that the compound formed during the polymerization has weak bond which can easily be broken during degradation process. This is evident from the char yield of these compounds. The 5% weight loss temperature is also very less compared to that of the pure monomer. From the DTG curve of polymerized compound, the polymers PVS and PVS 6 show single degradation step whereas other three blends show two stages of degradation. The maximum degradation takes place at around 400°C. The initial stages of degradation temperature for the compounds PVS 1, PVS 2, and PVS 6 are 304, 355, and 249°C, respectively. The polymer, PVS, is showing the highest char value compared to the blends. The observed char for the polymer from the blend is less indicative that the thermoset resulting from the thermal polymerization of a blend of PVS is having no sufficient thermally stable structural units.

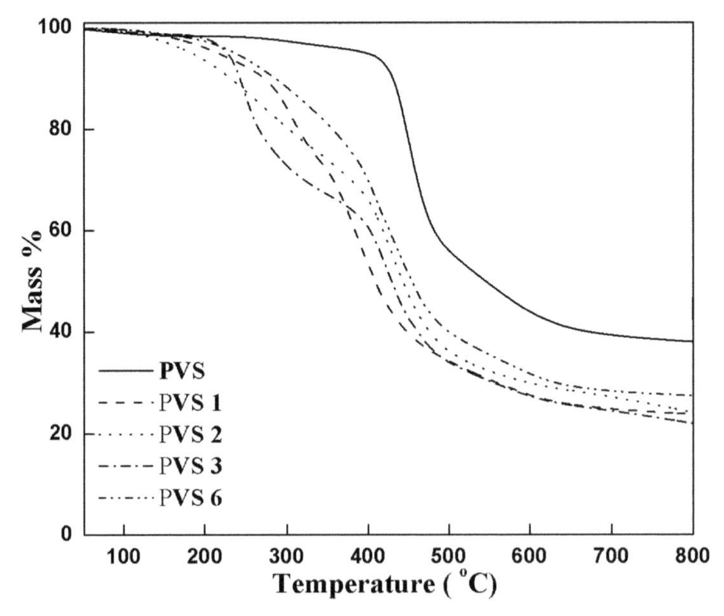

Figure 7.6 Thermogravimetric curves of the polymers (β = 20°C/min).

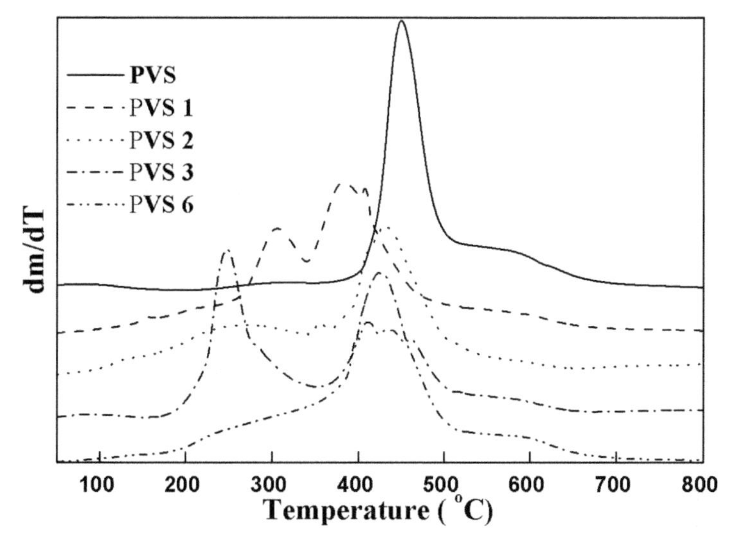

Figure 7.7 DTG curves of the polymers (β = 20°C/min).

Table 7.2 Hermogravimetric studies: The degradation parameters for the different isobismaleimides and its blends recorded at polymers 20°C/min

	Temperature(°C)						T_{max}(°C)		
Sample	T_5	T_{10}	T_{25}	T_{50}	T_i	T_e	T_{M1}	T_{M2}	Char Yield % at 800°C
PVS	397	431	424	546	363	680	–	450	38
PVS 1	215	273	335	409	202	665	304	381	24
PVS 2	210	231	343	443	202	638	355	430	24
PVS 3	228	243	288	429	164	646	249	425	22
PVS 6	237	286	382	454	179	700	–	428	27

7.4 Conclusion

The compounds' VS was prepared and a 1:1 mol ratio blend of this monomer, and different diphenols were also prepared and thermally polymerized. The FTIR studies of the polymers indicate that the polymerization takes place through the isoimide ring open and also the formation of copolymers. The blending of the compound VS with different diphenols will affect the onset and cure maxima expect VS 6. This may due to the influence of the structure of the compound. Based on the char residue, the polymer from the blend is comparatively less stable than polymers from pure monomers. The introduction of the diphenol makes the polymer very weak bond for the degradation; this results in the less stability of the compound.

Acknowledgments

The authors would like to acknowledge the Management, the Principal and the Dean of Kamaraj College of Engineering and Technology, S. P. G. C. Nagar, K. Vellakulam Post-625 701, India, and Thiagarajar College (Autonomous), Madurai – 625009, for providing all the facilities to do this work.

References

[1] Lin, S. C., and Pearce, E. M. (1994). *High Performance Thermosets: Chemistry, Properties, Applications*. New York, NY: Carl Hanser Verlag.
[2] Fink, J. K. (2005). *Reactive Polymers Fundamentals and Applications – A Concise Guide to Industrial Polymers*. New York, NY: William Andrew Publishing.
[3] Lee, H., and Neville, K. (1967). *Handbook of Epoxy Resin*. New York, NY: McGraw-Hill Inc.
[4] Fang, T., and Shimp, D. A. (1995). Polycyanate esters: science and applications. *Prog. Poly. Sci.* 20, 61–118.
[5] Fan, J., Hu, X., and Yue, C. Y. (2003). Dielectric properties of self-catalytic interpenetrating polymer network based on modified bismaleimide and cyanate ester resins. *J. Poly. Sci. Part B Poly. Phys.* 41, 1123–1134.
[6] Lakshmi, M., and Reddy, B. S. R. (2002). Synthesis and characterization of new epoxy and cyanate ester resins. *Eur. Poly. J.* 38, 795–801.
[7] Hamerton, I., and Hay, J. N. (1998). Recent developments in the chemistry of cyanate esters. *Poly. Int.* 47 465–473.
[8] Herr, D. E., Nikolic, N. A., and Schultz, R. A. (2001). Chemistries for high reliability in electronics assemblies. *High Perf. Poly.* 13, 79–100.
[9] Kopf, P. W., Wagner, E. R. (1973). Formation and cure of novolacs: NMR study of transient molecules. *J. Poly. Sci. Part A Poly. Chem.* 11, 939–960.
[10] Knop, A., and Pilato, L. A. (1985). *Phenolic Resin*. New York, NY: Springer-Verlag.
[11] Billmeyer, F. W. (1971). *Textbook of Polymer Science*. New York, NY: John Wiley & Sons.

[12] Yan, Y., Shi, X., Liu, J., Zhao, T and Yu, Y. (2002). Thermosetting resin system based on novolak and bismaleimide for resin-transfer molding. *J. Appl. Poly. Sci.* 83, 1651–1657.

[13] Culbertson, B. M., Tiba, O., Deviney, M. L., and Tufts, T. A. (1989). "Bisoxazoline Phenolic Resin Step-Growth Copolymerizations: New Systems for Electronic, Mold Making, and Resin Transfer Molding," in *Proceedings of the 34th International SAMPE Symposium*, Reno, NV.

[14] Kaji, M., and Endo, T. (1999). Synthesis of a novel epoxy resin containing naphthalene moiety and properties of its cured polymer with phenol novolac. *J. Poly. Sci. Part A: Poly. Chem.* 37, 3063–3069.

[15] Ogato, M., Kinjo, N and Kawata, T. (1993). Effects of crosslinking on physical properties of phenol–formaldehyde novolac cured epoxy resins. *J. Appl. Poly. Sci.* 48, 583–601.

[16] Tyberg, C. S., Bears, K., Sankarapandian, M., Shih, P., Loos, A. C., Dillard, D., et al. (1999). Tough, void-free, flame retardant phenolic matrix materials. *Construct. Build. Mater.* 13, 343–353.

[17] Han, S., Kim, W. G., Yoon, H. G., and Moon, T. J. (1998). Kinetic study of the effect of catalysts on the curing of biphenyl epoxy resin. *J. Appl. Poly. Sci.* 68, 1125–1137.

[18] Benedetto, A. T. D. (1987). Prediction of the glass transition temperature of polymers: a model based on the principle of corresponding states. *J. Poly. Sci. Part B Poly. Phys.* 25, 1949–1969.

[19] Han, S., Kim, W. G., Yoon, H. G., and Moon, T. J. (1997). A kinetic study of biphenyl type epoxy-xylok resin system with different kinds of catalysts. *Bull. Korean Chem. Soc.* 18, 119–1203.

[20] Han, S., Kim, W. G., Yoon, H. G., and Moon, T. J. (1998). Kinetic study of the effect of catalysts on the curing of biphenyl epoxy resin. *J. Appl. Poly. Sci.* 68, 1125–1137.

[21] Fainleib, A., Galy, J., Pascault, J. P., and Sue, H. J. (2001). Two ways of synthesis of polymer networks based on diglycidyl ether of bisphenol A, bisphenol A, and sulfanilamide: Kinetics study. *J. Appl. Poly. Sci.* 80, 580–591.

[22] Thomas, A. S., John, D. H. (1992). Thermosetting bis(isoimide) resin composition. US Patent No. 5079338.

[23] John, D. H. (1992). Thermosetting bis(isoimide) resin composition. US Patent No. 5082920.

[24] Maeda, H. K., Kunimune K. C. (1994). Process for producing poly-isoimide. US Patent No. 5294696.
[25] Stephen, A. E., Richard, G. H., Gregory J. A. (1998). Isomaleimides and polymers derived therefrom. US Patent No. 5744574.
[26] Stephen, M. D., Dennis, B. P., Jose, A. (2006). Osuna Maleimide compounds in liquid form. US Patent No. 7102015.
[27] Wankand, R. L., and Harper, J. D. (1988). Thermosetting bis(isoimide)-diphenol resin. US Patent 4732963.

8

Natural Fiber Based Bio-materials: A Review on Processing, Characterization and Applications

M. K. Gupta and R. K. Srivastava

Department of Mechanical Engineering, Motilal Nehru National Institute of Technology, Allahabad, India

Abstract

Nowadays, researchers and scientists have to overcome the challenge of increasing use of synthetic fibers to maintain environmental balance, which has opened a new approach of research toward natural fibers. Natural fiber-reinforced polymer composites (NFRPCs) include natural fibers (sisal, jute, banana, hemp, kenaf, coir, bamboo, curaua, etc.) as a reinforcement, and thermosets, thermoplastics, and bio-polymers as a matrix. In the recent decades, natural fibers are being used as reinforcement in place of glass and other synthetic fibers due to their benefits such as low cost, low density, availability in abundance, environmental friendliness, biodegradability, and high specific strength and modulus. In view of the above advantages, NFRPCs have been used in many applications like automotive, packaging, construction, household, toys, furniture, and so on. In the present review, an overview of the developments in the area of NFRPCs in terms of description and classification of the composites, fabrication methods, characterization of composites, and applications is given.

8.1 Composite Materials

A composite material can be defined as a material having two or more chemically distinct phases, which are separated by a distinct interface at the microscopic scale. The two phases consist of the discontinuous and

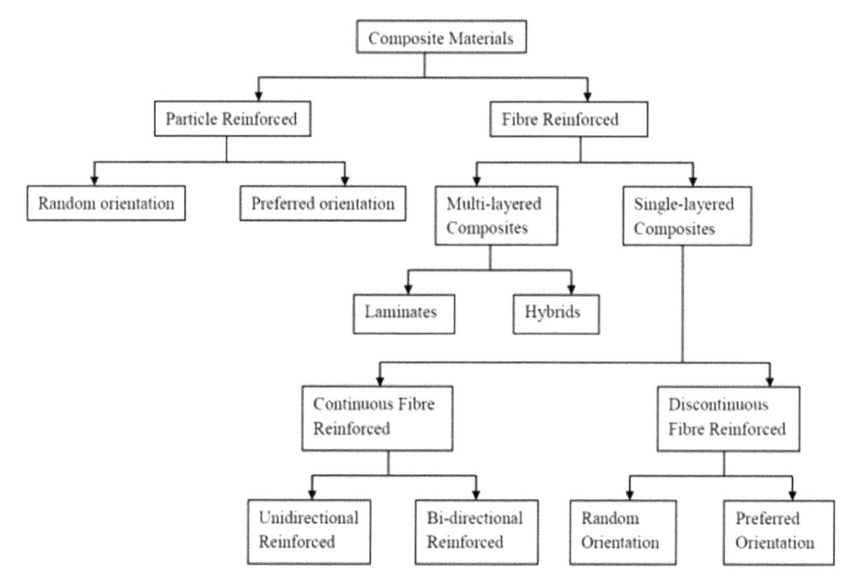

Figure 8.1 Classification of composites on basis of reinforcement.

the continuous phase, where the discontinuous phase is embedded into the continuous phase. The discontinuous phase is generally harder and stronger than the continuous phase. The discontinuous phase is called the reinforcement, whereas the continuous phase is termed as the matrix. The properties of composite materials depend on the properties of their constituents, their distribution, and the interaction between them. The shape, size, distribution and orientation of the reinforcement also govern the properties of the composite. The interaction between the reinforcement and matrix decides the mechanical strength of composites. This interaction mainly depends on the geometrical shape, size, orientation, and volume fraction of reinforcement. A brief classification of the composites on the basis of reinforcement is given in Figure 8.1. A brief definition and classification of these composites are given in the following subsection.

8.1.1 Particle Reinforced Composite

The matrix of such composites uses a particle as reinforcement where the shape of the particulates may be spherical, cubic, tetragonal, platelet, or of other regular or irregular geometry. However, while the particles are not very effective in improving the fracture resistance, they can improve the stiffness of the composite to some extent. The other properties of the matrix such as mechanical, thermal, and tribological properties can be improved by using the particle fillers.

8.1.2 Fiber-reinforced Composite

Fiber-reinforced composites are being preferred over particle-reinforced composites due to their industrial application because of high specific strength and stiffness. In fibrous composites, reinforcements are fibers whose length is very high compared to their cross-sectional dimensions. The fiber has significant effect in improving the fracture resistance of the matrix because of its long dimension to decrease the growth of incipient cracks normal to the reinforcement. Fibers cannot be used directly in engineering applications due to their small area of cross section. Therefore, they are reinforced into the matrix to form a fibrous composite. The matrix binds the fibers together, transfers the load to the fibers, and protects them against the environmental chemical effects.

A fiber-reinforced composite can be classified into two broad groups, i.e., continuous (long) fiber composites and discontinuous (short) fiber composites, whose details follow.

8.1.2.1 Continuous fiber composite

The continuous fiber composite consists of continuous or long form of fibers embedded in the matrix. A continuous fiber has very high length-diameter ratio. These fibers are normally stronger and stiffer than the bulk material matrix. On the basis of the orientation of fibers within the matrix, it is again subdivided into two categories: (a) unidirectional reinforcement and (b) bidirectional reinforcement. In case of unidirectional reinforcement, the fibers are oriented only in one direction, whereas in bidirectional reinforcement the fibers are oriented in two directions which may be either at right angles to one another or at some desired angle.

8.1.2.2 Discontinuous fiber composite

The discontinuous fiber composite consists of discontinuous or short fibers reinforced into the matrix. The short fiber-reinforced composites are the material of choice for large-scale production due to their low cost and ease of fabricating the complex parts. Further, the discontinuous fiber-reinforced composite is divided into two parts such as preferred oriented fiber composite and random oriented fiber composite. In the preferred oriented fiber composite, the fibers are oriented in prearranged directions, whereas in the latter type, the fibers are not in a predetermined orientation but are randomly oriented.

8.1.3 Laminate Composite

Laminate composites are those composites which are composed of more than one layer of fibers, held together by the matrix. Usually, these layers are kept in alternate arrangement for better bonding between reinforcement and matrix. According to the application of the composite, these laminates can have unidirectional or bidirectional orientations of fiber reinforcement. The man-made fibers are used in most of the applications of laminate composites due to their good combination of physico-mechanical and thermal behavior.

8.1.4 Flake Composite

Flakes are used in many industrial applications because they are not expensive to produce and are usually cheaper than fibers. Flakes are generally used in place of fibers due to their compatibility. The flake-reinforced composites have a higher theoretical modulus of elasticity than fiber-reinforced composites. Flakes have many advantages over fibers in structural applications. This composite provides uniform mechanical properties in the plane of the flakes. The other properties of these composites are also encouraging, such as high strength, high flexural modulus, high dielectric strength, high heat resistance, and low cost.

8.1.5 Hybrid Composite

Hybrid composites are those which have more than one reinforcement in a single matrix, or a single reinforcement with multiple matrices, or multiple reinforcements with multiple matrices. Therefore, the hybrid composites provide more flexibility than single fiber-reinforced composites in terms of mechanical properties. In hybridization, a combination of a high modulus fiber and low modulus fibers are preferred. The high modulus fiber provides the stiffness and load-bearing qualities, whereas the low modulus fiber makes the composite more damage tolerant and keeps the cost of the material low. The mechanical properties of a hybrid composite can be changed by varying the volume fraction or weight fraction and stacking sequence of different fibers. The hybrid composites are considered to be the more advanced composites than conventional fiber-reinforced polymer composites. The classification of composites based on matrix is given in the following subsection.

8.2 Classification Based on Matrix Materials

(i) Metal Matrix Composite (MMC)
(ii) Ceramic Matrix Composite (CMC)
(iii) Polymer Matrix Composite (PMC)

The details about these composites are given as follows.

8.2.1 Metal Matrix Composite

In MMC, aluminum, copper, magnesium, and titanium are used as matrix which is mainly useful in aerospace applications. These matrix materials require high modulus reinforcements. Therefore, alumina, boron and carbon fiber or silicon carbide which offers high mechanical strength is mostly used. The MMCs have properties as follows:

(i) High strength, fracture toughness and stiffness
(ii) Strength retention at higher temperatures
(iii) High transverse strength
(iv) Better electrical conductivity
(v) High thermal conductivity
(vi) Good erosion resistance, etc., as compared to organic matrices.

In spite of the several advantages listed above, the MMC also has some drawbacks such as higher densities and lower specific mechanical properties as compared to PMC. Further, MMCs require high energy for its fabrication.

8.2.2 Ceramic Matrix Composite

Ceramics are defined as solid materials which exhibit very strong ionic bonding in general and covalent bonding in a few cases. In CMC, ceramic materials such as alumina, silicon carbide, aluminum nitride, and silicon nitride are used as matrix, whereas short fibers or whiskers of silicon carbide and boron nitride are used as reinforcement. The main commercially available CMCs are carbon/carbon (C/C), carbon/silicon carbide (C/SiC), silicon carbide/silicon carbide (SiC/SiC) and alumina/alumina (Al_2O_3/Al_2O_3).

The advantages of CMCs are high melting point, good corrosion resistance, stability at elevated temperatures, high compressive strength, and high strength and high hardness. One of the main objectives in producing CMC is to increase the toughness. The brittleness of the ceramics makes the fabrication process complicated. Therefore, the ceramic material is used in powder form during processing.

8.2.3 Polymer Matrix Composite

In PMC, matrix materials are polymeric materials—either thermosets, thermoplastics, or bio-polymers. In general, the mechanical properties of polymers are not sufficient for structural applications as their strength and stiffness are low as compared to metals and ceramics. These limitations can be overcome by adding reinforcement such as fibers and particles into polymers. The PMCs are very popular due to low cost, low density, and easy processing. The processing of PMC does not require high pressure or high temperature. The equipments required for processing of PMC are of simple configuration and their operations are easier. For this reason, development of PMCs advanced rapidly and soon became very user-friendly for structural applications too. The main disadvantages of PMCs are low thermal resistance and high coefficient of thermal expansion.

8.3 Natural Fiber Reinforced Polymer Composites

Natural fiber-reinforced polymer composites include natural fibers, i.e., jute, banana, sisal, kenaf, hemp, pineapple leaf, bamboo, coir, and so on as reinforcement and polymers as matrix. For ecological and environmental considerations, natural fiber have been recommended as a better replacement to man-made fibers for the polymer matrix due to their advantages such as low cost, low density, availability in abundance, environmental friendliness, non-toxicity, high flexibility, renewability, biodegradability, relative non-abrasiveness, high specific strength, and ease of processing. On other hand, these fibers have some limitations also, i.e., low impact strength, poor interface between fiber and matrix, poor thermal stability, and high moisture absorption which can be controlled using the hybridization technique, chemical treatments, and the addition of nano-particles [1–7]. The use of NFRPCs has been reported in many applications such as automobile, packaging, aerospace, and construction industries [8–12].

8.3.1 Matrix

Thermosets, thermoplastics and bio-polymers are used as a matrix for NFRPCs. Thermosets are those which cannot be heat-softened but thermoplastics can be heat-softened, melted, and reshaped as many times as desired. Matrix is used to keep the fibers straight and also transfer of loads to fibers. The matrix also provides rigidity and shape, and protects the fibers from chemical and corrosion effect. Thermoset polymers are better than thermoplastics in mechanical properties, thermal stability, chemical resistance, and durability. Epoxy resin is the most used polymer matrix by researchers due to its high tensile strength, tensile modulus, and compressive strength. Thermoplastic matrix is also used by various researchers due to its low processing, design flexibility, and easy fabrication of complex parts. In thermoplastic-based composites, the temperature is restricted within 230°C to avoid degradation of natural fibers. However, thermoset-based composites can be easily processed at room temperature without degradation of fibers. The comparative study of the properties of thermosets and thermoplastics is shown in Tables 8.1 and 8.2 respectively. Table 8.1 shows that the epoxy has comparatively better mechanical properties as compared to the others.

8.3.2 Reinforcement

The main objective of reinforcement is to increase the mechanical strength of the matrix. In NFRPCs, natural fibers as reinforcements have been used in different shapes, sizes, and stacking sequences. The commonly acceptable classification of reinforcing materials is given in Figure 8.2. The properties such as low density, biodegradability, eco-friendliness, and high specific strength of natural fibers make them a better replacement for synthetic fiber for medium-strength applications. The mechanical properties of natural fibers are given in Table 8.3. The chemical composition of natural fibers plays an important role in the mechanical properties of natural fiber-reinforced polymer composite. The chemical compositions of natural fibers are given in Table 8.4.

8.3.3 Fabrication Methods

There are many well-established processing methods available for NFRPCs. All of these methods are specific to the materials which are being processed. The selection of fabrication method depends upon factors such as shape, cost, matrix materials, orientation of fibers, and the number of components required. The following methods are widely used by different researchers.

Table 8.1 Properties of thermosets [13]

Thermosets	Tensile Strength (MPa)	Tensile Modulus (GPa)	Flexural Strength (MPa)	Flexural Modulus (GPa)	T_g(°C)	Specific Gravity
Epoxies	55–130	2.7–4.1	110–150	3–4	170–300	1.2–1.3
Phenolics	50–60	4–7	80–135	2–4	175	1.2–1.3
Polyesters	34–105	2.1–3.5	70–110	2–4	130–160	1.1–1.4
Vinyl esters	73–81	3–3.5	130–140	3	–	1.1–1.3

Table 8.2 Properties of thermoplastics [13]

Thermoplastics	Density (g/cm^3)	T_g(°C)	Tensile Strength (MPa)	Tensile Modulus (GPa)
Polypropylene	0.910	−10	28.0	2.0
Polylactide (2002D)	1.24	47.0	56.3	3.6
Polylactide (4032D)	1.24	50.7	65.8	3.6
Nylon 11	–	–	30–70	–
Polyethylene (HDPE)	0.96	-110	26	1.4
Polyethylene (LLDPE)	0.93	–	14	0.45
Polyethylene (LDPE)	0.92	–	12	0.18
PVC	1.35	90	48	3.3
Polystyrenes	1.04	95	46	2.9
Acrylonitrile-butadiene-styrene	1.05	102	46	2.5
Polycarbonate	–	151	59.82	–

8.3.3.1 Hand lay-up

The hand lay-up method is the simplest way of processing the thermoset-based fibers composite. This method is suitable for all kinds of fibers. In this method, firstly a releasing agent is sprayed on the surface of the mold to avoid the sticking of the polymer to the surface. The thin plastic sheets are used at the top and the bottom of the mold to get a good surface finish of the product. The fibers are placed at the upper surface of the mold either in the form of a woven mat or in chopped form. Then mixture of thermosets resin and a suitable hardener (matrix) is poured on the fibers already placed in the mould. The matrix is uniformly spread with the help of a brush. The second layer of fibers is then placed on the matrix surface, and then a roller is used to remove the air and excess matrix. This process is repeated for each layer of fiber and matrix, till the required thickness is achieved. After placing the plastic sheet, a release agent is sprayed on the inner surface of the top mold which is then kept on the plastic sheet and then pressure is applied. After curing at room temperature, the mould is opened and the developed composite part is taken out.

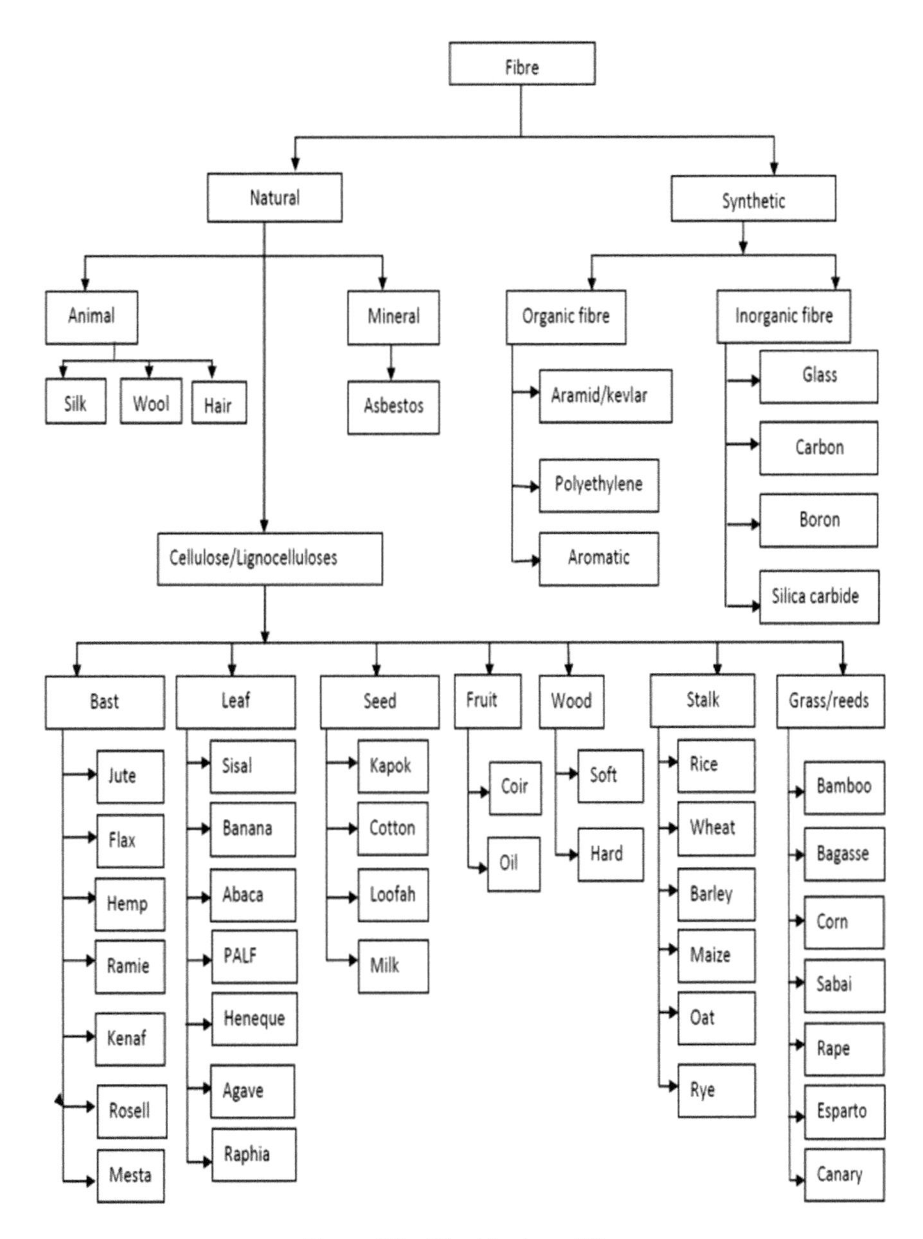

Figure 8.2 Classification of fibers.

Table 8.3 Physical and mechanical properties of natural fibers [13]

Fibers	Diameter (μm)	Density (g/cm^3)	Tensile Strength (MPa)	Tensile Modulus (GPa)	Elongation (%)
Alfa	–	1.4	247	21.5	1.96
Abaca	10–30	1.5	430–813	31.1–33.6	2.9
Jute	25–250	1.3–1.49	393–800	13–26.5	1.16–1.5
Sisal	50–200	1.34	610–710	9.4–22	2–3
Cotton	–	1.5–1.6	287–597	5.5–12.6	7.0–8.0
Kenaf	–	1.45	930	53	1.6
Wood	–	1.5	600–1020	18–40	4.4
Ramie	20–80	1.5	400–938	61.4–128	3.6–3.8
Coir	150–250	1.2	175	4–6	30
Flax	25	1.5	500–1500	27.6	2.7–3.2
Hemp	25–600	1.47	690	70	2.0–4.0
Pineapple	50	1.526	170–1627	60–82	2.4
Banana	100–250	0.8	161.8	8.5	2.0
Bagasse	49	–	96.24	6.42	4.03
Bamboo	88–125	800	441	35.9	1.3
Curaua	7–10	1.4	87–1150	11.8–96	1.3–4.9
PALF	20–80	0.8–1.6	180–1627	1.44–82.5	1.6–14.5
Coconut	–	1.1	140–225	3–5	25–40
Date palm	100–1000	–	135	4.6	3.6
Oil palm	174	0.7–1.55	206	3.567	4
Nettle	20	–	1594	87	2.11

Table 8.4 Chemical composition of natural fibers [13]

Natural Fiber	Cellulose (%)	Hemi Celluloses (%)	Lignin (%)	Pectin (%)	Moisture (%)
Flax	64.1–71.9	16.7–20.6	2.0–2.2	1.8–2.3	8–12
Hemp	70.2–74.4	17.9–22.4	3.7–5.7	0.9	6.2–12
Jute	61–71.5	12.0–20.4	11.8–13	0.2	12.5–13.7
Kenaf	31–57	21.5	8–19	3–5	–
Ramie	68.6–76.2	13.1–16.7	0.6–0.7	1.9	7.5–17
Sisal	65.8–78	8–14	10–14	0.8–10	10–22
Pineapple	70–82	–	5–12.7	–	11.8
Banana	63–64	10–19	5	–	10–12
Cotton	82.7–90	5.7	–	0–1	7.85–8.5
Coir	32–43	0.15–0.25	40–45	3–4	8
Abaca	56–63	15–17	7–9	7–9	–
Bamboo	26–43	30	21–31	–	–

The hand lay-up method has many advantages such as low setup cost, less lead time, easy processing, and fabrication of complex parts. This method also has some limitations such as non-suitability for large production, long curing time, difficulty in quality control, poor surface finish, wastage of materials, and variations in the quality of products.

8.3.3.2 Compression moulding

Compression moulding is a closed moulding process with high pressure application. In this method, two matched metal moulds are used to fabricate a composite product. The base plate is stationary while the upper plate is movable in a compression moulder. The reinforcement and the matrix are placed in the metallic mold and the whole assembly is kept between the compression molder. The required amount of heat and pressure depends upon the shape and size of the composites. The reinforcement and matrix are placed between the molder plates which flow due to application of pressure and heat. The curing of the composites takes place at room temperature. After the curing of the composite, the mold is opened and the composite product is taken out for further processing. This method is suitable for all kinds of polymers and fibers. The advantages of this process are higher fabrication speed, good surface finish, minimum materials wastage, low maintenance cost, and no development of residual stresses. Many parts, panels, and structures of the automobile industry are developed by using this method.

8.3.3.3 Injection moulding

The injection molding method is used for the formation of plastic parts with excellent dimensional accuracy. Products such as house ware, toys, automobile parts, furniture, packaging items, and medical disposal syringes are produced by this method. Injection moulding is a process of forming products by forcing molten plastic material under pressure into a mould where it is cooled, solidified, and subsequently released by opening the two halves of the mold. This process is suitable for all kinds of polymers and fibers. The main advantages of this method are higher production rate, minimum wastage of materials, and production of complex geometry. High setup cost is the main disadvantage of this method.

8.3.3.4 Pultrusion

This is a continuous process to manufacture products that have a constant cross-section such as rod stock, structural shapes, beams, channels, pipe, tubing, fishing rods, and golf club shafts. In this process, the continuous travelling of the reinforcement is impregnated with resin by passing it through a resin bath and then is pulled through a steel die. The steel die strengthens the saturated reinforcement, puts the shape of the stock, and controls the fiber/resin ratio. The die is heated to speedily cure the resin. The pultrusion process is suitable for thermoset-polymer composites reinforced by either synthetic or natural fibers.

The pultrusion method has many advantages such as being a low-cost automated system where human involvement is minimal, producing high quality products, high surface finish of the product compared to other composite processing methods, and high production rate as it is a continuous production process. Tapered, complex shapes and thin wall parts cannot be produced by this method.

8.3.3.5 Filament winding

Filament winding is the most common method to produce axi−symmetric composite parts such as pipes, tubes, tanks, cylinders, domes and spheres. In this process, fiber strands are unwound and passed continuously to the resin tank where fiber strands are impregnated completely with the resin. Now, these resin-impregnated strands are passed onto a rotating mandrel. These strands are wound around the mandrel in a controlled manner and in a specific fiber orientation. The matrix materials used in this method are epoxy, polyester, polyvinyl ester, and phenolic resin, and the reinforcements are glass, carbon, and aramid. This method is not suitable for a woven mat or to be stitched into a fabric form of fibers. The filament winding process has many advantages such as a high degree of uniformity in fiber distribution, minimal labor involvement, ability to change fiber orientation as desired, and lower cost of products as compared to other methods. Further, the size of components is not restricted. This method has some disadvantages such as relatively high capital investment, and the fact that very precise control over the mechanism is required for uniform distribution and orientation of fiber. It is not possible to produce the reverse geometry in this method.

8.3.4 Structure of Natural Fiber

The structure of natural fiber plays an important role in the achievement of mechanical properties of their prepared composite. Natural fibers are the cellulose fibers which consist of some helically wound cellulose microfibrils

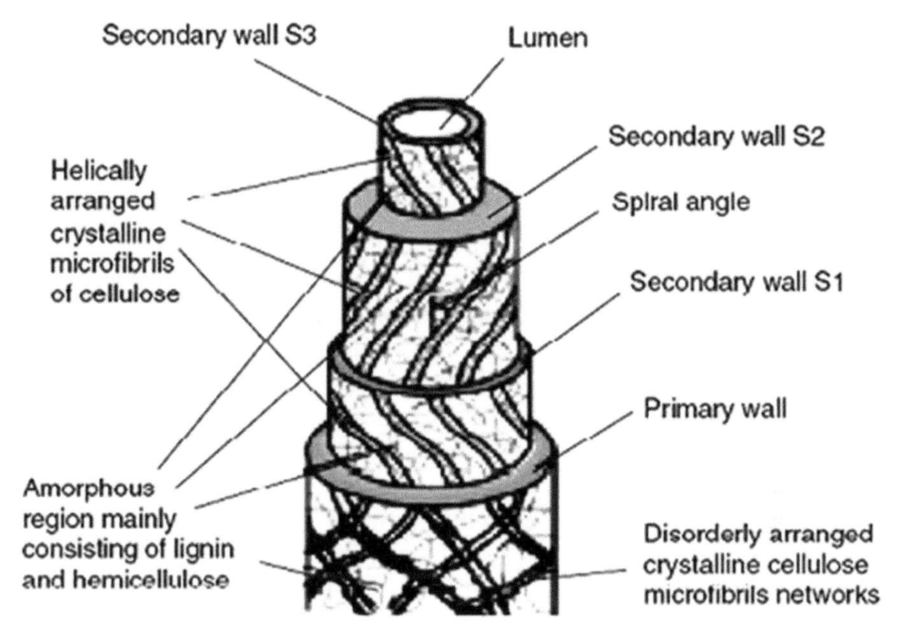

Figure 8.3 Structure of natural fiber.

and these are bound together by an amorphous lignin matrix. Lignin helps to keep the water in the fiber. It acts as a protection against biological attack and as a stiffener to give stems their resistance against gravitational forces and wind. Hemi-cellulose is found in the natural fibers and acts as a compatibilizer between cellulose and lignin. The cell wall in a fiber is not a homogenous membrane as shown in Figure 8.3. Each fiber consist of a complex, layered structure and has a thin primary wall—which is the first layer deposited during cell growth—encircling a secondary wall. The secondary wall consiss of three layers and the thick middle layer defines the mechanical properties of that fiber. The middle layer has a series of helically wound cellular microfibrils which are formed from long-chain cellulose molecules. The angle formed between the fiber axis and the microfibrils is called the microfibrillar angle and the value of the microfibrillar angle varies from one fiber to another. The diameter of these microfibrils is of the range of about 10–30 nm; they are made up of 30–100 cellulose molecules in extended chain conformation and provide the mechanical strength to the fiber.

8.4 Characterization

8.4.1 Mechanical Characterization

Mechanical characterization is a basic investigation of polymer-based fiber composites to study the strength, modulus, hardness, and stiffness. It was carried out in terms of tensile strength, tensile modulus, flexural strength, flexural modulus, impact strength, hardness, and compressive strength by many researchers. Mishra et al. [14] worked on the mechanical properties of epoxy-based bidirectional jute fiber composite. The jute composite was fabricated by the hand lay-up method with varying wt. % of fibers. It was reported that the composite with 48 wt. % fiber content had the maximum value of tensile strength, tensile modulus, flexural strength, flexural modulus and impact energy. Hossain et al. [15] evaluated the effect of fiber orientation on the tensile properties of jute fiber-reinforced epoxy laminated composite. The composite was prepared by a vacuum-assisted resin infiltration technique with varying stacking sequences, i.e., (0/0/0/0), (0/+45°/−45°/0) and (0/90°/90°/0) keeping constant 25 vol. % of fibers. It was concluded that the composite with stacking sequences 0/0/0/0 and 0/+45°/−45°/0 showed the higher values of tensile strength in the longitudinal direction than in the transverse direction. Gowda et al. [16] prepared jute fabric-reinforced polyester composites by hand lay-up method with constant 45 vol. % of fibers subjected to tensile, flexural, and impact tests. It was observed that the longitudinal tensile properties was five times higher than polyester resin and two times higher than transverse tensile properties. The flexural properties of the composite were observed to improve about two times more than polyester resin while a large improvement was observed in impact properties. Ray et al. [17] prepared the jute fiber-reinforced vinylester resin composites using pultrusion method with varying vol. % of fibers such as 8, 15, 23, 30, and 35. It was reported that the composite with 35 vol. % fiber content had the optimum mechanical properties as compared to all other composites. Kaewkuk et al. [18] studied the mechanical properties of sisal fiber-reinforced polypropylene composite. It was reported that the tensile strength and tensile modulus were found to increase with an increase in fiber content whereas impact strength and elongation-at-break was decreased. Prasad and Rao [19] carried out the mechanical properties of jowar, sisal and bamboo-reinforced polyester composite. They reported that the tensile strength of jowar composite was almost equal to that of the bamboo composite but 1.9 times more than that of the sisal composite. The flexural strength of the jowar composite was 4 and 35% greater than those of the bamboo and sisal composites respectively. Bakare et al. [20] developed and characterized the unidirectional sisal fibers reinforced rubber seed oil-based polyurethane composites, and observed that the tensile and flexural property of the composite increased with an increase in fiber loading up to 30 wt. %.

The effect of hybridization on the mechanical properties of the single natural fiber-reinforced polymer composite was also studied by many researchers. Shanmugam and Thiruchitrambalam [21] worked on the mechanical properties of unidirectional hybrid palm stalk/jute fiber-reinforced polyester composite. They reported that mechanical properties of palm fiber reinforced polyester composite were improved due to incorporation of jute fibres. Boopalan et al. [22] investigated the mechanical properties of the hybrid jute/banana fiber-reinforced epoxy composites. The hybrid composites are prepared by hand lay-up techniques with various percentage (100/0, 75/25, 50/50, 25/75, and 0/100) of jute/banana fibers by keeping constant 30 wt. % of total fiber content. It was concluded that the composite J50B50 shows 17, 4.3, and 35.5% greater tensile strength, flexural strength, and impact strength respectively than the jute fiber-reinforced epoxy composite. Venkateshwaran et al. [23] fabricated the banana/sisal-reinforced epoxy composite and studied its mechanical properties. The results showed that the composite S50B50 (50% sisal and 50% banana fiber) had the maximum value of tensile, flexural and impact strength.

8.4.2 Thermal Characterization

Thermal analysis has become an extremely widely used technique to determine the interfacial characteristics of NFRPCs. It was carried out using Dynamic Mechanical Analyzer (DMA), Thermogravimetric Analysis (TGA), Differential Scanning Calorimeter (DSC), and Fourier Transform Infrared Analysis (FTIR) by many researchers. Many researchers have reported studies on thermal analysis of natural fibers reinforced polymer composites. Pothan et al. [24] studied the dynamic mechanical properties of banana fiber-reinforced polyester composites. It was reported that the composite with 40% volume fraction fiber loading shows the maximum value of storage modulus whereas a lower value of loss modulus and damping parameters. Nair et al. [25] studied the thermal and dynamic mechanical properties of short sisal fiber-reinforced polystyrene composites. They reported that the sisal fiber-reinforced composite shows higher thermal stability than neat polystyrene and sisal fibers. The storage modulus decreased upon increasing the temperature and the glass transition temperature of the composite shifted toward lower temperature than neat polystyrene. Mohanthy et al. [26] studied the viscoelastic behavior of jute fiber-reinforced high-density polyethylene composites and observed that the storage modulus was found to increase with an increase in fiber loading up to 30%, whereas damping parameters was decreased compared to epoxy. Shinoj et al. [27] worked on the dynamic mechanical properties of the oil palm fiber linear low-density polyethylene biocomposite in terms of storage modulus, loss modulus and damping parameter. It was highlighted that the storage and loss modulus increased with an increase in fiber content whereas the values of the $Tan\delta$ peak decreased.

Boopalan et al. [22] studied the thermal properties of hybrid jute/banana fiber-reinforced epoxy composite. They suggested that the hybrid composite J50B50 (50% jute and 50% banana fiber) of 30 wt. % showed higher thermal stability than others composites. Jawaid et al. [28] also reported the studies on the thermal properties of hybrid oil palm/woven jute fiber-reinforced epoxy composite. It was found that the addition of woven jute fibers in the oil palm fiber-reinforced epoxy composite increased its thermal stability. Saw and Datta [29] studied the thermal properties of hybrid jute/bagasse fiber-reinforced epoxy composite. They suggested that the chemical-treated fibers composites showed better thermal stability than all the other composites as degradation temperature was shifted from 438°C to 475°C. Shanmugam and Thiruchitrambalam [21] studied the dynamic mechanical properties of alkali-treated hybrid palmyra palm leaf stalk/jute fiber-reinforced polyester composite. The results indicated that addition of jute fibers to palmyra palm leaf stalk fiber and alkali treatment of fibers had enhanced storage and loss modulus. The maximum damping behavior was observed for the composite with higher jute fiber content.

8.4.3 Water Absorption Properties

The water absorption properties of natural fiber-reinforced polymer composites were reported by different researchers. Singh et al. [30] studied the water absorption behavior of banana fiber-reinforced epoxy composites with different weight percentages of fiber content. It was suggested that the water absorption property was reduced due to addition of silica filler material with banana composite. Reis et al. [31] worked on the moisture absorption behavior of hemp fiber reinforced polypropylene composites and reported that the treated fiber composites showed decreased moisture absorption behavior compared to untreated fiber composite. Venkateshwaran et al. [23] investigated the water absorption properties of hybrid banana/sisal-reinforced epoxy composite. It was observed that the addition of sisal fiber in banana fiber reinforced-composite of up to 50 wt. % resulted in decreasing the water absorption properties. Dixit and Verma [32] carried out the water absorption behavior of coir/sisal/jute fiber-reinforced polyester composites. The water absorption test was performed in distilled water at different time intervals such as 24, 48, and 72 h. It was highlighted that the hybrid jute and sisal fiber-reinforced polyester composite gave the minimum water uptake compared to all other composites.

8.4.4 Tribological Behavior

Tribological behavior in terms of specific wear rate and the coefficient of frication of the NFRPCs were presented by different researchers and they found that incorporation of fibers into polymer matrix can improved its tribological properties. Bajpai et al. [33] studied the friction and wear properties of natural fibers (grewia optiva, nettle and sisal)-reinforced PLA composite. It was reported that incorporation of natural fibers into PLA improves the wear behavior of neat PLA. The friction coefficient and specific wear rate of composites were found to reduce by 10–44% and 70% respectively compared to neat PLA. Chand and Dwivedi [34] investigated the effect of maleic anhydride-grafted polypropylene as a coupling agent on the abrasive behavior of jute fiber-reinforced polypropylene composite. It was reported that use of a coupling agent gave better wear resistance than the jute polypropylene composite. Tayeb [35] worked on the tribological behavior of sugarcane and glass polyester composite. It was suggested that the sugarcane polyester composite showed a better degree of wear resistance and friction coefficient than the glass fiber-reinforced polyester composite. Deo and Acharya [36] studied the abrasive wear of the lantana camera fiber-reinforced epoxy composite. They reported that the abrasive wear loss increased on increasing the load, and optimum wear loss was found for the composite with 40 wt. % of fibers content.

8.5 Application of Natural Fiber Reinforced Polymer Composite

The applications of NFRPCs are increasing day by day in various applications such as automotive parts, packaging, building materials, sports, medicals, electrical industries, and so on. The major use of NFRPCs is in the packaging industry due to high specific strength and long durability. It is found that 42% of the natural fibers was used in the packaging industry, 20% in building and construction, 8% in automotive sector, and 30% in other applications.

Some selected applications of NFRPCs are as follows:

- *Building and construction industry*: Panels for partition and false ceiling, partition boards, walls, floors, window and door frames, roof tiles, mobile or pre-fabricated buildings which can be used in times of natural calamities such as floods, cyclones, earthquakes, etc.
- *Storage devices*: Post-boxes, grain storage silos, bio-gas containers, etc.
- *Furniture*: Chairs, tables, showers, bath units, etc.
- *Electric devices:* Electrical appliances, pipes, etc.
- *Everyday applications:* Lampshades, suitcases, helmets, etc.
- *Transportation:* Automobile and railway coach interiors, boats, etc.
- *Toys:* Soft toys, dolls and many more
- *Sports:* Bicycle frames, tennis rackets, etc.
- *Packaging:* In all kinds of packaging
- *Kitchen:* Cupboards, counter tops, kitchen cabinets, etc.

8.6 Conclusion

In the present study, the fabrication methods, properties, applications, and characterization in terms of mechanical properties, thermal properties, water absorption behavior, and tribological behavior of natural fiber-reinforced polymer composites are addressed and the following conclusions could be drawn:

- Researchers and scientists have a lot of expectations from biodegradable natural fibers to reduce in the near future the environmental burden created due to huge consumption of synthetic fibers.
- Mechanical properties such as tensility, flexure, and impact were found to increase with an increase in fiber loading, hybridization, and chemical treatments.
- Dynamic mechanical properties and thermal properties were also found to increase due to the effect of hybridization and chemical treatments.
- The limitations of natural fiber-reinforced polymer composites were improved due to the effect of hybridization and chemical treatments.

References

[1] Alam, T., Kumar, A., Gupta, M. K., Srivastava, R. K., and Singh, H. (2014). Thermal characterization and fracture toughness of sisal fibre reinforced polymer composite. *I. J. Scienti. Eng. Technol.* 3, 1071–1073.

[2] Gupta, M. K., and Srivastava, R. K. (2015). Effect of sisal fibre loading on dynamic mechanical analysis and water absorption behaviour of jute fibre epoxy composite. *Mater. Today Proc.* 2, 2909–2917.

[3] Nair, K. C. M., Thomas, S., and Groeninckx, G. (2001). Thermal and dynamic mechanical analysis of polysterine composites reinforced with short sisal fibre. *Compos. Sci. Technol.* 61, 2519–2529.

[4] Gupta, M. K., and Srivastava, R. K. (2015). Mechanical and water absorption properties of hybrid sisal/glass fibre reinforced epoxy composite. *Am. J. Polym. Sci. Eng.* 3, 208–219.

[5] Gupta, M. K., and Srivastava, R. K. (2015). Effect of sisal fibre loading on wear and friction properties of jute fibre reinforced epoxy composite. *Am. J. Polym. Sci. Eng.* 3, 198–207.

[6] Idicula, M., Malhotra, S. K., Josheph, K., and Thomas, S. (2005). Dynamic mechanical analysis of randomly oriented intimately mixed short banana/sisal fibre reinforced polyester composites. *Compos. Sci. Technol.* 65, 1077–1087.

[7] Mourya, H., Gupta, M. K., Srivastava, R. K., and Singh, H. (2015). Study on the mechanical properties of epoxy composite using short sisal fibre. *Mater. Today* 2, 1347–1355.

[8] Bisaria, H., Gupta, M. K., Sandilya, P., and Srivastava, R. K. (2015). Effect of fibre length on mechanical properties of randomly oriented short jute fibre reinforced epoxy composite. *Mater. Today* 2, 1193–1199.

[9] Gupta, M. K. (2016). Dynamic mechanical and thermal analysis hybrid jute/sisal fibre reinforced epoxy composite. *J. Mater. Des. App.* DOI: 10.1177/1464420716646398.

[10] Sahari, J., Sapuan, S. M., Zainudin, E. S., and Maleque, M. A. (2013). Mechanical and thermal properties of environmentally friendly composites derived from sugar palm tree. *Mater. Des.* 52, 285–289.

[11] Ranganna, H., Karthikeyan, N., Nikhilmurth, Y. V., and Kumar, S. R. (2012). Mechanical & thermal properties of epoxy based hybrid composites reinforced with sisal/glass fibres. *I. J. Fib. Text. Res.* 2, 26–29.

[12] Kumar, A., Gupta, M. K., Srivastava, R. K., and Singh, H. (2014). Viscoelastic properties and fracture toughness of hybrid polymer composite using banana/sisal fibre. *I. J. Adv. Res. Sci. Eng.* 3, 1–9.

[13] Gupta, M. K., and Srivastava, R. K. (2016). Mechanical properties of hybrid fibres reinforced polymer composite: a Review. *Polym. Plast. Technol. Eng.* 55, 626–642.

[14] Mishra, V., and Biswas, S. (2013). Physical and mechanical properties of bi-directional jute fiber epoxy composites. *Procedia. Eng.* 51, 561–566.

[15] Hossain, M. R., Islam, M. A., Vuurea, A. V., and Verpoest, I. (2013). Effect of fibre orientation on the tensile properties of jute epoxy laminated composite. *J. Scient. Res.* 5, 43–54.

[16] Gowda, T. M., Naidu, A. C. B., and Chhaya, R. (1999). Some mechanical properties of untreated jute fabric-reinforced polyester composites. *Compos. Part A* 30, 277–284.

[17] Ray, D., Sarkar, B. K., Rana, A. K., and Bose, N. R. (2001). The mechanical properties of vinylester resin matrix composites reinforced with alkali-treated jute fibres. *Compos. Part A* 32, 119–127.

[18] Kaewkuk, S., Sutapun, W., and Jarukumjorn, K. (2013). Effects of interfacial modification and fibre content on physical properties of sisal fiber/polypropylene composites. *Compos. Part B* 45, 544–549.

[19] Prasad, A. V. R., and Rao, K. M. (2011). Mechanical properties of natural fibre reinforced polyester composites: jowar, sisal and bamboo. *Mater. Des.* 32, 4658–4663.

[20] Bakare, I. O., Okieimen, F. E., Pavithran, C., Khalil, H. P. S. A., and Brahmakumar, M. (2010). Mechanical and thermal properties of sisal fibre-reinforced rubber seed oil-based polyurethane composites. *Mater. Des.* 31, 4274–4280.

[21] Shanmugam, D., and Thiruchitrambalam, M. (2013). Static and dynamic mechanical properties of alkali treated unidirectional continuous palmyra palm leaf stalk fibre/jute fibre reinforced hybrid polyester composites. *Mater. Des.* 97, 533–542.

[22] Boopalan, M., Niranjana, M., and Umapathy, M. J. (2013). Study on the mechanical properties and thermal properties of jute and banana fibre reinforced epoxy hybrid composites. *Compos. Part B* 51, 54–57.

[23] Venkateshwaran, N., ElayaPerumal, A., Alavudeen, A., and Thiruchitrambalam, M. (2011). Mechanical and water absorption behaviour of banana/sisal reinforced hybrid composites. *Mater. Des.* 32, 4017–4021.

[24] Pothan, L. A., Oommen, Z., and Thomas, S. (2003). Dynamic mechanical analysis of banana fibre reinforced polyester composites. *Compos. Sci. Technol.* 63, 283–293.

[25] Nair, K. C. M., Thomas, S., and Groeninckx, G. (2001). Thermal and dynamic mechanical analysis of polystyrene composites reinforced with short sisal fibres. *Compos. Sci. Technol.* 61, 2519–2529.

[26] Mohanty, S., Verma, S. K., and Nayak, S. K. (2006). Dynamic mechanical and thermal properties of MAPE treated jute/HDPE composites. *Compos. Sci. Technol.* 66, 538–547.

[27] Shinoj, S., Visvanathan, R. S., Panigrahi, S., and Varadharaju, N. (2011). Dynamic mechanical properties of oil palm fibre (OPF)-linear low density polyethylene (LLDPE) biocomposites and study of fibre-matrix interactions. *Biosyst. Eng.* 109, 99–107.

[28] Jawaid, M., Abdul Khalil, H. P. S., and Alattas, O. S. (2012). Woven hybrid bicomposites: dynamic mechanical and thermal properties. *Compos. Part A* 43, 288–293.

[29] Saw, S. K., and Datta, C. (2009). Thermomechanical properties of jute/bagasse hybrid fibre reinforced epoxy thermoset composites. *Bioresources* 4, 1455–1476.

[30] Singh, V. K., Gope, P. C., and Chauhan, S. (2012). Mechanical behavior of banana fiber based hybrid bio composites. *J. Mater. Environ. Sci.* 3,185–194.

[31] Reis, P. N. B., Ferreira, J. A. M., Antunes, F. V., and Costa, J. D. M. (2007). Flexural behaviour of hybrid laminated composites. *Compos. Part A* 38, 1612–1620.

[32] Dixit, S., and Verma, P. (2012). The effect of hybridization on mechanical behaviour of coir/sisal/jute fibers reinforced polyester composite material. *Res. J. Chem. Sci.* 2, 91–93.

[33] Bajpai, P. K., Singh, I., and Madaan, J. (2013). Tribological behavior of natural fibre reinforced PLA composites. *Wear* 297, 829–840.

[34] Chand, N., and Dwivedi, U. K. (2006). Effect of coupling agent on abrasive wear behaviour of chopped jute fibre reinforced polypropylene composites. *Wear.* 261, 1057–1063.

[35] Tayeb, N. S. M. (2008). A study on the potential of sugarcane fibres/polyester composite for tribological applications. *Wear* 265, 223–235.

[36] Deo, C., and Acharya, S. K. (2010). Effects on fibre content on abrasive wear of lantana camara fibre reinforced polymer matrix. *Indian J. Eng. Mater. Sci.* 17, 219–223.

9

Tribological Performance of Polymer Composite Materials

Raghvendra Kumar Mishra[1,2,3] and Sajith T. Abdulrahman[2]

[1]Director, BSM Solar and Environmental Solution, A-348, Awas Vikas Colony, Unnao, UP, 261001, India
[2]International Interuniversity centre for Nanoscience and Nanotechnology, Kerala, India
[3]Indian Institute of Space Science and Technology, India

Abstract

Improving the tribological properties of the mating surfaces gained more attention in recent days on considering the factor conservation of energy. Polymer composites are used as structural materials in aerospace, automobile, and medical fields to enhance the friction and wear properties especially in the areas where the fluid lubricants cannot be used. Over the decades, the research was focused on the factors affecting the tribological performance by incorporating fillers. In this chapter, we studied the factors affecting the tribological performance of the mating polymer surfaces. The role of nature and size of the fillers were examined here in detail. Moreover, the advanced polymer nanocomposites which can be used for future tribological applications were discussed here.

9.1 Introduction

Tribology is the consideration of materials' deterioration, friction together with lubrication. It also accounts the way of interaction between mating surfaces in the relative movement of the manufactured systems, which covers bearing structure as well as lubrication. Tribology is not much old technology, but actually it is a complicated, multidisciplinary technology undertaking in which developments are taken by collaborative endeavors of scientists from various areas such as physical engineering, production, materials technology, as well as manufacturing, chemical composition and its engineering, purgatives, computations, biomedical science and its mechanism, engineering, etc. Tribology is a word of mechanical engineers when the components are confronted with relative motion underneath various circumstances. Tribological science addresses friction, wear, and lubrication sciences. From a commercial point of perspective, a number of components are confronted with such tribological loading during their service period, and the majority of failures happen as a result of the high mass removal through the serves [1, 2]. Both friction and wear are not materials characteristics. They are just reactions to a particular tribological system which generally consists of a bearing driver shaft in addition to lubricant pairing together with a variety of aspects, some of the tribosystems such as support, joining material, bearing, gear, seals, clutch, and brakes. The scheme for tribosystem is mentioned in Figure 9.1.

The most important obstacle is the fact that friction and wear values will not be conveniently distributed from one system to some other. Evaluations between calculated values will only be workable whenever similar tribological systems are considered. The tribological characteristic of materials might be possible only for particular purposes depending on modeling and simulation examination offered for the working circumstances. The common stress contains the technological as well as actual physical load parameters which include load, sliding speed, as well as the contact period of time together during motion and, in addition, the temperature situations stressing the system structure. The system structure is dependent upon the character outline of the components such as the base, contacting body together with surrounding environment.

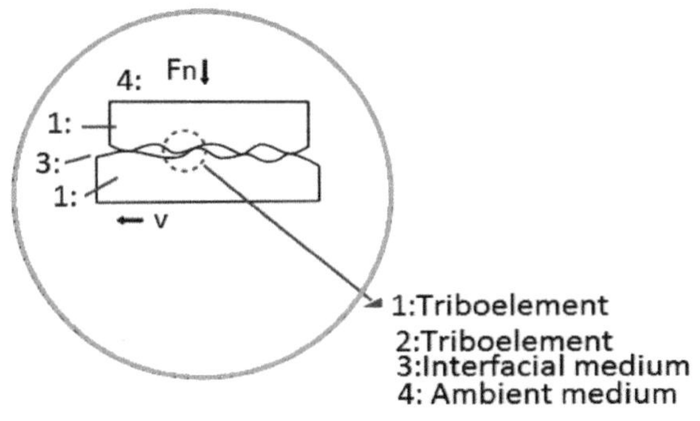

Figure 9.1 The scheme for the tribological subsystem.

Due to these reasons, identifying the tribological overall performance of materials turns out to be a vital concern when we look at the material selection for new component design and fundamental mechanical characteristics. There are lots of variables that affect the tribological behavior of materials, for example, functioning variables (applied load, sliding distance, and sliding velocity), contact conditions (dry/wet), contact mechanisms (point, line, and area), and morphology of materials. In today's decade, numerous materials have now been developed such polymers and metals composites; in addition, they found their applications in a lot of sectors. Biopolymers and polymer composites are also attractive research areas in today's decade [3, 4]. The tribological research field concentrates on the potential for using reinforcements for different composites such as 1) composite aging strategy has positive effects on the adhesive wear abilities of polymeric composites; 2) tribological properties of aged composite in numerous solutions are highly influenced by the viscosity of the solution; and 3) sliding distance has noticeable influence on the tribological properties of the composite aged in water and salt water. Tribology of polymers may be caused by two principal mechanisms, deformation and adhesion. The deformation mechanism includes total dissipation of energy at the contact area. The adhesion component is important for almost all types of the frictions regarding the polymer. This is a consequence of breaking of weak bonding forces between polymer chains in the material. Wear damages the surface functioning of polymers and can affect their functioning [5–7]. It is really thought that the coefficient of friction may be decreased and wear resistance may be enhanced by picking material by assessing bond strength and sharing and also the tear of rubbing materials within and over the contact place [8–11]. Mica has great mechanical, thermal, electrical, and chemical characteristics and not often present in any kinds of products. Strengthening of thermoplastics and polymer blends by mica happens to be the main topic of a number of studies because of its special combination of characteristics [12–15]. An array of research reports have been carried out as the influence of test conditions, environment on the friction, and wear behavior of polymer and polymer composites. It was absolutely stated that the friction coefficient of polymers compared to metals decreases upon increasing the load; at the same time, polymers indicate an increase in friction coefficient upon increasing the load [16–18]. A wide range of polymers can be viewed to competitive materials for tribological purposes due to their lower friction coefficients, effective damping properties, and self-lubricating capability. The enhancement of mechanical and/or tribological characteristics of polymers by the inclusion of particulate filler materials happens to be extensively examined underneath severe friction environment. In the case of heavy load, neat polymer composites are not reliable for antiwear and friction reduction [19–22].

The nanoparticle-enhanced polymer composites could be the most advanced types of materials, because of high strength and modulus at surprisingly low amounts of nanoparticles loading in the blends and composites [3, 23]. Nanocomposites are appropriate for traditional polymer processing, compared with the traditional fiber-reinforced composites. If the fillers or fiber possess dimensions regarding the order of nanometers, even a small quantity will improve the properties when compared to traditional composites [24]. Special characteristics of polymer nanocomposites are related to the substantial filler surface area-to-volume ratio, which leads to considerable enhancement in interfacial features of contact within the polymers and particles [25–27]. The huge increase in interfacial features makes contact with considerable fraction of polymer segments to have interaction directly using the filler particles at very low particle quantities. Mechanical and tribological properties of polymer nanocomposites are not quite straightforward to improve the specific properties of the components, but it is possible to determine by many factors and synergetic interactions. Incorporation of nanomaterials into polymer matrix is very complicated due to dispersion, distribution problem of nanomaterials, and also these aspects can have extremely important aspects to manipulate the performance of a certain system [23, 27–29]. Tribology is naturally complex with no dominant laws for dry sliding friction or wear. In addition to high tech in polymeric nanocomposites, tribology contains various qualitative descriptors of crucial system parameters, for example, bulk mechanical properties, particle dispersion, morphology, composition, and chemistry [30–32].

The concept tribology, taken from the ancient Greek vocabulary “tribos” which means rubbing, is considered as interdisciplinary discipline combined with technology associated with friction, wear, and lubrication. The tribology area is focused on building and then enhancing the material surfaces properties which may have minimal friction as well as higher wear protection for technology purposes. The frequently used materials in tribology range between metals, alloys to ceramics, solid lubricants, and polymers together with composites. Tribological performance of polymers is analyzed in the mid-20th century to the modern day. Significantly, numerous studies have already been performed on the tribology of polymers as well as its utilization of polymers for tribological requirements raising and also expanding significantly in recent years [30, 33].

Wear is referred to as damage to a solid surface that often will involve intensifying loss of material which is because of relative movement between surface as well as a contacting solid, liquid, or even gas. The modifications in surface of materials occur from physical forces, temperature, environment, as well as types of chemical reactions. Polymers because of their particular configuration together with mechanical characteristic tend to be more concerned to these types of conditions [34, 35]. Wear can be categorized in various manners. The normal classifications of wear refer to the lubricated or non-lubricated wear as per the presence or the absence of effective lubricants. An alternative way is simply to categorize wear on the principle of the basic mechanism which is in operating condition such as adhesion, abrasion, fatigue, and corrosion mechanism. Abrasion wear takes place between the pair of surfaces that have rigid and hard particles or sometimes hard asperities which are forced against as well as move along a solid surface contact [33]. As a consequence of hardness variations between surfaces, abrasive wear is identified on both of the surfaces. (To explain, both of the surfaces have been exposed to abrasive wear.) It is the rate on which the materials' abrade is dependent on the properties of each one surface, the existence of abrasives between surfaces, the speed of contact, as well as other environmental situations. Abrasive wear leads to the much softer material getting taken off the path followed by the asperity throughout the movements of the harder surface [36]. Abrasion is usually classified based on the kinds of contacts, in addition to contact condition [37, 38]. The kinds of contacts consist of two-body as well as three-body wear. Two-body takes place if an abrasive slides along a surface; in addition, the three-body wear takes place if an abrasive is arrested between one surface and the other [39, 40].

The performance of tribological system depends on the material properties (e.g., hardness, adhesion behavior, structure of materials, deformation behavior, and surface energy), geometric properties of materials (e.g., topography, contact ratio, and dimension), interfacial liquid and gaseous medium (e.g., humidity, component, viscosity, compatibility, and molecular structure), interfacial solid particle medium (e.g., abrasive particles, wear particles, solid lubricant, and dispersion), environmental condition, and finally type of load, motion, and temperature. This tribological system is composed of the collective stress/operational inputs, system structure, and the functional and loss outputs. The collective stress includes the technical and physical load parameters including load, sliding speed, and duration along with the movement and temperature conditions stressing the system structure. The system structure is determined by the property profiles of the substantial elements including the base, opposing body and ambient, and the intermediate medium. Friction is the force of opposition to movement between a pair of systems in touch. Friction may be referred to as the macroscopic point by the straightforward rules of friction from physicists Guillaume Amontons as well as Charles-Augustin de Coulomb. Those physicists discovered a linear relationship between resultant friction forces with employed normal load. Depending on this, a dimensionless primary parameter may be come out of, referred to as the coefficient of friction. It is actually described by the ratio of the resultant friction force and employed normal force. Nevertheless, the exact mechanism of sliding friction takes place at a microscopic scale; this means that tribological concepts on friction additionally include the topography of the surface areas. The mechanisms liable for the energy transforming operation in the close to surface area include things like tribological stress (energy introduction), energy exchange and also energy dissipation by thermal progression, energy discharge, and energy dissipation.

Wear is understood to be the irreversible material loss of any interaction surfaces. The actual physical and chemical parameters which change the material properties together with structural and frictional properties were considered as wear mechanisms. All these wear mechanisms consist of surface area fatigue (fracture), abrasive (grooves), adhesion (materials' exchange), and tribochemical reaction (particles). Friction and wear mechanisms are highly influenced by the structure of the tribological body along with the generated combined stress. Friction and wear mechanisms are not prone to happen in an isolated fashion, but actually via a superposition of mechanisms which is difficult to quantify and as a consequence manage. This superposition takes place in tribo-technical approaches in non-detectable ratios and also in ratios that differ across time period and also location, so that it is very hard to estimate friction in addition to wear details in a tribo-contact. This is the reason that tribological testing is highly essential for estimating tribological tendencies. In this context, we need to analyze fully grasp tribologically measured information and also mechanism-oriented study; we must have entire understanding of the playing mechanisms in a tribo-contact. Tribological examination permits us to obtain details about tribo-performance of components to generate completely new and consequently much better products designs. It is easy to focus on blends of materials to attain improved tribological features. Tribological results of the tests together with surface area results through analytical techniques can help in the approximation the tribo-performance comprising friction together with wear, malfunction mechanisms, kinetics of exchange films of current materials, or even innovative prototypes dependent on numerous facts and effects. These data facilitate us to realize factors which effect tribological performance such as the consequences of a variety of material combinations which include filler, filler concentration, synergetic results of fillers, and material design along with the effect of supplementary aspects of the system structure. For optimizing tribological performances of the surfaces, an insight of identifying the crucial points affecting the tribo-system is necessary. Subsequently, determining methods to develop effectiveness by lowering wear for optimizing the material pairings results in lowering of the friction and wear intensity by choosing and utilizing the suitable lubricants. The scope of the tribological unit is of key significance in bearing choice. The number of factors will responsible for collective stress (e.g., character of the load, character of the movement, temperature ranges, and time period aspect), the mating companion (e.g., materials, such as physical arrangement and chemical attributes), geometrical characteristics such as the contact ratio and also topography (roughness, isotropy, and anisotropy), the interfacial media and also its right character profile, the environment medium and also its properties, and thermal conductivity of the development. This chapter provides an insight into tribological performance of polymer nanocomposites and also similar kinds of nanocomposites suitable for these applications.

9.2 Tribological Characterization Techniques for Polymer Composites

The friction coefficient and wear resistance are not genuine material characteristics; these charcerstics are surface characteristics and it varies from the system to system when the materials are in motion. More often than not, it is really of the main issue to produce polymer composites which have low friction and low wear characteristics underneath dry sliding condition in opposition to smooth metallic counterparts. Relationship among the wear of polymer composites and operational guidelines is desirable to have better comprehending concerning the wear behavior [20, 41–44]. There have been many studies in to the impact of test conditions, contact geometry, types of fillers, and surrounding environment, friction, and wear tendency of polymer composites. In previous scientific studies, the tribological characteristics of polyamide (PA66), ultrahigh molecular weight polyethylene (UHMWPE), polyoxymethylene (POM), polyphenylene (PPS), and their composites have been described [45–48]. Considerable improvement regarding the wear resistance of fiber-reinforced epoxy composites was achieved, and it was also noticed that the wear rate reduced significantly with the inclusion of fibers

[49, 50]. It was introduced an observational analysis of the friction and wear tendency of carbon-epoxy and glass-epoxy composites. The analysis revealed that a rise in the sliding speed or a decline in the load results in decreasing the coefficient of friction and therefore the sliding wear characteristics of carbon-epoxy samples were more improved compared to glass-epoxy. Researchers noticed that the inclusion of glass fibers into poly(ether ether ketone) (PEEK) leads to increased wear rates under various kinds of loadings [51]. The variations in the sliding speed and load on the wear of a glass-epoxy composite were studied and observed that the wear loss improved with increasing applied load and sliding speed [52]. The previous studies revealed that glass-epoxy composites loaded with the hard powders of tungsten carbide (WC) and tantalum niobium carbide showed a lower volume loss and a lower particular wear rate [53–55]. Their tribological behavior was revealed to the types of materials included and it was absolutely detected the wear of the composites was additionally affected by the degree of load and the sliding distance, and finally the sliding speed. Wear tests have been carried out on a unique track of the disk. The specific wear rate can be measured by the following equation, which indicates the relationship between mass loss, density, sliding distance, and load.

The specific wear rate is expressed as

$$K_s = \frac{\frac{\Delta M}{\rho LD} mm^3}{Nm} K_s = \frac{\frac{\Delta M}{\rho LD} mm^3}{Nm}$$

where M = mass loss during test (g), ρ = density of composites (g/cm^3), L = Applied load (N), and D = sliding distance (m); the schematic setup for wear test is mentioned in Figure 9.2.

The majority of mechanical equipment include the following—automatic washers possess pulleys to revolve the drum and cams to use switches; computer printers possess gears, cams, and pulleys to supply paper and regulation printing heads; and photocopiers and automatic teller machines possess practically lots of such components in each system. These applications call for low priced, low-to-zero maintenance, non-lubricated, and first and foremost simplified designs. For those points, thermoplastic polymer-based materials can be the crystal clear option for this application. Injection-molded thermoplastic polymer gears could be easily manufactured in very vast quantities at low price, which is extremely rapid process considering the product development efficiency. Components and efficiency of polymeric gears have been examined by a wide range of people; the way these components' wear has displayed had gained a lot of attention [57, 58]. Dry Sliding Wear Test

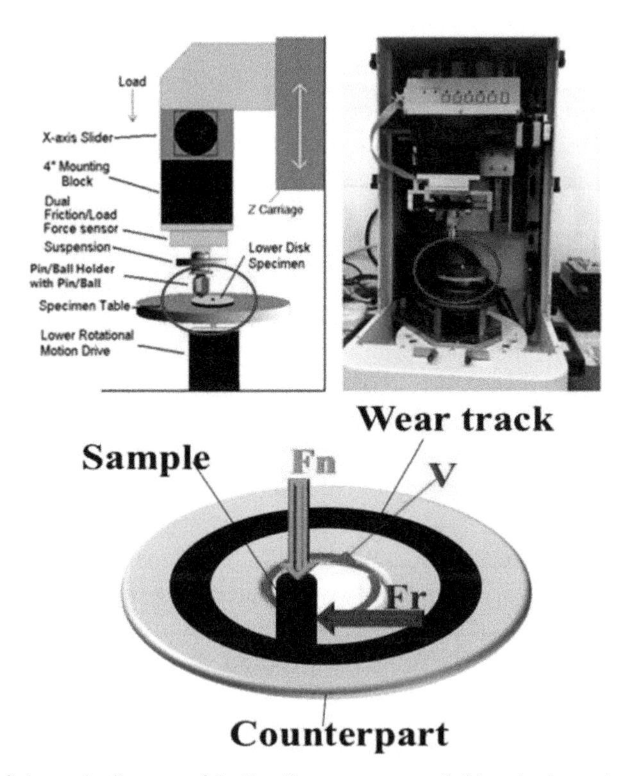

Figure 9.2 Schematic diagram of the P-o-D test apparatus (left) and schematics of the P-o-D tribological tests (right) [56].

A pin-on-disk setup (ASTM G99 standard, Make: DUCOM Instruments, Bangalore) can be used for dry sliding wear tests [59]. If the highest s/n ratio is noticed in the tested samples, it indicates the lowest wear, and the minimum s/n ratio suggests the utmost wear when we look at the sample [33]. The filler content is one of the significant factors in composites, followed by sliding velocity and normal load although the sliding distance gets the least or very little significance on wear rate of the particulate loaded composites [60, 61].

9.3 Preparation of Polymer Nanocomposites

Polymer nanocomposites tend to be hybrid organic and inorganic materials with one or more dimensions associated with dispersion phase significantly less than 100 nm. Because of the nanoscale dimensions, nanocomposites show exceptional mechanical and tribological properties [24, 62, 63]. The polymer nanocomposites could be prepared making use of the following ways—melt blending, in situ polymerization, solution casting, etc. Melt blending comprises the heating and melt blending polymer granules at high temperatures and bellows the degradation temperature, subsequently including the filler and blending by utilizing a high-speed mixer to produce the uniform distribution filler within polymers. In situ polymerization includes the penetration of filler when we look at the polymer monomer or monomer solution. Solution casting includes the production of composites by using suitable solvent; polymer and filler were dissolved in the solution and subsequently casted [64–66]. Extrusion and injection molding have been used as the most frequently options for preparing polymer nanocomposites. Processing parameters and surface modification of the nanoparticles show an important role in dispersion of nanoparticles in polymer matrix. The inclusion of nanofillers has a tendency to raise the tensile strength of the prepared nanocomposites. Wear maps are most useful in determining a variety of wear mechanisms associated and the wear transition [67–71]. The engineering plastic has commonly been utilized in automobile for outside parts of the body for attractive and aerodynamic necessities. ABS plastics are typically found in refrigerator lining automotive components and highway safety appliances. Polyamide plastics are most frequently found in bearings, gears, cams, and belt reinforcement, and carbon fiber-reinforced polymer composites are most favored in aerospace purposes including jaguars, Tornados, Boeing, and harriers. Carbon fiber-reinforced PEEK has been utilized in developing helicopter blades for lowering the overall weight. The field of tribology focuses on the style and design, friction, wear, as well as lubrication of interacting surfaces in comparative movement. Polymer composite parts are entirely utilized significantly for tribological applications recently. A lot of advancements of modern polymers composites contain incorporated nanofillers in consideration that reinforcing agents, leading to the word "polymeric nanocomposites," in general, it is often showed that these types of fillers of small dimensions (compared to the traditional micrometer-sized materials or even elements) may also lead to amazing enhancements in the friction in addition to wear properties of collectively bulk materials and coatings. Much like the bulk mechanical feedback, the tribological features of polymers are tremendously based upon the consequences of relative speed of the interacting surfaces, normal load, and the surroundings. Consequently, to take care of these kinds of influences and also for a better control over the results, polymers' properties are adjusted by introducing suitable fillers. Hence, these are almost always utilized in the form of composite or even, at the best mixed manner with the best possible composition for the predominantly friction and also wear performances.

9.4 Tribology Study of Different Polymer Nanocomposites

The initial work outs on polymer tribology possibly began with the sliding friction tests on rubbers and also elastomeric-based materials. Additionally, focus on some other polymers such as thermosets together with thermoplastics resulted in the growth and development of the two-term model of friction. The two-term model recommends that the frictional force is the result of the interfacial as well as cohesive works, which is carried out on the surface of the polymer substance. This is exactly provided that the counter face is adequately difficult in comparison with the polymer mating surface and also goes through merely gentle or even absolutely no elastic deformation.

The interfacial frictional work is the response to adhesive interactions, so the scope to owning this element undoubtedly depends on variables, for example, the stiffness of the polymer, molecular arrangement, glass transition temperature, as well as crystallinity of the polymer, surface roughness of the counterface in addition to chemical–electrostatic relations between the counterfaces along with the polymer. As an illustration, an elastomeric solid, having their own glass transition temperature below the room temperature and as a consequence for the reason that it is extremely soft, can have quite high adhesive element resulting in higher friction. Beyond interfacial work is the involvement of the cohesive term, which happens to be on account of the plowing movements of the asperities of the harder counterface into the polymer. The energy essential to the plowing motion would depend mainly upon the tensile strength along with the elongation before fracture (or even toughness) of the polymer and then the geometric outlines (height as well as cutting angle) of the asperities on the counterface. The elastic hysteresis can be another aspect, which is usually correlated to the cohesive terminology of polymers that display huge viscoelastic strains, for example, with regard to rubbers and elastomers. The effects of the interracial as well as cohesive stress are determined by the existing interface as well as surrounding temperatures, which also accounts for the rate of relative velocity of the material. Pressure possesses certain impact on the interfacial friction mainly because normal contact pressure has a tendency to change the shear strength of the interface layer.

Moreover, it is seen that as the contact pressure raises, the shear stress rise linearly resulting in higher friction. The friction is tremendously improved by the members of the class of polymers, viz., elastomers, thermosets, and thermoplastics (semicrystalline and amorphous). Semicrystalline linear thermoplastic can provide the most reasonable coefficient of friction, while elastomers and rubbers display significant values. This is due to the molecular structure of the linear polymers that assists molecules stretch quite easily in the direction of shear providing minimum frictional resistance. The unavoidable effect of friction in a sliding contact is wear. Wear of polymers is a complicated procedure. Based upon recognition, wear of a polymer sliding against a hard counter face are referred to as interfacial, cohesive, abrasive, adhesive, chemical wear, and so on. It is recognized that polymer wear can be tremendously affected by the kind (elastomeric, amorphous, and also semicrystalline) of the polymer. Generally, substantial tensile strength along with higher elongation at fail enhances wear resistance in a polymer. Consequently, a few of the linear thermoplastic polymers with semicrystalline microstructure carry out much better in wear resistance compared to thermosets or amorphous thermoplastics. These studies have been in series with the idea for polymers which indicates surface hardness is not a regulating reason for wear resistance. In actual fact, high hardness of a polymer can be harmful for wear resistance in dry sliding against hard counterface as hardness normally comes with low toughness for polymers. A higher extent of elongation at failure of a polymer signifies that the shear stress in a sliding can be significantly decreased because of substantial plastic deformation of the polymer within an extremely narrow layer near the interface. This interfacial layer suits just about all the energy dissipation processes, and the bulk of the polymer goes through a bare minimum of deformation or wear. Frictional heat produced at the interface is the leading problem to higher wear life of the polymer. The UHMWPE and PEEK are the most important polymers which can be utilized in tribological purposes without having reinforcement. Due to outstanding biocompatibility as well as wear resistance of UHMWPE, UHMWPE has made set of comprehensive application in the form of bearing material for synthetic human joints. PEEK has become a well-known polymer as matrix for certain relatively new composites with the purpose of making wear-resistant materials. PEEK can be used as a high-temperature polymer and has a tendency to display low wear rate, while the coefficient of friction is relatively higher (around 0.3). Nylons are further tribological materials, which exhibit lower friction and also lower wear. Poly(tetrafluoroether) (PTFE) is a linear fluorocarbon, which usually exhibits really low friction coefficient compared to a number of other thermoplastics; however, it shows higher wear rate because of its actual exceptional features of slippage in the crystalline configuration of the molecular bond structure. In the case of composite, PTFE acts as a superb solid lubricant because of lower friction character. Amorphous thermoplastics similar to PMMA and poly(styrene) exhibit higher coefficients of friction as well as higher wear rates. Amorphous thermoplastics similar to that PMMA and poly(styrene) cannot well apply in a wear test. Thermosets are usually employed in the way of composites like fiber strengthening which may significantly decrease wear. Fiber strengthening enhances the material's resistance to subsurface crack formation and propagation providing decreased plowing by the counterface asperities or even fatigue cracking. Interface friction is also fully optimized by introducing an appropriate fraction of a solid.

There is certainly a necessity to tailor the majority of polymers by appropriate filler which could decrease the wear rate together with the design necessity, which may improve or even reduce the coefficient of friction. The filler is capable of doing several different tasks based upon the range of the matrix and filler. A few of these tasks are reinforcing of the matrix (higher load supporting capability), change in the subsurface crack-arresting capability (much better toughness) and lubricating influence at the interface by reducing the shear stress along with the improvement of the thermal conductivity of the polymer. The whole facts of the tribology of polymer composites are often rather complicated in order to define the description. A self-lubricating polymer, for example, PTFE, could be made wear retardant by reinforcing the main part with the hard and perhaps strong filler materials just like particles of ceramics/metals and also an appropriate strong fiber (carbon, aramid, and also glass fibers). The drawback to utilizing fillers (particularly the particle kind) is the fact that the composite materials can become fairly much less tough compared to the pristinepolymer and therefore influence wear by fatigue; nevertheless, it could be prevented by appropriate optimization of the mechanical and also tribological characters. A fiber will help in lowering the wear of the composite after few running-in time. Polymer nanocomposite is reliant for its actual much better mechanical qualities on the exorbitantly high interface region between the filler (nanoparticles or even nanofibers) along with the polymer matrix. Substantial interface results in a much better bonding between the composites' component in addition to consequently much better durability together with toughness properties over unfilled polymer or even conventional polymer composites. The particular wear rates associated with these nanocomposites are usually discussed within the range of several orders lesser than the specific wear rates of the virgin polymer, once the best possible weight percent of the nanoparticles is formulated. The effect of inclusion of carbon fiber on the wear properties of polyphenylene sulfide/poly (tetrafluoroethene) has been reported, and it was found that the inclusion of carbon fibers decreases the coefficient of friction of the composites, when the carbon fibers are quiet low, then carbon fibers are come out from the matrix, and friction coefficient increases, while it becomes difficult to break the hard CF particles which in turn lead to reduction in friction coefficient at higher content of carbon fibers [72, 73]. The incorporation of zirconia (ZrO_2) particles into the polyimide polymer matrix seems to reduce the friction coefficient of the composites. However, once atomic oxygen is exposed, the friction coefficient of the composites relatively raises that could be caused by the generation of ZrO_2-rich layer [74]. Wear rate and friction coefficient of nano ZrO_2-based PEEK composites enhance with a rise in load; it is a result of high plastic deformation and scuffing mechanism [75, 76]. It is examined in which the inclusion of nanopowder brings down the friction coefficient. The increment in pressure raises the wear rate in the composites. Nevertheless, the nanofiller-reinforced nanocomposites offer a lesser extent raise in wear rate with a raise in pressure in comparison with composites filled up with traditional fillers [77, 78]. The inclusion of WS2 nanoparticles into the PEEK leads to decreased wear rate. The introduction of WS2 (Fullerene shaped) displayed 10% less wear rate in comparison with neat PEEK, although WS2 (needle shaped) particles lower the wear rate about by 60% [79–81]; however, the wear rate improves using the inclusion of carbon nanotube (CNT) and graphite nanoparticles. The adding up of CNT enhances the wear rate of composites by means of 20%, and the wear rate of graphite nanoparticles-reinforced composites is pretty much three times more compared to neat PEEK. These modifications are typically as a result of hardness associated with the reinforcing materials [79, 82–84], as mentioned in Table 9.1.

Table 9.1 Morphology of various polymer composites

Materials	Morphology	Remarks	Ref.
Transmission electron microscopy (TEM) microstructure of the graphite nanosheets	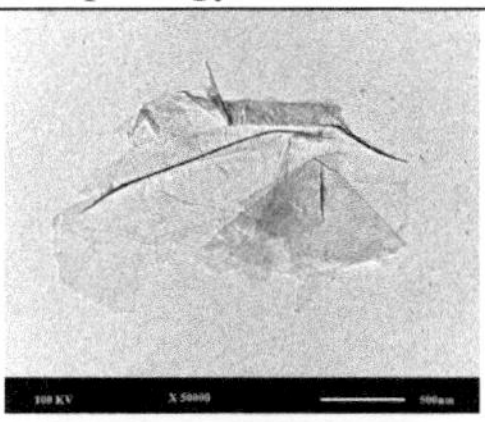	Graphite nanosheets were around in the shape of a thin film with a couple of ripples within the plane.	[56]
TEM microstructure of the surrounded graphite nanosheets		After surface alteration, unique morphological adjustments similar to rougher surface, without any effect on the diameter of the graphite nanosheet, along with the size still ranged from 2 to 4 μm.	[56]
Fractured surfaces of the poly (ether ether ketone) (PEEK) composites (PEEK/GNS = 90/10)	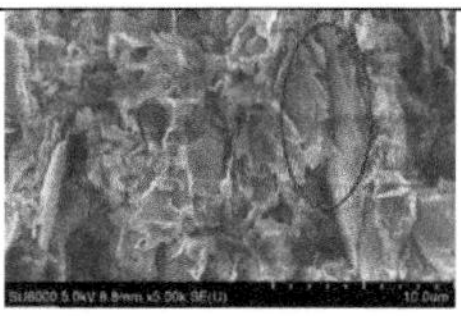	The graphite nanosheets were aggregated and also produced bundles inside the PEEK matrix	[56]
Fractured surfaces of the PEEK composites (PEEK/GNS = 80/20)		The GNSs were aggregated and formed bundles inside the PEEK matrix	[56]
Fractured surfaces of the PEEK composites (PEEK/wrapped GNS = 90/10)	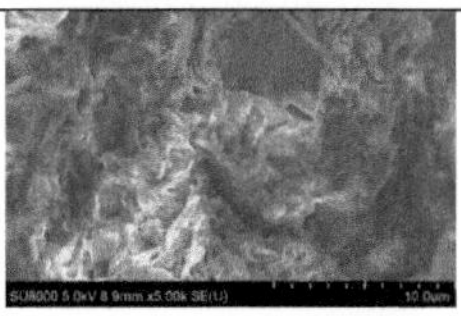	PES was formulated into the system, the GNS dispersion was enhanced, and also outstanding interfacial compatibility between the PEEK matrix and PES.	[56]
Fractured surfaces of the PEEK composites (PEEK/wrapped GNS = 80/20)	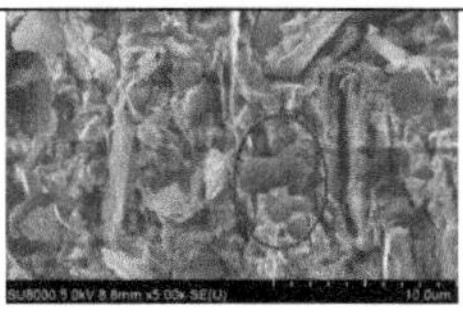	Certain aggregations of PES/GNS within the PEEK matrix.	[56]

(Continued)

Table 9.1 Continued

Materials	Morphology	Remarks	Ref.
TEM micrograph for the distribution of the nanoalumina particle in the PEEK matrix (alumina content 0.8 vol. %)	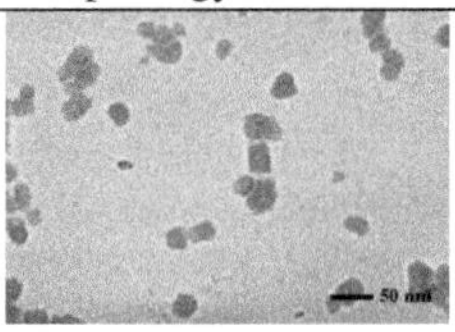	Spread separately, inconstant, semihomogeneously dispersed nanoparticles	[85]
TEM micrograph for the distribution of the nanoalumina particle in the PEEK matrix (alumina content 1.6 vol. %)	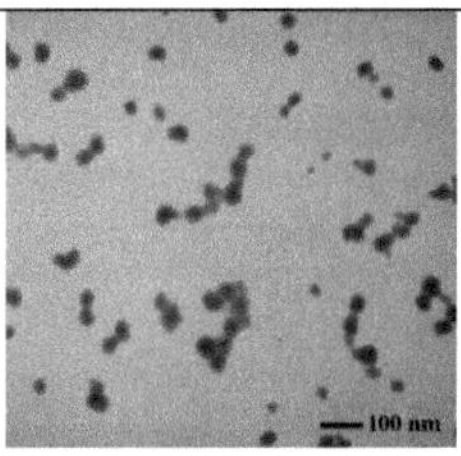	Scattered individually, irregular, semihomogeneously dispersed nanoparticles.	[85]
TEM of photoes of nano-Al_2O_3/epoxy compos-ites(20% original nano-Al_2O_3)	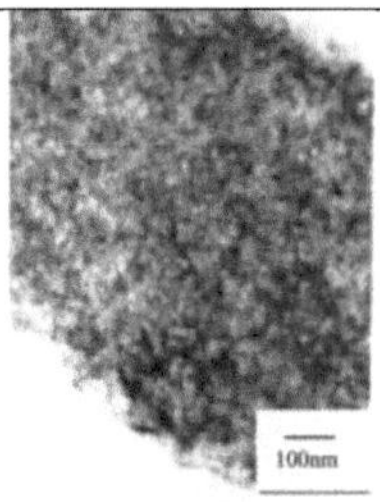	20% original nano-Al_2O_3, diameter of conglomeration about 500 nm,	[86]
TEM of photoes of nano-Al_2O_3/epoxy composites(5% modified nano-Al_2O_3)	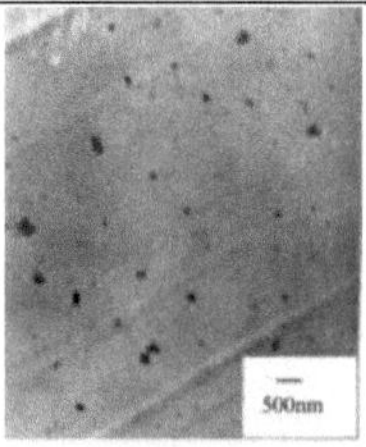	5% tailored nano-Al_2O_3, nicely separate, and also impregnated by the epoxy resin, substantial with the content of nano-Al_2O_3 raised	[86]
TEM of photoes of nano-Al_2O_3/epoxy composites (20% modified nano-Al_2O_3)	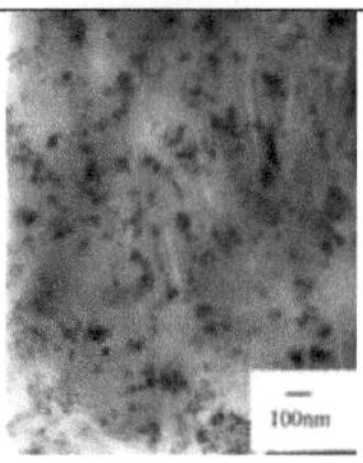	20% tailored nano-Al_2O_3, epoxy resin has a good capability to wet the functionalized nano-Al_2O_3, substantial with the contents of nano-Al_2O_3 raised	[86]

Table 9.1 Continued

Materials	Morphology	Remarks	Ref.
TEM of photoes of nano-Al_2O_3/epoxy composites (40% modified nano-Al_2O_3)	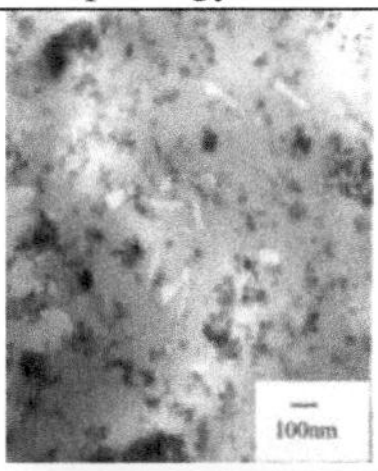	40% altered nano-Al_2O_3, epoxy resin has a highly effective ability to wet the functionalized nano-Al_2O_3, intense with the content of nano-Al_2O_3 increased	[86]
Short carbon fibers (SCFs)/PEEK composites(80 vol% PEEK/20 vol% SCFs)	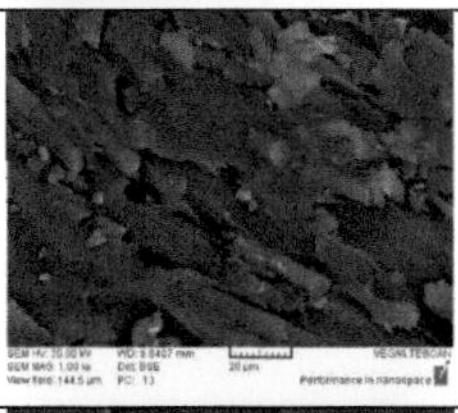	CFs in several sizes, SCFs cut down into small-size pieces due to the high shear, and SCFs in various lengths.	[87]
Short carbon fibers(20 vol%)/PEEK (80 vol%)compound (98 wt%) and 2 wt% nano-SiO_2	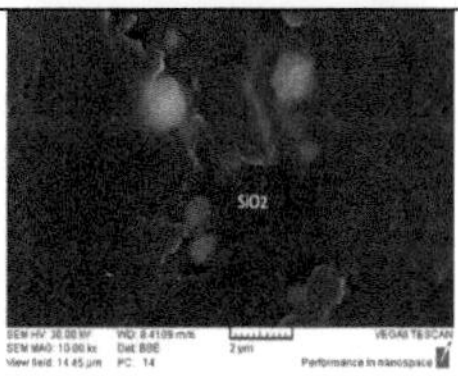	Nano-SiO_2 particles tend to agglomerate to decrease their area to volume ratio	[87]
Microwave-synthesized reduced graphene oxide (MRG)		Transparent and layered structure	[88]
boron–nitrogen-co-doped reduced graphene oxide(B–N-MRG)	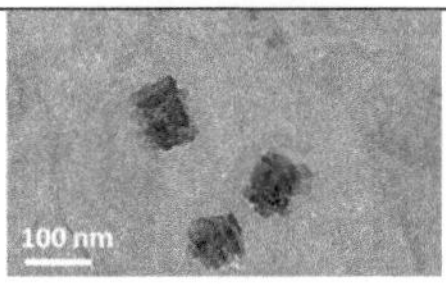	Transparent and layered structure	[88]
TiO_2-reinforced boron–nitrogen-codoped reduced graphene oxide (TiO_2–B–N-MRG)	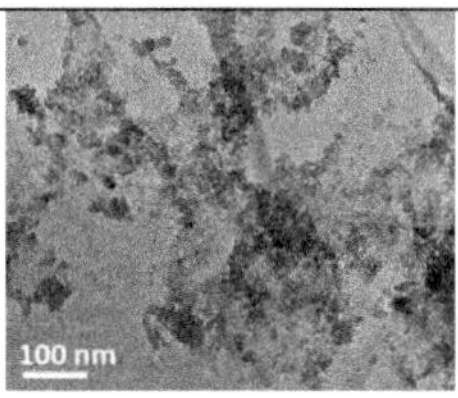	Transparent and layered structure, exhibiting TiO_2 nanoparticles on the surface of doped graphene	[88]

(*Continued*)

Table 9.1 Continued

Materials	Morphology	Remarks	Ref.
Microwave-synthesized reduced graphene oxide	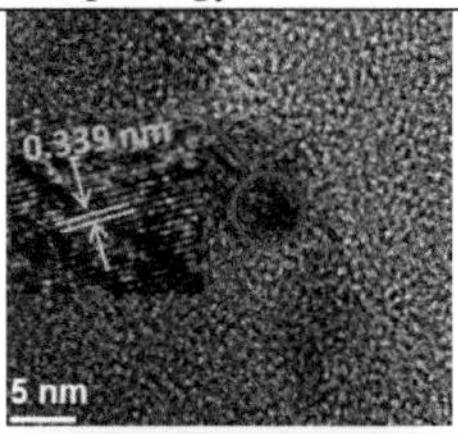	Lattice spacing of graphene sheets in 0.339 nm	[88]
Boron-doped reduced graphene oxide	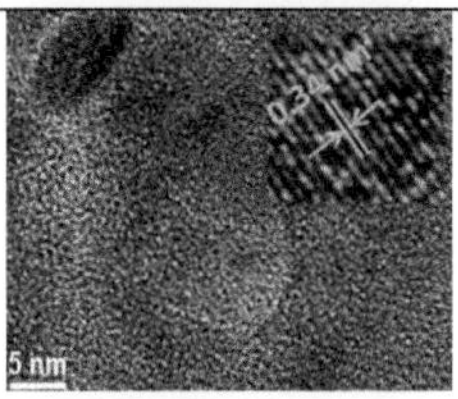	Lattice spacing of graphene sheets in 0.34 nm	[88]
TiO_2-reinforced boron–nitrogen-codoped reduced graphene oxide	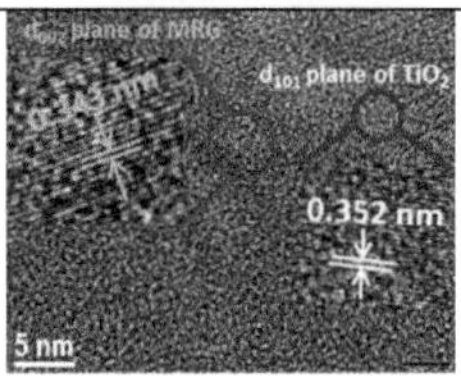	TiO_2 nanoparticles on the surface of doped graphene	[88]

Table 9.2 Mechanical properties of various polymer composites

Materials	Properties	Remarks	Ref.
Tensile strength of poly (ether ether ketone) (PEEK)/GNS and PEEK/wrapped GNS composites	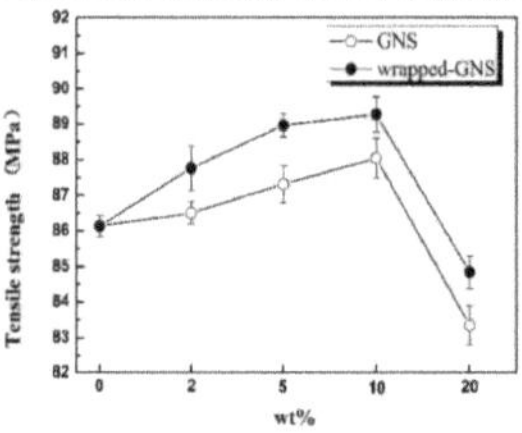	Improvement in tensile strength of the PEEK/wrapped GNS	[56]
Elongation at break of the PEEK/GNS and PEEK/wrapped GNS composites	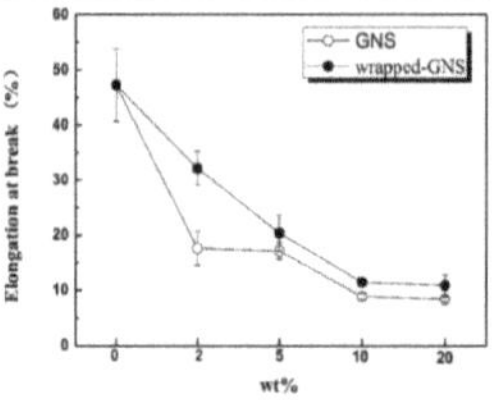	Improvement in elongation at break of the PEEK/wrapped GNS	[56]

(*Continued*)

Table 9.2 Continued

Materials	Properties	Remarks	Ref.
Flexural strength of the PEEK/GNS and PEEK/wrapped GNS composites	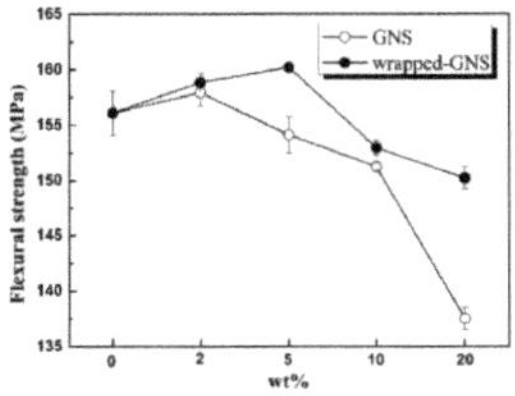	Improvement in flexural strength of the PEEK/wrapped GNS	[56]
Impact strength of the epoxy/nano-Al_2O_3 nanocomposites	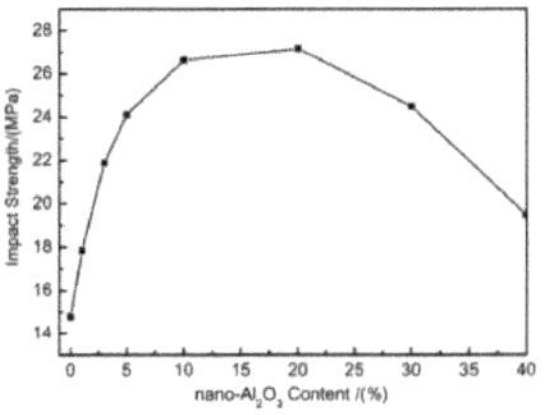	Improvement because of the huge interfacial area, the better interaction at interface, and microcracks are excessive, and the plastic deformation is oversized at higher content of nanofillers, which grows into the crack and declines the properties	[86]
Elastic modulus PEEK Pure PEEK PS PEEK (80vol %)/ SCFs (20vol %) (100wt %) PS1S PEEK (80vol %) /SCFs (20vol %) (99wt %) / Nano-SiO_2 (1wt%) PS1.5S PEEK (80vol %)/ SCFs (20vol %) (98.5wt %)/nano-SiO_2 (1.5wt%) PS2S PEEK (80vol %)/ SCFs (20vol %) (9wt %)/ nano-SiO_2 (2wt %)	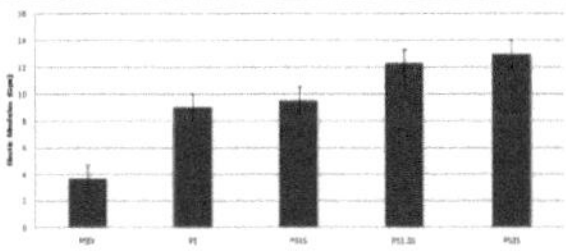	Inclusion of CFs into the polymer matrix mostly increases the elastic modulus, and incorporation of nano-SiO_2 into the conventional composite results in a second improvement in mechanical properties of the composite.	[87]

Table 9.3 Tribological properties

Materials	Properties	Remarks	Ref.
Friction coefficient of PEEK, PEEK/GNS and PEEK/wrapped GNS composites	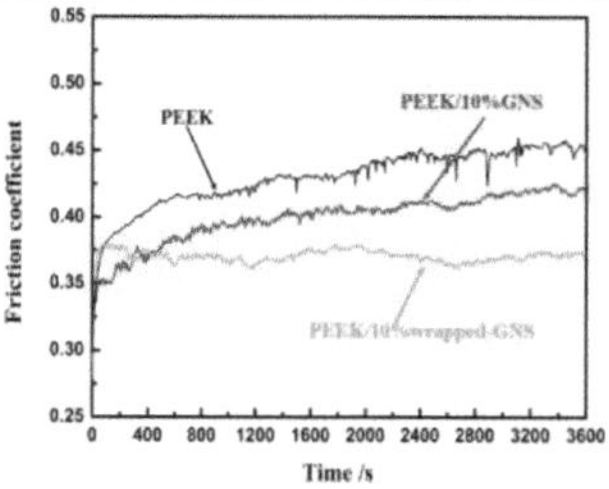	The friction coefficient improved remarkably at the beginning and fluctuated slightly at the steady stage.	[56]
Wear rate values of PEEK/GNS and PEEK/wrapped GNS	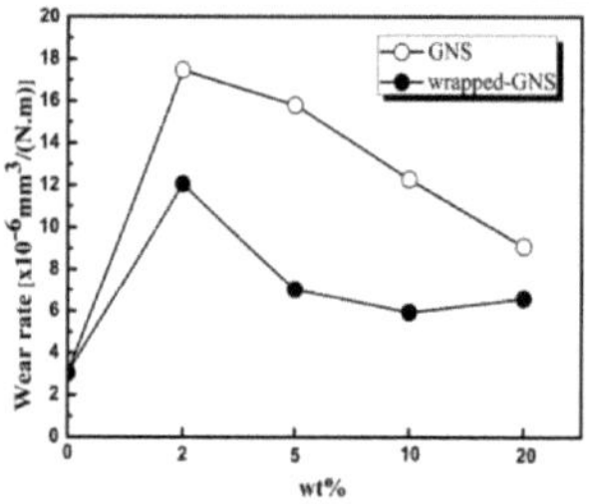	Wear rates of all of the types of composites were higher than that of pure PEEK due to spalling of PEEK.	[56]
Friction coefficients of PEEK/ SCF/PTFE/ graphite and PEEK/SCF/PTFE/ graphite/nano-SiO_2	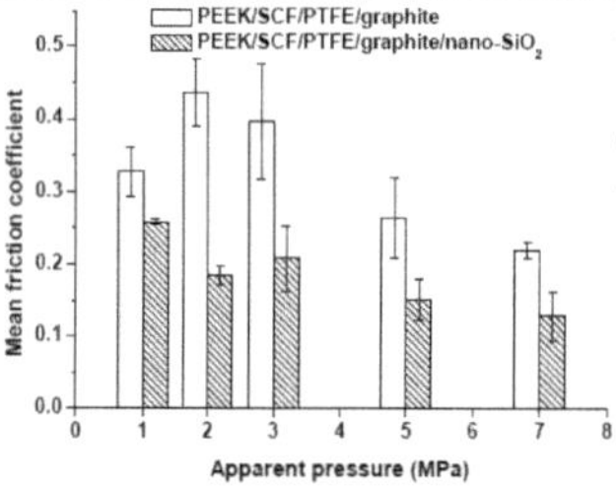	Nano-SiO_2 significantly decreases the friction coefficients.	[89]
Sliding velocity of PEEK/SCF/PTFE/ graphite and PEEK/SCF/PTFE/ graphite/nano-SiO_2 at 1 m/s and 5 MPa.	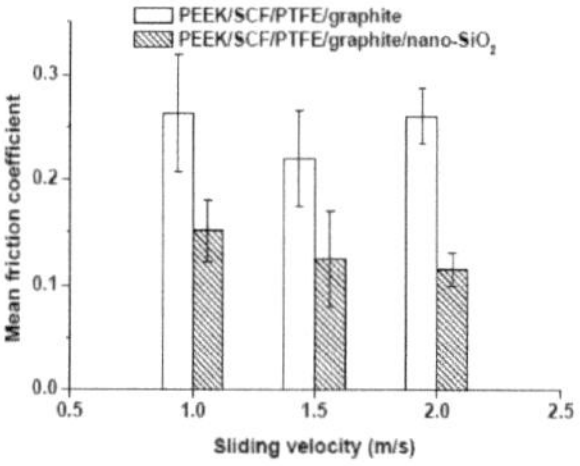	Nanoparticles remarkably reduce the friction coefficients	[89]
Effects of apparent pressure on the specific wear rates of PEEK/ SCF/PTFE/ graphite and PEEK/SCF/PTFE/ graphite/nano-SiO_2	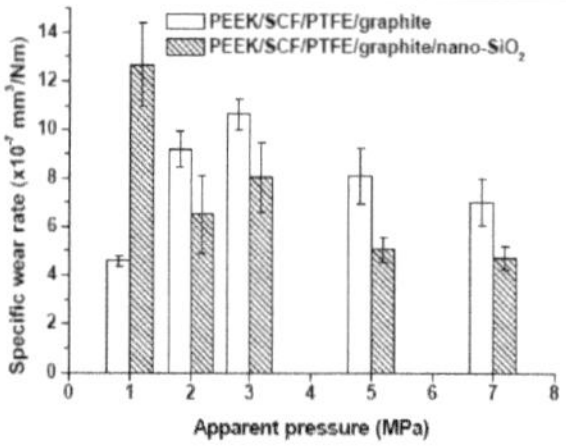	Positive effect of nano-SiO_2 on the wear resistance is more pronounced at high pressure.	[89]

(*Continued*)

Table 9.3 Continued

Materials	Properties	Remarks	Ref.
Effects of sliding velocity on the specific wear rates of PEEK/SCF/PTFE/graphite and PEEK/SCF/PTFE/graphite/nano-SiO_2.	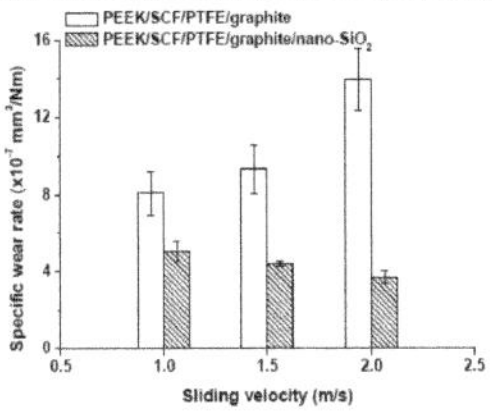	Positive effect of nano-SiO_2 on the wear resistance is more pronounced at high velocities.	[89]
Variation in the friction coefficient of steel surface supplemented with paraffin oil and blends of various graphene-based hybrid nanomaterials in paraffin oil with time. * nitrogen-doped reduced graphene oxide (N-MRG)		Significant reduction in the COFFICENT OF FRICTION for the blends of graphene-based nanomaterials	[88]
Hardness PEEK — Pure PEEK PS — PEEK (80vol %) /SCFs (20vol %) (100wt %) PS1S — PEEK (80vol %) /SCFs (20vol %) (99wt %)/nano-SiO_2 (1wt%) PS1.5S — PEEK (80vol %) /SCFs (20vol %) (98.5wt %)/nano-SiO_2 (1.5wt %) PS2S — PEEK (80vol %) /SCFs (20vol %) (9wt %)/nano-SiO_2 (2wt %)	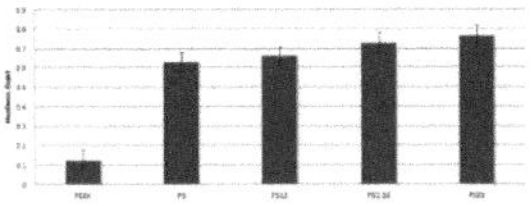	Hardness is evidently enhanced by the addition of the SCFs and SiO_2 nanoparticles	[87]

The quantity of the nanomaterials is a critical parameter for various characteristics of the polymer matrix. The surface treatment of the nanomaterials is usually a key factor for enhancing the interfacial adhesion between the nanomaterials and also polymer matrices [90]. The same quantity of surface-modified nanomaterials has great degree of dispersion in the polymer matrix in comparison with unmodified nanomaterials dispersed in the polymer matrix. The nanoparticle dimensions, particle–matrix interface adhesion, as well as adding quantity emphatically have an effect on the mechanical characteristics of nanoparticle-reinforced polymer composites [90, 91]. The nanoparticles promptly improve Young's modulus of polymer matrices that is on account of considerably higher stiffness of nanoparticles when compared with pure polymer. The stress transfer functional mechanism acts as a significant role in the nanoparticle-reinforced polymer nanocomposites [92, 93]. The highly effective interfacial adhesion efficiently enhances the stress transfer functional mechanism from particles to polymer matrix, leading to an improvement in the strength of the polymer nanocomposites. In consideration that the effective interfacial adhesion functional mechanism thoroughly enhances the mechanical characteristics of the nanocomposites, consequently, that allows to obtain much better interfacial adhesion, the surfaces of the nanoparticles are to be altered with appropriate coupling agents. The experts noticed less strong interface between the nanoparticles with polymer, which contributes to brittle-to-ductile transformation at room temperature [90, 97].

Polymer nanocomposites with metallic, metallic oxides, and carbon nanoparticles in the form of reinforcements are more widely used to produce the modern gadgets in the digital as well as optoelectronic products [92, 98]. The inclusion of metallic as well as metallic oxide nanoparticles in the polymer matrices possesses modernized electrical characteristics, and it has received possible a tremendous consideration.The essential electrical characteristics of the polymer nanocomposites are actually analyzed with results of size, shape, and quantity of the fillers [99, 100]. The enhancement of dielectric characteristics of nanocomposites primarily varies according to the (i) the huge surface area of nanoparticles, which produces significant interaction with the polymer matrix; (ii) altering the polymer morphology as a result of the surface of nanoparticles; (iii) dimensions' consequence; (iv) charge distribution between the nanoparticles and the matrix; and also (v) scattering effect [93, 101, 102]. The interface between the polymers along with the particles includes a significant role in changing the dielectric characteristics of the nanocomposites. It is exactly recognized that the metal oxide nanoparticles carry very high surface energy. Generally, the unmodified nanoparticles tend to agglomeration in the composites, which leads to decreased dielectric characteristics [103].

The blending of conducting metallic nanoparticles in the polymer matrix improved the dielectric constant of the nanocomposite. The metal-nanoparticle incorporated nanocomposites have turned out to be electrically conductive adhesives in electronic packaging applications [103]. The nanocomposite coatings have a considerable performance in the corrosion hindrance in comparison to traditional coatings [104, 105]. The many other metallic and also metallic oxides, for example, Ag, SiO_2, Fe_3O_4, clay, metal oxide nanoparticles and nanotubes, and also ZnO, are actually employed as a filler material for hybrid polymer coatings [106–108]. Polymer membranes are commonly used in separation and also filtration purposes because of their higher stability and performance, less energy needs, as well as simple operation [109–111]. The significant issue with membranes is fouling, which in turn causes a decrease in flux and a rise in the operating expenditure of the system. To be able to solve this challenge, the hydrophilic nanoparticles are utilized in the polymeric membranes to considerably improve the pure water flux and the fouling resistance. Silica, titania, alumina, and also ZnO nanoparticles are often used in the form of filler materials in the ultrafiltration polymer membranes as a result of higher thermal durability as well as substantial surface area [112–114]. Biodegradable polymer nanocomposites are frequently used in the biomedical purposes such as tissue architecture, bone repair and replacement, dental treatments, and regulated drug delivery. Biodegradable polymer nanocomposites consist of the materials that can deteriorate and are continually absorbed and/or disposed of by the body, whether degradation is induced mainly through hydrolysis or negotiated by metabolic means [115, 116]. The polymer nanocomposite when it comes to biomedical use involves specific policies such as, for example, biodegradability, biocompatibility, well mechanical characteristics, and, in certain instances, inventive needs. Because of this, types of polymer matrix and nanomaterials are really valuable when it comes to fabrication of polymer nanocomposites in biomedical opportunities [117]. Ag and Au nanoparticles are popularly employed filler materials as the product of the bio-nanocomposites because of their strong antibacterial characteristic thoroughly as low toxicity [118]. The nanofiber-strengthened polymer nanocomposites provide amazing characteristics on the basis of the dimensions, dispersion, and distribution of the fibrous material [119, 120]. The dispersion regarding the nanofiber and interfacial adhesion in the particle–matrix interface acts as an essential factor in deciding the properties regarding the polymer, which is certainly nanofiber-reinforced [121]. If the dispersion is poor, it can reduce the physical and mechanical properties of the polymer matrix. A number of metal, metal oxide, and carbon fibers can be used for the formulation of fiber-based polymer nanocomposites. In the last few years, natural fibers have obtained considerable interest in the area of polymer composites as a result of benefits such as decreased tool wear, low priced, lower density per unit volume, and appropriate specific strength. The renewable, sustainable, and degradable types of the natural fibers are ideal for making use of nanofillers in polymer composites. A number of natural fibers such as, for example, sisal, jute, wood fibers, and kenaf, have now been utilized as fillers, when we look at the polymer nanocomposites. Commonly, the natural fibers are generally hydrophilic and still have a relatively inferior fiber/matrix interactions, water resistance, and reasonably poor durability; to prevent this concern, the surface of the natural fibers should be dealt with using inorganic reagents or organic coupling agents [122]. Metal and metal oxide fibrous nanomaterials have now been commonly applied as filler materials for polymer nanocomposites and express effective properties, which are electrical properties due to the greater aspect ratio, ultimately contributing to lower percolation thresholds [123]. Polymer nanocomposites along with higher dielectric constant (κ) materials have obtained an escalating attention for many different uses such as energy storage, high-κ capacitors, electroactive devices, as well as gate dielectrics [124]. The metal oxide wires such as, for example, TiO_2, ZnO, and ZrO_2 have special optical and electrical properties, exceptional chemical stability, and higher dielectric constant characteristics. Table 9.4 show the various types of nanomaterials.

Table 9.4 An overview on nanomaterials

Classification of Nanomaterials	Example	Remarks
OD nanomaterials	Nanoparticles	Spherical, cubes, and polygonal shapes with nanodimensions are found.
	Nanoclusters	A cluster of nanoparticles and a narrow size distribution
	Fullerenes	Hollow carbon molecules form a closed cage

(*Continued*)

Table 9.4 Continued

Classification of Nanomaterials	Example	Remarks
	Quantum dots	Tiny particles or nanocrystals of a semiconducting material with diameters in the range of 2–10 nanometers.
	Nano Onions and nanopowder	Nested carbon spheres and powdered with particles having sizes under 100 nm, respectively.
1D Nanomaterials	Nanowires	Nanometric width dimensions and exhibiting aspect ratios (the ratio between the length and the width) of 1,000 or more.
	Nanorods	1D nanostructure with nanometric width dimensions and exhibiting aspect ratios 3–5.
	Nanotubes	Tubes with diameters in the nanoscale.
	Nanofibers	Diameter of less than 100 nm
2D Nanomaterials	Nanosheets	Thickness in a scale ranging from 1 to 100 nm
	Nanowalls	A nanoscale wall
	Nanofilms	A film of nanoscale thickness
	Nanolayers	Large group of extended two-dimensional nanoobjects of different physical and chemical nature
3D Nanomaterials	Nanofoams	Porous materials (foams) containing a significant population of pores with diameters less than 100 nm.
	Nanocomposites	Reinforced by nanoscale particles or nanostructures

9.4.1 Metallic Nanoparticles-based Polymer Nanocomposites

Fillers can alter the physical, rheological, chemical resistance, thermal, optical, and electrical properties of a polymeric component. Simultaneously, tribology properties of polymers can be heighten with the inclusion of metallic fillers, short fibers, and some others [125, 126].

The particles in these applications are for the most part of micrometer size. The utilization of nanofillers in system has been drawing attention from materials scientists, technologists, and industrialists [127, 128]. Nowadays, improvement of their wear properties by surfaces modification of the polymer composite gains significance [129, 130]. The primary reason of that is an ever increasing number of mechanical parts, for example, bushings, cams, gears, rollers, wheels, brakes, transports, and sliding shoes, are produced from short/continuous fiber impregnated composites, where friction and wear are key parameters taken into consideration. Nowadays, with the expectation of enhancing the tribological characteristics of the composites, different sorts of nanoparticle-filled polymer composites are produced and their properties are studied [131].

It has been noted that the inorganic fillers upgrade the mechanical properties as well as the tribological properties of polymers. Studies has revealed that the nanoparticles were proved to have some special advantages over micrometer particles for tribological applications [132, 133]. In spite of the fact that the reinforcement effect upon the stiffness, toughness, and wear performance of the composites can be dictated by the properties of the composite constituents, these properties depend emphatically on the microstructure spoken to by the filler size, shape, and the homogeneity of particle distribution [134, 135]. Metal-based materials are predominantly made out of metals (e.g., silver, gold, and copper nanoparticles) [136, 137]. Another highlight of the metal nanoparticles is their optical properties, which are not quite the same as those of their mass partner. This is because of an impact called limited surface plasmon reverberation (LSPR). A unique metal nanoparticle, which mean gold, silver, platinum, and palladium nanoparticles, demonstrates LSPR [138]. The surface of these nanoparticles can be functionalized with various substance and biochemical atoms empowering particular official of natural particles, making them helpful in sensors. As of now, plasmonic segments are being researched for future applications in malignancy treatment, sunlight-based cells, waveguides, optical interconnection, camera light-emitting diodes (LEDs), OLEDs, and then some [139, 140]. At the point when light hits a metal surface (of any size), a portion of the light wave engenders along the metal surface offering ascend to a surface plasmon—a gathering of surface conduction electrons that spread toward a path parallel to the metal/dielectric (or metal/vacuum) interface. At the point when a plasmon is created in a regular mass metal, electrons can move unreservedly in the material and no impact is enrolled. On account of nanoparticles, the surface plasmon is restricted in space, so it sways forward and backward synchronizedly in a little space, and the impact is called localized surface plasmon resonance. The vitality of the localized surface plasmon resonance relies upon the molecule shape, measure, piece, between molecule dividing, and dielectric condition [141]. Polymer composites are materials where a polymer is loaded with an inorganic manufactured as well as regular compound so as to expand a few properties, for example, warm resistance or mechanical quality, or to diminish different properties, for example, electrical conductivity for gasses like oxygen or water vapor [142, 143]. Materials with synergistic properties are utilized to plan composites with custom-fitted attributes. For example, high-modulus yet weak carbon filaments are added to low-modulus polymers to make a hardened lightweight polymer composite with some level of durability. Some Ni particles go into existing free volume spaces in vulcanized materials—in this way improving mechanical properties, including the dynamic storage modulus E' [144].

9.4.2 Nanometal Oxide-based Polymer Nanocomposites

Metal oxide-based polymer composites consist of titanium dioxide, zinc dioxide, and indium tin oxide. Titanium nanoparticles (30–50 nm) have optical and catalytic properties; they can absorb ultraviolet (UV) light [145]. Nano-TiO_2 is reasonable for transparent coatings and for new life style sunscreens, which are portrayed by a high protective factor, however, transparent appearance [146]. They can be employed for catalysis applications, for example, photocatalytic cleaning of water and air to disintegrate natural toxins. Thin films of TiO_2 are utilized on windows to receive the self-cleaning properties on the glass. Zinc oxide (ZnO) has some comparative properties to TiO_2 (i.e., it can be utilized for UV protection, in creams or coatings) [147, 148]. Contrasted with TiO_2, it has a weaker photocatalytic effect. By controlling development conditions of ZnO and TiO_2, a variety of nanostrctured can be produced such as nanoscale fibers, sphere, rings, and so on [149, 150]. For example, Zinc oxide fibers can be utilized to make adaptable, compact power sources that could be incorporated into material. Indium tin oxide (ITO) is a semiconducting material whose fundamental properties is the combination of electrical conductivity and optical properties [151, 152]. It is broadly utilized in the form of a thin-film; which can be used for preparation of touch screens, LEDs, semiconducting sensors, and so forth. Indium tin oxide is an infrared safeguard and is as of now utilized as a protecting cover on window glass. Among different metal oxide fillers, nanosized zinc oxide (ZnO), zirconium oxide (ZrO_2), and cerium oxide (CeO_2), indium tin oxide fillers-based polymer nanocomposites have gained important consideration in various sectors [153, 154]. A few investigations have been directed on the blend and structure property of metal oxide nanoparticles. Many examinations showed that the superb properties of nanocomposites rely upon the particle distribution. Many techniques, for example, in situ polymerization, intercalation strategy, sol–gel technique etc. have been utilized to prepared the nanocomposites [155, 156]. In every one of these strategies, in situ polymerization has been viewed as a common because it is less expensive, practicability, and processibble. The nanoparticles introduce high propensity to agglomerate affected by Vander Wall's attraction. Agglomeration of these particles in the polymer can result in diminished interfacial adhesion between matrix and nanoparticles. Non-uniform distribution of the particles causes the performance of nanocomposites. Numerous techniques can be utilized to introduce the nanoparticles inside polymers including ultrasonication bath, probe sonicator, high shear blending, ball milling, high -speed stirrer, and so on.

9.4.3 Nanoclay-based Polymer Nanocomposites

Nanoclay-based nanomaterials are normally found as platelets, stacked from a couple to one thousand sheets. A solitary sheet of montmorillonite (MMT) was accounted to have an in-plane Young's modulus from 178 to 265 GPa. Montmorillonite is the most broadly utilized earth nanofiller, sandwiched between two silicate layers of an octahedral sheet of alumina [157–159]. The nanometer-scale sheets of aluminosilicates have 1–5 nm thickness and 100–500 nm in distance across. Polymer–nanoclay composites can be grouped into three sorts—immiscible composites; intercalated nanocomposites; and exfoliated nanocomposites [160–162]. The wide-angle X-ray diffraction pattern for the composite is the same as that for the organo-clay powder. Clay are modified through cation exchange, silane group, and adsorption of polar polymers [163–165]. Cation exchange relies upon precious particle size, the pH, and the sort of replaceable particles. The measure of cations that can be exchange relies upon the measure of replaceable locales and the structure of the silicate [166–168]. Organosilanes contain natural moiety where a covalent bond can be made between the silicate filler and the polymer lattice. The particle exchange responses with cationic surfactants incorporate essential, auxiliary, tertiary, and quaternary alkylammonium or alkylphosphonium cations [169, 170]. These surfactants can give useful bunches on the clay that respond with the polymer lattice or sometimes start the polymerization of monomers to enhance the quality of the interface between the inorganic and the polymer lattice. The expanded modulus and effect properties have been found in the nylon and clay composites because of the development of ionic bonds between the amine bunches on the nylon and clay sheets. These interchangeable cations influence interconnection with natural particles, which control the morphology (intercalation and exfoliation).

9.4.4 Carbon Nanotube-based Polymer Nanocomposites

Carbon nanotubes-based polymer composites has been using in various sector such as automobile, aerospace, electronics because carbon nano tube hold excellent physical, chemical and thermal properties. High Young's moduli within the direction of the nanotube's axis, electrical conductivities are found, conductivity fluctuate from insulating to bimetallic. Carbon nanotubes have tensile strengths 20 times that of high-strength alloys. This means that utilize the reinforcing capability of the nanotube (CNT) and maximize the mechanical properties of the composites is important consideration, however, interfacial adhesion is a critical factor. If the interfacial region is stronger than the matrix, composites can offer excellent performance. However, if the surface region is weaker, debonding is occurred at the interface. The extent of interaction would depend upon the surface properties carbon nanotubes as well the dispersion and distribution of carbon nanotube within the matrix. To extend the interaction between the polymer and carbon nanotube, polymers can also be grafted to CNTs. There are two methods—the primary, usually named "grafting to" approach, involves preformed polymer chains reacting with the surface of nanotubes. The addition of CNTs might increase the glass transition, melting, crystalinity and thermal decomposition temperatures of the polymer matrix because of their constraint of polymer chains by the carbon nanotube. Enhancements on the dispersion and alignment of CNTs during a chemical compound matrix might decrease the percolation threshold worth. Polymer/CNT composites might even be accustomed defend human eyes, optical parts, optical sensors, and optical switching.

9.4.5 Graphene-based Polymer Nanocomposites

Carbon-based chemical compound composites comprises of carbon with graphene structures. Graphene nanosheet with Young's modulus of 1 TPa and supreme strength of 130 criteria is one among the well-known strongest materials. It has a highly specific space of 2600 m^2/g, terribly high electrical conduction (6000 S/cm), thermal conduction ($\sim$5000 W/mK), and so on. CNT bundles act as stress concentration points inside the polymer matrix and it adverse affect the mechanical properties of the polymer nanocomposites [171–173]. Graphite consists of the many graphene sheets and stack together by van der Waals forces [174–176]. The graphene oxide (GO) is a form of graphene, it possesses polar functionalities such as carbonyl, hydroxyl, epoxides, and carboxyl groups. The epoxide and hydroxyl radical groups are set within the basal plane of the graphene sheets whereas the carbonyl and carboxyl groups' are found at the sides. The groups within the graphene chemical compound and some treatment reduces the van der Waals forces between the graphene oxide layer and create exfoliated form, which help to penetrate water molecules as well as polymer chain water molecules into the gallery space between layer of the platelets. Underneath the influence of the sonication and reduction in van der Waals force, exfoliation of GO is performed. The element functionalities produce electricity repulsion that stops the reaggregation of the exfoliated graphene oxide in water [177–179]. Graphene oxide is electrically insulating because of the disruption of the C structure throughout the oxidization method. This can be reversed by reducing thegraphene oxide. However, the reduction of GO dispersion without stabilizer causes precipitation ofCarbon particles because of the fast and irreversible aggregation of graphene sheets. Hence, before reduction, the surface of GO sheets must be changed by covalent or non-covalent functionalization. Reducing agents that are typically used for the reduction of organic ketones, chemical group acids, and epoxy useful groups are often used for the reduction of pure GO or functionalized GO. Non-covalent functionalization of graphene has been achieved via π–π stacking or van der Waals interactions. Such ways stop any structural harm that affects the electronic properties [177, 180–182]. Graft and block copolymers are used as compatibilizers for polymer/CNT composites. Here one chain of the copolymers interacts with the carbon nanofillers by π–π interactions while the other chain is miscible with the matrix polymer. This ends up in well-distributed CNTs within the polymer matrix. If dispersion and distribution of graphene in polymer matrix are achieved, it can improve the mechanical and tribology performance of composites. The interfacial problem is main concern for producing effect composites and their application in wear as well as friction performance. However, It has been found that graphene composites is sutiable candidate for developing the composites with excellent tribology performance, it is due to the excellent thermal, electrical, mechanical properties of graphene.

9.4.6 Fullerenes-based Polymer Nanocomposites

Fullerenes-based polymer composites is referred to the polymer consisting carbon with fullerenes structures. As a result of the outstanding physical properties and chemical properties of C60,fullerene-based composites are expected considerably improved properties such as mechanical , chemical, thermal performance of polymer in comparison with neat polymer; the content of the C60s within the matrix is crucial factor for increasing the mechanical properties the nanocomposite [183–185]. The broadband dielectric spectroscopic analysis is one in all the foremost economical tools for finding out molecular relaxations of polymers. It covers a broad frequency vary, permitting activity of various relaxation processes at the same time, together with secondary and even entire chain relaxation processes underneath favorable circumstances. The doping of atomic number 6 (C60) with a compound solution could be a comparatively a replacement approach to reinforce the optical and electrical properties of compound films. Macromolecule – containing Fullerene has been used to develop simple processible semiconductor diode. Fullerene tends to aggregates and troublesome to handle, therefore, organic modifications are needed to resolve these types of issues. Because, composites containing fullerenes have great potential for versatile applications such as photoconductors, and electrodes in Li batteries, electro-optical structures in nanoelectronics, non-linear optics, and electrical devices [186–189]. Reinforcing ability of carbonous fillers is especially concerning mechanical performance and electrical conduction improvement for compound composite applications. Atomic number 6 has been used as an additive in many thermoplastic polymers together with synthetic resin, polypropene, polyamide, and poly(ethylene-co-acrylic acid). Once C60 is hooked up to a compound chain, properties of the compound are modified and a few new properties can be seen. At a similar time, the atomic number 6 compound is also soluble in common solvents. Through controlled modification of C60, polymers with totally different structures are often obtained, resembling side-chain polymers, main-chain polymers, asteroid polymers, etc. With the event of fullerene organic chemistry, several low-molecular organic reactions are often utilized in the compound modification of fullerene [188, 190, 191]. By dominating the useful groups in polymers and reaction conditions, several well-defined fullerene polymers, resembling side-chain fullerene polymers, main-chain atomic number 6 polymers, nerve fiber fullerene polymers, asteroid fullerene polymers, atomic number 6 end-capped polymers, and so on, are ready. At a similar time, several living polymerization ways are introduced into the preparation of the compound fullerenes. By anionic chemical action, well-defined six-arm asteroid fullerene compounds are often obtained (the iniferter technique permits well-defined arms for the star fullerene polymer to be synthesized). With ATRP, mono- or bi-substituted fullerene compounds with well-controlled polymer chain length are often ready. So it is currently doable to manage the design of fullerene polymers exactly by ways that represent an honest platform for the sensible applications of atomic number 6 derivatives. Doping fullerene into compound systems has found additional and additional uses in the preparation of electronic and optical materials. Several conducting polymers doped with C60 and its derivatives are studied for various application [188, 192–194].

9.4.7 Nanodiamonds-based Polymer Nanocomposites

Nanodiamonds-based polymer composites contain carbon with nanodiamond (ND) structures. Nanodiamond created by detonation synthesis in massive industrial quantities is ideal nanofiller for several composite applications. It is manufactured in the form of a tiny and uniform particles of $\sim$5 nm in diameter that have high surface area, tailorable surface chemistry [195–197]. It has distinctive properties, as well as superior thermal conduction, hardness, and Young's modulus, high electrical resistance, and a high refraction index than bulk diamond. To purify ND detonation soot, two subsequent purification steps area unit is applied. First, air reaction approach is applied, leaving the precise control of the sp2/sp3 carbon relation. This study suggests tempering of ND containing detonation soot at a temperature of 430 °C for 5 hours in air to get the most sp^2 carbon. Associate acid treatment is applied to get rid of non-carbon impurities similar to metals, metal oxides, and different impurities originating from the detonation chamber or the explosives used for the synthesis itself. Common acid treatments contain mixtures of various acids and water [198–201]. Besides removing impurities, acid treatments additionally introduce chemical element containing useful groups which has been be used for any surface functionalization of NDs. The kinds and contents of groups are important aspects for NDs. To an oversized growth, it depends on the history of the sample, i.e., on how it had been created, purified, modified, etc. Therefore, careful qualitative and quantitative characterization of NDs is of outermost importance for the successful development of its applications. For best performance of NDs in its applications, it is of outermost importance to have acceptable surface chemistry. The range of accessible chemical modification techniques are additionally divided into gas and wet chemistry treatment. They are non-toxic that makes them compatible with medical applications such as drug delivery, bio imaging, and tissue engineering, and additionally as super molecule mimics and a filler material for nanocomposites. The first particle sizes, the prevailing industrial diamond nanoparticles are classified into three groups of products, i.e., nanocrystalline particles, ultrananocrystalline particles, and diamondoids. The sizes of the nanocrystalline particles are within the vary of tens of micrometers whereas the sizes of the ultrananocrystalline particles are with in nanometers [197, 202–204]. The diamondoids (well-outlined H-terminated molecular forms) contain many tens of carbon atoms. Nanocrystalline diamond particles occur within the kinds of isolated monocrystalline or crystalline particles. Their sizes are within many tens of nanometers. The process includes purification, grinding, and grading of the powder. Compared with the varieties of ND, monocrystalline particles obtained by the grinding technique have rather sharp edges. Inside the category of ND materials, the purest type is monocrystalline natural ND powder. The processing of diamond particles obtained by shock-wave method (De Pont method) yields polycrystalline ND particles have a size of~20 to 25 nm [204–206]. This sort of ND has not been used for bio applications since it has a high content of impurities. There are many kinds of ultrananocrystalline diamond particles based on the detonation [197, 204–206]. For the synthesis of ND, three ways had been commercialized. The most in-depth and huge scale production technique is de Pont technique, wherein carbon precursors (e.g., coal, graphite, and carbon black) are transformed into diamond in a capsule by the application of circular shock wave (~140 GPa) generated outside the capsule. The carbon precursors are placed in a tube that is successively placed in the outer tube lined with a driving tube. The free space within the tube is full of the explosive [201, 207, 208]. The circular shock waves generated by the ignition of explosive transforms sp2 carbon material into ND particles. A mix of metallic powder (Al, Cu, and Ni) and graphite (6–10%) is employed to forestall graphitization of diamond. The second technique for bulk scale production of ND relies on the detonation of carbon containing material with explosives. During this technique, diamond synthesis takes place inside the carbon containing material. A diamond isometric section no more than 20 nm is created in inert atmosphere whereas in air atmosphere, the landslide particle size is smaller (8 nm) [201, 207, 209]. The yield of the diamond is regarding 3.4% of mass of explosives or 17% of the mass of the initial carbon. The third technique is explosion technique for diamond synthesis and ends up in diamond clusters from explosive utilized as a precursor material. The range of impurities are additionally found within the diamond soot similar to metal, non-diamond carbon, and concrete rubbish from the reaction chamber, the purification steps contains removal of magnetic impurities by mechanical and magnetic separation from larger rubbish through sieving. Afterward, the raw soot is refined by acid treatment. The treatment not solely reduces the metal content but additionally oxidizes the non-diamond carbon in the soot. It is well established that the reaction of sp2 carbon is far easier than that of diamond carbon [201, 210, 211]. The sp2 carbons are removed and therefore the material is received with enrich in diamond content. The artificial diamond created by this technique contains 100–300 ppm of nitrogen vacancy (N-V). The nitrogen vacancy (N-V) color defect centers area unit is accountable for the emission of visible radiation, and thus artificial diamonds are widely employed in cellular models as imaging agents. Chemical vapor deposition technique is the important technique for crystalline. Furthermore, as single-crystalline diamond films, the diamond substrate is needed in every case. Chemical vapor deposition (CVD) ways are oft used for coated skinny films. There are several varieties of the CVD technique that rely upon the activation of precursor gases similar to microwave plasma-increased CVD technique that uses microwave energy to activate the gases. It is foremost used ways; however, it is expensive to line up and run. RF plasma CVD technique is employed to activate the gases. In DC plasma CVD technique, direct power is employed. In high-frequency CVD, the activation of the precursor gases is achieved by heating transition metal component, e.g., tungsten/tantalum or molybdenum. It is easy way, more cost-effective ways, using multiple filaments permits massive deposition of the ND films. Diamond films cab be obtained by carbon vapors deposition on diamond/non-diamond substrate [212–214]. The grain size of the nanocrystalline film ranges from 5 nm to 100 nm. The dimensions have been controlled by growth of precursors (hydrogen wealthy and carbon poor precursors), surface temperature, biasing voltage, and film pressure [204, 215–218]. Ultrananocrystalline diamond films with a grain size of 3–5 nm are typically found consisting argon and H poor condition. In spite of distinction in grain size, two diamond films additionally show distinction in their sp2 carbon content. Nanocrystalline diamond film possess sp2 carbon content up to 50%, whereas in ultrananocrystalline diamond film, the sp2 carbon content is 2–5% [204, 219–222]. Thermoplastics can be reinforced with ND even without surface functionalization and at low concentration. To boost its tribological properties, various studies have been performed. In this manner, Improvement in wear resistance and reduction in friction constant were calculated. The link between wear performance and ND agglomerate size was additionally investigated by another study. The smaller agglomerate was found to decrease the damage whereas the larger agglomerate resulted in magnified wear properties in the blend composites. A significant challenge that limits the applications of polymers is the low thermal stability and melting temperature [223–227]. But the reinforcement of ND in polymer enhances the thermal properties. Diamond has excellent thermal conduction; in addition, in introduction of nano diamond in polymer matrix can improve the thermal conductivity. This increase in thermal conduction is dependent on the interface of the polymer matrix and nanodiamonds (nanofillers because interface plays an important role in reducing surface phonon scattering. It is noteworthy that the thermal stability of the polymers can be effectively improved by fictionalization of NDs [228–231]. Nanocomposites have also distinctive optical properties that have been considered as prime interest for varied applications. Once nanodiamons is introduced in polymers, materials with excellent index of refraction could also be achieved. Nanofillers such as nanodimaonds additionally have an effect on the visual look of polymer films, e.g., surface finishing and packaging applications. Currently, in cosmetics, sun light protecting nanofillers are used. Nanodiamond has excellent optical properties, e.g., high index of refraction and capability of blocking light in UV radiation region. The absorption efficiency was shown to be a function of concentration, surface functionalities, and agglomerate size. Photonic crystals are patterned by ND agglomerates, wherein interparticles' distance variations lead to variation of colors in ND/oil suspensions. These nanocomposites have nice potential for ultraviolet radiation scratch resistance and protecting coatings. Nanodiamonds possess nice potential for engineering new nanocomposites for medicine applications thanks to its wealthy surface chemistry, biocompatibility, and non-toxicity. Principally perishable polymers are developed with poor mechanical properties that greatly limit their applications. Electromagnetic compatibility is another property necessary for electronic devices to figure out the functionality. It is essential to protect electronic devices like integrated circuits, processors, information processing equipment, medicine devices, and telecommunication units from electromagnetic interference (EMI). Nanocomposites reinforced with acceptable nanofillers have smart shielding impact against EMI. Onion-like carbon has metallic conducting properties; onion-like carbon is generated by ND tempering (sp3 carbon is regenerated into sp2 carbon). By the tempering time and temperature, the ratio of sp2/sp3 carbon was controlled. The great electromagnetic absorption properties were shown by Onion-like carbon attributable to its high electron conduction and bigger defective inner shells.

9.5 Conclusion

Section 9.1 explains the basic mechanism and the factors, which affects the tribological behavior of the mating surfaces. Several factors (e.g., Preparation method, materials, and morphology of materials) influence the tribological behavior. Various preparation methods and characterization techniques have great importance in the final performance. Tribological properties along with the mechanical and morphological results have been discussed in Section 9.2. The results indicate that the graphene- and silica-based nanomaterials' incorporation into the polymer material has improved the tribological properties. And finally, Section 9.3 explained the polymer nanomaterials, which are used for tribological application in future.

References

[1] Bhushan, B. (2013). *Introduction to Tribology*, 2nd Edn. Hoboken, NJ: John Wiley & Sons
[2] Bhushan, B. (2001). *Modern Tribology Handbook*. Boca Raton, FL: CRC Press.
[3] Karingamanna Jayanarayanan, N. R., and Mishra, R. K. (2017). *Thermal and Rheological Measurement Techniques for Nanomaterials Characterization*. Amsterdam: Elsevier, 123–157.
[4] Remya, V. R., Deepak Patil, Abitha, V. K., Rane, A. V., and Mishra, R. K. (2016). Biobased materials for polyurethane dispersions. *Chem. Int.* 2, 158–167.
[5] Heinrich, G., and Klüppel, M. (2008). Rubber friction, tread deformation and tire traction. *Wear* 265, 1052–1060.
[6] Burton, Z., and Bhushan, B. (2005). Hydrophobicity, adhesion, and friction properties of nanopatterned polymers and scale dependence for micro- and nanoelectromechanical systems. *Nano Lett.* 5, 1607–1613.

[7] Zhang Newby, B., and Chaudhury, M. K. (1998). Friction in adhesion. *Langmuir* 14, 4865–4872.

[8] Lopez Arteaga, I., Busturia, J. M., and Nijmeijer, H. (2004). Energy dissipation of a friction damper. *J. Sound Vib.* 278, 539–561.

[9] Carbone, G., and Mangialardi, L. (2004). Adhesion and friction of an elastic half-space in contact with a slightly wavy rigid surface. *J. Mech. Phys. Solids* 52, 1267–1287.

[10] Tomlinson, S. E., Lewis, R., and Carré, M. J. (2009). The effect of normal force and roughness on friction in human finger contact. *Wear* 267, 1311–1318.

[11] Greenwood, J. A., and Williamson, J. B. P. (1966). Contact of Nominally Flat Surfaces. *Proc. R. Soc. A Math. Phys. Eng. Sci.* 295, 300–319.

[12] Ozkoc, G., Bayram, G., and Bayramli, E. (2004). Effects of polyamide 6 incorporation to the short glass fiber reinforced ABS composites: An interfacial approach. *Polymer* 45, 8957–8966.

[13] Teixeira, D., Giovanela, M., Gonella, L. B., and Crespo, J. S. (2013). Influence of flow restriction on the microstructure and mechanical properties of long glass fiber-reinforced polyamide 6.6 composites for automotive applications. *Mater. Des.* 47, 287–294.

[14] Fu, S. Y., Lauke, B., Li, R. K. Y., and Mai, Y. W. (2005). Effects of PA6,6/PP ratio on the mechanical properties of short glass fiber reinforced and rubber-toughened polyamide 6,6/polypropylene blends. *Compos. Part B Eng.* 37, 182–190.

[15] Unal, H., and Mimaroglu, A. (2004). Influence of Filler Addition on the Mechanical Properties of Nylon-6 Polymer. *J. Reinf. Plast. Compos.* 23, 461–469.

[16] Unal, H., Mimaroglu, A., Kadioglu, U., and Ekiz, H. (2004). Sliding friction and wear behaviour of polytetrafluoroethylene and its composites under dry conditions. *Mater. Des.* 25, 239–245.

[17] Unal, H., Mimaroglu, A., and Arda, T. (2006). Friction and wear performance of some thermoplastic polymers and polymer composites against unsaturated polyester. *Appl. Surf. Sci.* 252, 8139–8146.

[18] Liu, Y., and Schaefer, J. A. (2006). The sliding friction of thermoplastic polymer composites tested at low speeds. *Wear* 261, 568–577.

[19] Dorri Moghadam, A., Omrani, E., Menezes, P. L., and Rohatgi, P. K. (2015). Mechanical and tribological properties of self-lubricating metal matrix nanocomposites reinforced by carbon nanotubes (CNTs) and graphene - a review. *Compos. Part B Eng.* 77, 402–420.

[20] Ronkainen, H., Varjus, S., and Holmberg, K. (1998). Friction and wear properties in dry, water- and oil-lubricated DLC against alumina and DLC against steel contacts. *Wear* 222, 120–128.

[21] Yazawa, S., Minami, I., and Prakash, B. (2014). Reducing friction and wear of tribological systems through hybrid tribofilm consisting of coating and lubricants. *Lubricants* 2, 90–112.

[22] Liu, H., and Xue, Q. (1996). The tribological properties of TZP-graphite self-lubricating ceramics. *Wear* 198, 143–149.

[23] Thomas, S., Thomas, R., Zachariah, A. K., and Mishra, R. K. (2017). *Thermal and Rheological Measurement Techniques for Nanomaterials Characterization*. Amsterdam: Elsevier.

[24] Raghvendra KM, Sravanthi, L. (2017). Fabrication techniques of micro/nano fibres based nonwoven composites: a review. *Mod. Chem. Appl.* 5:206.

[25] Mishra, R. K., Cherusseri, J. Bishnoi, A., and Thomas, S. (2017). *Spectroscopic Methods for Nanomaterials Characterization*. Amsterdam: Elsevier, 369–415.

[26] Chirayil, C. J., Abraham, J., Mishra, R. K., George, S. C., and Thomas, S. (2017). *Thermal and Rheological Measurement Techniques for Nanomaterials Characterization.* Amsterdam: Elsevier, 1–36.

[27] Sravanthi, L., Valapa, R. B., Mishra, R. K., Pugazhenthi, G., and Thomas, S. (2017). *Thermal and Rheological Measurement Techniques for Nanomaterials Characterization.* Amsterdam: Elsevier, 67–108.

[28] Mishra, R. K., Zachariah, A K., and Thomas, S. (2017). "Energy-dispersive X-ray spectroscopy techniques for nanomaterial," in *Microscopy Methods in Nanomaterials Characterization,* eds S. Thomas, R. Thomas, A. Zachariah, and R. K. Mishra (Amsterdam: Elsevier), 383–405.

[29] Sabu, T., Thomas, R., Zachariah, A., and Mishra, R. K. (2017). *Microscopy Methods in Nanomaterials Characterization.* Amsterdam: Elsevier.

[30] Rohatgi, P. K., Tabandeh-Khorshid, M., Omrani, E., Lovell, M. R., and Menezes, P. L. (2013). "Tribology for scientists and engineers," in *Tribology for Scientists and Engineers: From Basics to Advanced Concepts*, eds P. Menezes, M. Nosonovsky, P. I. Sudeep, V. K. Satish, and R. L. Michael (Berlin: Springer).

[31] Samyn, P., Schoukens, G., Quintelier, J., and De Baets, P. (2006). Friction, wear and material transfer of sintered polyimides sliding against various steel and diamond-like carbon coated surfaces. *Tribol. Int.* 39, 575–589.

[32] Bhimaraj, P. et al. (2005). Effect of matrix morphology on the wear and friction behavior of alumina nanoparticle/poly(ethylene) terephthalate composites. *Wear* 258, 1437–1443.

[33] Kato, K., and Adachi, K. (2001). "Wear mechanisms," in *Modern Tribology Handbook*, Vol. 1, ed. B. Bhushan (New York, NY: CRC Press).

[34] Bayer, R. G. (2004). Glossary of wear mechanisms, related terms, and phenomena. *Eng. Des. Wear Second Ed Revis. Expand.* 9, 367–374.

[35] Cenna, A. A., Williams, K. C., and Jones, M. G. (2011). Analysis of impact energy factors in ductile materials using single particle impact tests on gas gun. *Tribol. Int.* 44, 1920–1925.

[36] Williams, J. A., and Dwyer-Joyce, R. S. (2001). Contact Between Solid Surfaces. *Mod. Tribol. Handb.* 1, 121–162.

[37] Pettigrew, T. F., Tropp, L. R., Wagner, U., and Christ, O. (2011). Recent advances in intergroup contact theory. *Int. J. Intercul. Relat.* 35, 271–280.

[38] Schallamach, A. (1958). Friction and abrasion of rubber. *Wear* 1, 384–417.

[39] Harsha, A. P., Tewari, U. S., and Venkatraman, B. (2003). Three-body abrasive wear behaviour of polyaryletherketone composites. *Wear* 254, 680–692.

[40] Moore, M. A. (1974). A review of two-body abrasive wear. *Wear* 27, 1–17.

[41] Natarajan, N., Vijayarangan, S., and Rajendran, I. (2006). Wear behaviour of A356/25SiCp aluminium matrix composites sliding against automobile friction material. *Wear* 261, 812–822.

[42] Qin, Q. D., Zhao, Y. G., and Zhou, W. (2008). Dry sliding wear behavior of Mg2Si/Al composites against automobile friction material. *Wear* 264, 654–661.

[43] Straffelini, G. et al. (2016). Wear behavior of a low metallic friction material dry sliding against a cast iron disc: Role of the heat-treatment of the disc. *Wear* 348–349, 10–16.

[44] Uyyuru, R. K., Surappa, M. K., and Brusethaug, S. (2007). Tribological behavior of Al-Si-SiCp composites/automobile brake pad system under dry sliding conditions. *Tribol. Int.* 40, 365–373.

[45] Chen, Z., Li, T., Yang, Y., Liu, X., and Lv, R. (2004). Mechanical and tribological properties of PA/PPS blends. *Wear* 257, 696–707.

[46] Zoo, Y. S., An, J. W., Lim, D. P., and Lim, D. S. (2004). Effect of carbon nanotube addition on tribological behavior of UHMWPE. *Tribol. Lett.* 16, 305–310.

[47] Kobayashi, M., Koide, T., and Hyon, S. H. (2014). Tribological characteristics of polyethylene glycol (PEG) as a lubricant for wear resistance of ultra-high-molecular-weight polyethylene (UHMWPE) in artificial knee join. *J. Mech. Behav. Biomed. Mater.* 38, 33–38.

[48] Chand, N., Dwivedi, U. K., and Sharma, M. K. (2007). Development and tribological behaviour of UHMWPE filled epoxy gradient composites. *Wear* 262, 184–190.

[49] Wan, Y. Z. et al. (2006). Friction and wear behavior of three-dimensional braided carbon fiber/epoxy composites under dry sliding conditions. *Wear* 260, 933–941.

[50] Sharma, M., Rao, I. M., and Bijwe, J. (2009). Influence of orientation of long fibers in carbon fiber-polyetherimide composites on mechanical and tribological properties. *Wear* 267, 839–845.

[51] Harsha, A. P., and Tewari, U. S. (2004). Tribological Studies on Glass Fiber Reinforced Polyetherketone Composites. *J. Reinf. Plast. Compos.* 23, 65–82.

[52] Farias, M. C. M., Souza, R. M., Sinatora, A., and Tanaka, D. K. (2007). The influence of applied load, sliding velocity and martensitic transformation on the unlubricated sliding wear of austenitic stainless steels. *Wear* 263, 773–781.

[53] Zhu, W. et al. (2012). Nanocrystalline tungsten carbide (WC) synthesis/characterization and its possible application as a PEM fuel cell catalyst support. *Electrochim. Acta* 61, 198–206.

[54] Mohan, N., Mahesha, C. R., Shivarudraiah, Mathivanan, N. R., and Shivamurthy, B. (2013). Dry sliding wear behaviour of Ta/NbC filled glass-epoxy composites at elevated temperatures. *Procedia Eng.* 64, 1166–1172.

[55] Katiyar, P. K., Singh, P. K., Singh, R., and Kumar, A. L. (2016). Modes of failure of cemented tungsten carbide tool bits (WC/Co): a study of wear parts. *Int. J. Refract. Met. Hard Mater.* 54, 27–38.

[56] Jiang, W. et al. (2015). Poly(ether ether ketone)/wrapped graphite nanosheets with poly(ether sulfone) composites: preparation, mechanical properties, and tribological behavior. *J. Appl. Polym. Sci.* 132:41728.

[57] Imrek, H. (2009). Performance improvement method for Nylon 6 spur gears. *Tribol. Int.* 42, 503–510.

[58] Nagrial, M., Rizk, J., and Hellany, A. (2012). "Design and performance of magnetic gears using rare-earth permanent magnets," in *Proceedings of the 2012 International Conference on Robotics and Artificial Intelligence,* Rawalpindi, 86–90.

[59] ASTM (2010). ASTM G99: Standard Test Method for Wear Testing with a Pin-on-Disk Apparatus. *ASTM Stand.* G99, 1–5.

[60] Unal, H., and Mimaroglu, A. (2012). Friction and wear performance of polyamide 6 and graphite and wax polyamide 6 composites under dry sliding conditions. *Wear* 289, 132–137.

[61] Anjum, N., Ajit Prasad, S. L., and Suresha, B. (2013). Role of silicon dioxide filler on mechanical and dry sliding wear behaviour of glass-epoxy composites. *Adv. Tribol.* doi:10.1155/2013/324952

[62] Abitha, VK, VR Remya,. Vasudeo Rane, A., Jadhav, S., and Mishra, R. K. (2016). Influence of hybrid fillers on morphological and mechanical properties of carboxylated nitrile butadiene rubber composites. *J. Mater. Sci. Eng. with Adv. Technol.* 13, 13–27.

[63] Karande, R. D., Abitha, V., Rane, A. V., and Mishra, R. K. (2015). Preparation of polylactide from synthesized lactic acid and effect of reaction parameters on conversion. *J. Mater. Sci. Eng. with Adv. Technol.* 12, 1–37.

[64] Boussia, A. C., Vouyiouka, S. N., and Papaspyrides, C. D. (2012). Applying the traditional solution melt polymerization for the in situ intercalation of polyamide 6.6-clay nanocomposites. *Macromol. Mater. Eng.* 297, 68–74.

[65] Ahmad, M. B., Gharayebi, Y., Salit, M. S., Hussein, M. Z., and Shameli, K. (2011). Comparison of in situ polymerization and solution-dispersion techniques in the preparation of Polyimide/ Montmorillonite (MMT) Nanocomposites. *Int. J. Mol. Sci.* 12, 6040–6050.

[66] Briesenick, D., and Bremser, W. (2015). Synthesis of polyamide-imide-montmorillonite-nanocomposites via new approach of in situ polymerization and solvent casting. *Prog. Org. Coat.* 82, 26–32.

[67] Sengupta, R. et al. (2007). A short review on rubber/clay nanocomposites with emphasis on mechanical properties. *Engineering* 47, 21–25.

[68] Vlachopoulos, J., and Strutt, D. (2003). Polymer processing. *Mater. Sci. Technol.* 19, 1161–1169.

[69] Rousseaux, D. D. J. et al. (2011). Water-assisted extrusion of polypropylene/clay nanocomposites: a comprehensive study. *Polymer* 52, 443–451.
[70] Vlachopoulos, J., and Strutt, D. (2003). "The role of rheology in polymer extrusion," in *Proceedings of the New Technology for Extrusion Conference, Milan,* 1–26.
[71] Wang, J. (2012). "PVT properties of polymers for injection molding," in *Some Critical Issues for Injection Molding*, ed. J. Wang (Rijeka: InTech), 3–30.
[72] Zhang, G., Rasheva, Z., and Schlarb, A. (2010). K. Friction and wear variations of short carbon fiber (SCF)/PTFE/graphite (10 vol.%) filled PEEK: effects of fiber orientation and nominal contact pressure. *Wear* 268, 893–899.
[73] Jiang, Z., Gyurova, L. A., Schlarb, A. K., Friedrich, K., and Zhang, Z. (2008). Study on friction and wear behavior of polyphenylene sulfide composites reinforced by short carbon fibers and sub-micro TiO_2 particles. *Compos. Sci. Technol.* 68, 734–742.
[74] Lv, M., Wang, Q., Wang, T., and Liang, Y. (2015). Effects of atomic oxygen exposure on the tribological performance of ZrO_2-reinforced polyimide nanocomposites for low earth orbit space applications. *Compos. Part B Eng.* 77, 215–222.
[75] Ajayi, O. O., Lorenzo-Martin, C., Erck, R. A., and Fenske, G. R. (2011). Scuffing mechanism of near-surface material during lubricated severe sliding contact. *Wear* 271, 1750–1753.
[76] Unal, H., and Mimaroglu, A. (2006). Friction and wear characteristics of PEEK and its composite under water lubrication. *J. Reinf. Plast. Compos.* 25, 1659–1667.
[77] Kim, H. J., and Kim, D. E. (2009). Nano-scale friction: a review. *Int. J. Precis. Eng. Manuf.* 10, 141–151.
[78] Sharma, S., Bijwe, J., and Kumar, M. (2013). Comparison between nano- and micro-sized copper particles as fillers in NAO Friction Materials Regular Paper. *Nanomater. Nanotechnol.* 3, 1–9.
[79] Kalin, M., Zalaznik, M., and Novak, S. (2015). Wear and friction behaviour of poly-ether-ether-ketone (PEEK) filled with graphene, WS2 and CNT nanoparticles. *Wear* 332–333, 855–862.
[80] Rapoport, L. et al. (2005). Behavior of fullerene-like WS2 nanoparticles under severe contact conditions. *Wear* 259, 703–707.

[81] Wang, A. H. et al. (2008). Ni-based alloy/submicron WS2 self-lubricating composite coating synthesized by Nd:YAG laser cladding. *Mater. Sci. Eng.* A475, 312–318.
[82] Lu, D., Jiang, Y., and Zhou, R. (2013). Wear performance of nano-Al_2O_3 particles and CNTs reinforced magnesium matrix composites by friction stir processing. *Wear* 305, 286–290.
[83] Lim, D. S., You, D. H., Choi, H. J., Lim, S. H., and Jang, H. (2005). Effect of CNT distribution on tribological behavior of alumina-CNT composites. *Wear* 259, 539–544.
[84] Ray, D., and Gnanamoorthy, R. (2007). Friction and wear behavior of vinylester resin matrix composites filled with fly ash particles. *J. Reinf. Plast. Compos.* 26, 5–13.
[85] Kuo, M. C., Huang, J. C., and Chen, M. (2006). Non-isothermal crystallization kinetic behavior of alumina nanoparticle filled poly(ether ether ketone). *Mater. Chem. Phys.* 99, 258–268.
[86] Zheng, Y. P., Zheng, J. X., Li, Q., Chen, W., and Zheng, X. (2009). The influence of high content nano-Al_2O_3 on the properties of epoxy resin composites. *Polym. Plast. Technol. Eng.* 48, 384–388.
[87] Barkoula, N. M., Alcock, B., Cabrera, N. O., and Peijs, T. (2008). Fatigue properties of highly oriented polypropylene tapes and all-polypropylene composites. *Polym. Polym. Compos.* 16, 101–113.
[88] Jaiswal, V. et al. (2016). Synthesis, characterization, and tribological evaluation of TiO_2-reinforced boron and nitrogen co-doped reduced graphene oxide based hybrid nanomaterials as efficient antiwear lubricant additives. *ACS Appl. Mater. Interface* 8, 11698–11710.
[89] Zhang, G., Schehl, M., and Burkhart, T. (2016). *Effect of Low-Loading Nanoparticles on the Tribological Property of Short Carbon Fiber (SCF)/PTFE/Graphite Reinforced PEEK*. Kaiserslautern: University of Kaiserslautern, 11–20.
[90] Fu, S. Y., Feng, X. Q., Lauke, B., and Mai, Y. W. (2008). Effects of particle size, particle/matrix interface adhesion and particle loading on mechanical properties of particulate-polymer composites. *Compos. Part B Eng.* 39, 933–961.
[91] Ferdous, S. F., Sarker, M. F., and Adnan, A. (2013). Role of nanoparticle dispersion and filler-matrix interface on the matrix dominated failure of rigid C60-PE nanocomposites: A molecular dynamics simulation study. *Polymer* 54, 2565–2576.
[92] Tjong, S. C. (2006). Structural and mechanical properties of polymer nanocomposites. *Mater. Sci. Eng. R Rep.* 53, 73–197.

[93] Rozenberg, B. A., and Tenne, R. (2008). Polymer-assisted fabrication of nanoparticles and nanocomposites. *Prog. Poly. Sci.* 33, 40–112.

[94] Xie, Y., Hill, C. A. S., Xiao, Z., Militz, H., and Mai, C. (2010). Silane coupling agents used for natural fiber/polymer composites: a review. *Comp. Part A Appl. Sci. Manufact.* 41, 806–819.

[95] Mortezaei, M., Famili, M. H. N., and Kokabi, M. (2011). The role of interfacial interactions on the glass-transition and viscoelastic properties of silica/polystyrene nanocomposite. *Compos. Sci. Technol.* 71, 1039–1045.

[96] Gumbsch, P., Taeri-Baghbadrani, S., Brunner, D., Sigle, W., and Rühle, M. (2001). Plasticity and an inverse brittle-to-ductile transition in strontium titanate. *Phys. Rev. Lett.* 87, 85505.

[97] Tönnies, D., Maaβ, R., and Volkert, C. A. (2014). Room temperature homogeneous ductility of micrometer-sized metallic glass. *Adv. Mater.* 26, 5715–5721.

[98] Zhang, K., Park, B. J., Fang, F. F., and Choi, H. J. (2009). Sonochemical preparation of polymer nanocomposites. *Molecules* 14, 2095–2110.

[99] Sarkar, S., Guibal, E., Quignard, F., and SenGupta, A. K. (2012). Polymer-supported metals and metal oxide nanoparticles: synthesis, characterization, and applications. *J. Nanoparticle Res.* 14:715.

[100] Sau, T. K., Rogach, A. L., Jackel, F., Klar, T. A., and Feldmann, J. (2010). Properties and applications of colloidal nonspherical noble metal nanoparticles. *Adv. Mater.* 22, 1805–1825.

[101] Rahman, I. A., and Padavettan, V. (2012). Synthesis of Silica nanoparticles by Sol-Gel: Size-dependent properties, surface modification, and applications in silica-polymer nanocompositesa review. *J. Nanomater.* 2012:15.

[102] Yang, T.-I., Brown, R. N. C., Kempel, L. C., and Kofinas, P. (2008). Magneto-dielectric properties of polymer–nanocomposites. *J. Magn. Magn. Mater.* 320, 2714–2720.

[103] Siddabattuni, S., Schuman, T. P., and Dogan, F. (2013). Dielectric properties of polymer-particle nanocomposites influenced by electronic nature of filler surfaces. *ACS Appl. Mater. Interface* 5, 1917–1927.

[104] Musil, J. (2000). Hard and superhard nanocomposite coatings. *Surf. Coat. Technol.* 125, 322–330.

[105] Erdemir, A., and Voevodin, A. A. (2010). "Nano composite coatings for servere applications," in *Handbook of Deposition Technologies for*

Films and Coatings, ed. P. M. Martin (Boston, MA: William Andrew Publishing), 679–715.

[106] Rahman, M. M., Ahammad, A. J. S., Jin, J. H., Ahn, S. J., and Lee, J. J. (2010). A comprehensive review of glucose biosensors based on nanostructured metal-oxides. *Sensors* 10, 4855–4886.

[107] Hu, H., Wang, Z., and Pan, L. (2010). Synthesis of monodisperse Fe_33O_4@silica core-shell microspheres and their application for removal of heavy metal ions from water. *J. Alloys Compd.* 492, 656–661.

[108] Rahman, O. U., Kashif, M., and Ahmad, S. (2015). Nanoferrite dispersed waterborne epoxy-acrylate: Anticorrosive nanocomposite coatings. *Prog. Org. Coat.* 80, 77–86.

[109] Peighambardoust, S. J., Rowshanzamir, S., and Amjadi, M. (2010). Review of the proton exchange membranes for fuel cell applications. *Int. J. Hydrog. Energy* 35, 9349–9384.

[110] Smitha, B., Sridhar, S., and Khan, A. A. (2005). Solid polymer electrolyte membranes for fuel cell applications – A review. *J. Memb. Sci.* 259, 10–26.

[111] Cong, H., Radosz, M., Towler, B. F., and Shen, Y. (2007). Polymer-inorganic nanocomposite membranes for gas separation. *Separat. Purif. Technol.* 55, 281–291.

[112] Zularisam, A. W., Ismail, A. F., Salim, M. R., Sakinah, M., and Ozaki, H. (2007). The effects of natural organic matter (NOM) fractions on fouling characteristics and flux recovery of ultrafiltration membranes. *Desalination* 212, 191–208.

[113] Ng, L. Y., Mohammad, A. W., Leo, C. P., and Hilal, N. (2013). Polymeric membranes incorporated with metal/metal oxide nanoparticles: a comprehensive review. *Desalination* 308, 15–33.

[114] Jönsson, C., and Jönsson, A. S. (1995). Influence of the membrane material on the adsorptive fouling of ultrafiltration membranes. *J. Memb. Sci.* 108, 79–87.

[115] Nair, L. S., and Laurencin, C. T. (2007). Biodegradable polymers as biomaterials. *Prog. Polym. Sci.* 32, 762–798.

[116] Armentano, I., Dottori, M., Fortunati, E., Mattioli, S., and Kenny, J. M. (2010). Biodegradable polymer matrix nanocomposites for tissue engineering: a review. *Polym. Degrad. Stabil.* 95, 2126–2146.

[117] Peter, S. J., Miller, M. J., Yasko, A. W., Yaszemski, M. J., and Mikos, A. G. (1998). Polymer concepts in tissue engineering. *J. Biomed. Mater. Res.* 43, 422–427.

[118] Li, T. et al. (2010). Comparative toxicity study of Ag, Au, and Ag-Au bimetallic nanoparticles on Daphnia magna. *Anal. Bioanal. Chem.* 398, 689–700.

[119] Barick, A. K., and Tripathy, D. K. (2010). Effect of nanofiber on material properties of vapor-grown carbon nanofiber reinforced thermoplastic polyurethane (TPU/CNF) nanocomposites prepared by melt compounding. *Compos. Part A Appl. Sci. Manuf.* 41, 1471–1482.

[120] Luo, Z. P., and Koo, J. H. (2008). Quantitative study of the dispersion degree in carbon nanofiber/polymer and carbon nanotube/polymer nanocomposites. *Mater. Lett.* 62, 3493–3496.

[121] Al-Saleh, M. H., and Sundararaj, U. (2011). Review of the mechanical properties of carbon nanofiber/polymer composites. *Compos. Part A Appl. Sci. Manuf.* 42, 2126–2142.

[122] Corrales, F., et al. (2007). Chemical modification of jute fibers for the production of green-composites. *J. Hazard. Mater.* 144, 730–735.

[123] Yu, Y., et al. (2013). Influence of filler waviness and aspect ratio on the percolation threshold of carbon nanomaterials reinforced polymer nanocomposites. *J. Mater. Sci.* 48, 5727–5732.

[124] Yousefi, N. et al. (2014). Highly aligned graphene/polymer nanocomposites with excellent dielectric properties for high-performance electromagnetic interference shielding. *Adv. Mater.* 26, 5480–5487.

[125] Aly, A. A. (2012). Friction and wear of polymer composites filled by nano-particles: a review. *World J. Nano Sci. Eng.* 2, 32–39.

[126] Bijwe, J., Logani, C. M., and Tewari, U. S. (1990). Influence of fillers and fibre reinforcement on abrasive wear resistance of some polymeric composites. *Wear* 138, 77–92.

[127] Wei, L., Hu, N., and Zhang, Y. (2010). Synthesis of polymer-mesoporous silica nanocomposites. *Materials* 3, 4066–4079.

[128] Kaskel, S. (2006). Functional inorganic nanofillers for transparent polymers. in *VDI Berichte* 2006, 57–60. doi:10.1039/b608177k

[129] Spitalsky, Z., Tasis, D., Papagelis, K., and Galiotis, C. (2010). Carbon nanotube-polymer composites: chemistry, processing, mechanical and electrical properties. *Prog. Polym. Sci.* 35, 357–401.

[130] Findik, F. (2014). Latest progress on tribological properties of industrial materials. *Mater. Design* 57, 218–244.

[131] Schmidt, G., and Malwitz, M. M. (2003). Properties of polymer-nanoparticle composites. *Curr. Opin. Colloid Interface Sci.* 8, 103–108.

[132] YU, H., et al. (2008). Tribological properties and lubricating mechanisms of Cu nanoparticles in lubricant. *Trans. Nonferrous Met. Soc. China* 18, 636–641.

[133] Kumar, M., and Bijwe, J. (2013). Optimized selection of metallic fillers for best combination of performance properties of friction materials: a comprehensive study. *Wear* 303, 569–583.

[134] Li, Y., Swartz, M. L., Phillips, R. W., Moore, B. K., and Roberts, T. A. (1985). Effect of filler content and size on properties of composites. *J. Dent. Res.* 64, 1396–401.

[135] Wetzel, B., Haupert, F., Friedrich, K., Zhang, M. Q., and Rong, M. Z. (2002). Impact and wear resistance of polymer nanocomposites at low filler content. *Polym. Eng. Sci.* 42, 1919–1927.

[136] Rai, M., Yadav, A., and Gade, A. (2009). Silver nanoparticles as a new generation of antimicrobials. *Biotechnol. Adv.* 27, 76–83.

[137] Rodríguez-González, B., Burrows, A., Watanabe, M., Kiely, C. J., and Liz Marzán, L. M. (2005). Multishell bimetallic AuAg nanoparticles: synthesis, structure and optical properties. *J. Mater. Chem.* 15;1755.

[138] Johnson, P. B., and Christy, R. W. (1972). Optical constants of the noble metals. *Phys. Rev. B* 6, 4370–4379.

[139] Anker, J. N. et al. (2008). Biosensing with plasmonic nanosensors. *Nat. Mater.* 7, 442–453.

[140] El-Ansary, A., and Faddah, L. M. (2010). Nanoparticles as biochemical sensors. *Nanotechnol. Sci. Appl.* 3, 65–76.

[141] Dahlin, A. B. et al. (2012). Electrochemical plasmonic sensors. *Anal. Bioanal. Chem.* 402, 1773–1784.

[142] Du, J. H., Bai, J., and Cheng, H. M. (2007). The present status and key problems of carbon nanotube based polymer composites. *Express Polym. Lett.* 1, 253–273.

[143] Balazs, A. C., Emrick, T., and Russell, T. P. (2006). Nanoparticle polymer composites: where two small worlds meet. *Science* 314, 1107–1110.

[144] Kim, K. H., Ong, J. L., and Okuno, O. (2002). The effect of filler loading and morphology on the mechanical properties of contemporary composites. *J. Prosthet. Dent.* 87, 642–649.

[145] Newman, M. D., Stotland, M., and Ellis, J. I. (2009). The safety of nanosized particles in titanium dioxide- and zinc oxide-based sunscreens. *J. Am. Acad. Dermatol.* 61, 685–692.

[146] Faure, B. et al. (2013). Dispersion and surface functionalization of oxide nanoparticles for transparent photocatalytic and UV-protecting coatings and sunscreens. *Sci. Technol. Adv. Mater.* 14:23001.

[147] Fujishima, A., Zhang, X., and Tryk, D. A. (2008). TiO_2 photocatalysis and related surface phenomena. *Surf. Sci. Rep.* 63, 515–582.

[148] Zhao, X., Zhao, Q., Yu, J., and Liu, B. (2008). Development of multifunctional photoactive self-cleaning glasses. *J. Non. Cryst. Solids* 354, 1424–1430.

[149] Ismail, A. A. et al. (2013). TiO_2 decoration of graphene layers for highly efficient photocatalyst: impact of calcination at different gas atmosphere on photocatalytic efficiency. *Appl. Catal. B Environ.* 129, 62–70.

[150] Guo, M. Y. et al. (2011). ZnO and TiO_2 1D nanostructures for photocatalytic applications. *J. Alloys Compd.* 509, 1328–1332.

[151] Pan, Z. W., Dai, Z. R., and Wang, Z. L. (2001). Nanobelts of semiconducting oxides. *Science* 291, 1947–1949.

[152] Ellmer, K. (2001). Resistivity of polycrystalline zinc oxide films: current status and physical limit. *J. Phys. D. Appl. Phys.* 34, 3097–3108.

[153] Zhong Lin, W. (2004). Zinc oxide nanostructures: growth, properties and applications. *J. Phys. Condens. Matter* 16, R829–R858.

[154] Kolodziejczak-Radzimska, A., and Jesionowski, T. (2014). Zinc oxide-from synthesis to application: a review. *Materials* 7, 2833–2881.

[155] Madaeni, S. S., Zinadini, S., and Vatanpour, V. (2011). A new approach to improve antifouling property of PVDF membrane using in situ polymerization of PAA functionalized TiO_2 nanoparticles. *J. Memb. Sci.* 380, 155–162.

[156] Mirkin, C. A., Letsinger, R. L., Mucic, R. C., and Storhoff, J. J. (1996). A DNA-based method for rationally assembling nanoparticles into macroscopic materials. *Nature* 382, 607–609.

[157] Xie, D. F. et al. (2013). Elaboration and properties of plasticised chitosan-based exfoliated nano-biocomposites. *Polymer* 54, 3654–3662.

[158] Das, A., Costa, F. R., Wagenknecht, U., and Heinrich, G. (2008). Nanocomposites based on chloroprene rubber: effect of chemical nature and organic modification of nanoclay on the vulcanizate properties. *Eur. Polym. J.* 44, 3456–3465.

[159] Kojima, Y. et al. (1993). One-pot synthesis of nylon 6-clay hybrid. *J. Polym. Sci. Part A Polym. Chem.* 31, 1755–1758.

[160] Sinha Ray, S., and Okamoto, M. (2003). Polymer/layered silicate nanocomposites: a review from preparation to processing. *Prog. Polym. Sci.* 28, 1539–1641.

[161] Paul, D. R., and Robeson, L. M. (2008). Polymer nanotechnology: nanocomposites. *Polymer* 49, 3187–3204.

[162] Fischer, H. (2003). Polymer nanocomposites: from fundamental research to specific applications. *Mater. Sci. Eng. C* 23, 763–772.

[163] He, H., Ma, Y., Zhu, J., Yuan, P., and Qing, Y. (2010). Organoclays prepared from montmorillonites with different cation exchange capacity and surfactant configuration. *Appl. Clay Sci.* 48, 67–72.

[164] Herrera, N. N., Letoffe, J. M., Putaux, J. L., David, L., and Bourgeat-Lami, E. (2004). Aqueous dispersions of silane-functionalized laponite clay platelets. A first step toward the elaboration of water-based polymer/clay nanocomposites. *Langmuir* 20, 1564–1571.

[165] Speakman, S. A. (1900). Basics of X-ray powder diffraction. *Basics X-Ray Powder Diffr.* 7, 1–97.

[166] Mansour, N., Christopoulos, C., and Tremblay, R. (2011). Experimental validation of replaceable shear links for eccentrically braced steel frames. *J. Struct. Eng.* 137, 1141–1152.

[167] Alexandre, M., and Dubois, P. (2000). Polymer-layered silicate nanocomposites: preparation, properties and uses of a new class of materials. *Mater. Sci. Eng. R Rep.* 28, 1–63.

[168] Ketterings, Q. M., Reid, S., and Rao, R. (2007). Cation exchange capacity (CEC). *Cornell Univ. Coop. Ext.* 24, 1–2. doi: 10.1080/01431160305010

[169] Wang, D., and Buriak, J. M. (2005). Electrochemically driven organic monolayer formation on silicon surfaces using alkylammonium and alkylphosphonium reagents. *Surf. Sci.* 590, 154–161.

[170] Marras, S. I., Tsimpliaraki, A., Zuburtikudis, I., and Panayiotou, C. (2007). Thermal and colloidal behavior of amine-treated clays: the role of amphiphilic organic cation concentration. *J. Colloid Interface Sci.* 315, 520–527.

[171] Stankovich, S. et al. (2006). Graphene-based composite materials. *Nature* 442, 282–286.

[172] Huang, X., Qi, X., Boey, F., and Zhang, H. (2012). Graphene-based composites. *Chem. Soc. Rev.* 41, 666–86.

[173] Dragoman, M., and Dragoman, D. (2009). Graphene-based quantum electronics. *Prog. Quantum Electron.* 33, 165–214.

[174] Marcano, D. C. et al. (2010). Improved synthesis of graphene oxide. *ACS Nano* 4, 4806–4814.

[175] Georgakilas, V. et al. (2016). Noncovalent functionalization of graphene and graphene oxide for energy materials, biosensing, catalytic, and biomedical applications. *Chem. Rev.* 116, 5464–5519.

[176] Liu, J., Tang, J., and Gooding, J. J. (2012). Strategies for chemical modification of graphene and applications of chemically modified graphene. *J. Mater. Chem.* 22, 12435–12452.

[177] Bagri, A. et al. (2010). Structural evolution during the reduction of chemically derived graphene oxide. *Nat. Chem.* 2, 581–587.

[178] Georgakilas, V. et al. (2012). Functionalization of graphene: covalent and non-covalent approaches, derivatives and applications. *Chem. Rev.* 112, 6156–6214.

[179] Hernandez, Y. et al. (2008). High-yield production of graphene by liquid-phase exfoliation of graphite. *Nat. Nanotechnol.* 3, 563–568.

[180] Bourlinos, A. B. et al. (2003). Graphite oxide: chemical reduction to graphite and surface modification with primary aliphatic amines and amino acids. *Langmuir* 19, 6050–6055.

[181] Gilje, S. et al. (2008). Processable aqueous dispersions of graphene nanosheets. *Nat. Nanotechnol.* 3, 101–105.

[182] Sreeprasad, T. S., and Berry, V. (2013). How do the electrical properties of graphene change with its functionalization? *Small* 9, 341–350.

[183] Brusatin, G., and Innocenzi, P. (2001). Fullerenes in Sol-Gel materials. *J. Solgel Sci. Technol.* 22, 189–204.

[184] Roy, N., Sengupta, R., and Bhowmick, A. K. (2012). Modifications of carbon for polymer composites and nanocomposites. *Prog. Polym. Sci.* 37, 781–819.

[185] Thompson, B. C., and Fréchet, J. M. J. (2008). Polymer-fullerene composite solar cells. *Angew. Chem. Int. Ed. Engl.* 47, 58–77.

[186] Dai, L. M. (1999). Advanced syntheses and microfabrications of conjugated polymers, C-60-containing polymers and carbon nanotubes for optoelectronic applications. *Pol. Adv. Technol.* 10, 357–420.

[187] Diederich, F., and Gómez-López, M. (1999). Supramolecular fullerene chemistry. *Chem. Soc. Rev.* 28, 263–277.

[188] Giacalone, F., and Martín, N. (2006). Fullerene polymers: synthesis and properties. *Chem. Rev.* 106, 5136–5190.

[189] Avouris, P., Chen, Z., and Perebeinos, V. (2007). Carbon-based electronics. *Nat. Nanotechnol.* 2, 605–615.

[190] Vroman, I., and Tighzert, L. (2009). Biodegradable polymers. *Materials* 2, 307–344.

[191] Troshin, P. A. et al. (2009). Material solubility-photovoltaic performance relationship in the design of novel fullerene derivatives for bulk heterojunction solar cells. *Adv. Funct. Mater.* 19, 779–788.

[192] Dai, L., Lu, J., Matthews, B., and Mau, A. W. H. (1998). Doping of conducting polymers by sulfonated fullerene derivatives and dendrimers. *J. Phys. Chem. B* 102, 4049–4053.

[193] Zhang, W. B. et al. (2008). Clicking fullerene with polymers: synthesis of [60]fullerene end-capped polystyrene. *Macromolecules* 41, 515–517.

[194] Markov, D. E., Amsterdam, E., Blom, P. W. M., Sieval, A. B., and Hummelen, J. C. (2005). Accurate measurement of the exciton diffusion length in a conjugated polymer using a heterostructure with a side-chain cross-linked fullerene layer. *J. Phys. Chem. A* 109, 5266–5274.

[195] Turner, S. et al. (2009). Determination of size, morphology, and nitrogen impurity location in treated detonation nanodiamond by transmission electron microscopy. *Adv. Funct. Mater.* 19, 2116–2124.

[196] Gibson, N. et al. (2009). Colloidal stability of modified nanodiamond particles. *Diam. Relat. Mater.* 18, 620–626.

[197] Smith, B. R., Gruber, D., and Plakhotnik, T. (2010). The effects of surface oxidation on luminescence of nano diamonds. *Diam. Relat. Mater.* 19, 314–318.

[198] Neitzel, I., Mochalin, V., Knoke, I., Palmese, G. R., and Gogotsi, Y. (2011). Mechanical properties of epoxy composites with high contents of nanodiamond. *Compos. Sci. Technol.* 71, 710–716.

[199] Pichot, V. et al. (2008). An efficient purification method for detonation nanodiamonds. *Diam. Relat. Mater.* 17, 13–22.

[200] Neitzel, I., Mochalin, V., Knoke, I., Palmese, G. R., and Gogotsi, Y. (2011). Mechanical properties of epoxy composites with high contents of nanodiamond. *Compos. Sci. Technol.* 71, 710–716.

[201] Huang, Q. et al. (2014). Nanotwinned diamond with unprecedented hardness and stability. *Nature* 510, 250–253.

[202] Peng, B. L., Dhar, N., Liu, H. L., and Tam, K. C. (2011). Chemistry and applications of nanocrystalline cellulose and its derivatives: a nanotechnology perspective. *Can. J. Chem. Eng.* 89, 1191–1206.

[203] Yang, K., Feng, L., Shi, X., and Liu, Z. (2013). Nano-graphene in biomedicine: theranostic applications. *Chem. Soc. Rev.* 42, 530–547.
[204] Williams, O. A. (2011). Nanocrystalline diamond. *Diam. Relat. Mater.* 20, 621–640.
[205] Zhang, Z., Keys, A. S., Chen, T., and Glotzer, S. C. (2005). Self-assembly of patchy particles into diamond structures through molecular mimicry. *Langmuir* 21, 11547–11551.
[206] Filik, J. et al. (2006). Raman spectroscopy of nanocrystalline diamond: An ab initio approach. *Phys. Rev. B* 74:035423.
[207] Le Guillou, C., Rouzaud, J. N., Remusat, L., Jambon, A., and Bourot-Denise, M. (2010). Structures, origin and evolution of various carbon phases in the ureilite Northwest Africa 4742 compared with laboratory-shocked graphite. *Geochim. Cosmochim. Acta* 74, 4167–4185.
[208] Occelli, F., Loubeyre, P., and LeToullec, R. (2003). Properties of diamond under hydrostatic pressures up to 140 GPa. *Nat. Mater.* 2, 151–154.
[209] Robertson, J. (2002). Diamond-like amorphous carbon. *Mater. Sci. Eng. R Rep.* 37, 129–281.
[210] Ferrari, A. C., and Robertson, J. (2004). Raman spectroscopy of amorphous, nanostructured, diamond-like carbon, and nanodiamond. *Philos. Trans. A Math. Phys. Eng. Sci.* 362, 2477–2512.
[211] Reinke, P., and Oelhafen, P. (1999). Electronic properties of diamond/nondiamond carbon heterostructures. *Phys. Rev. B* 60, 15772–15781.
[212] Emslie, D. J. H., Chadha, P., and Price, J. S. (2013). Metal ALD and pulsed CVD: fundamental reactions and links with solution chemistry. *Coord. Chem. Rev.* 257, 3282–3296.
[213] Wrobel, A. M. et al. (2009). Growth mechanism and chemical structure of amorphous hydrogenated silicon carbide (a-SiC:H) films formed by remote hydrogen microwave plasma CVD from a triethylsilane precursor: Part 1. *Chem. Vap. Depos.* 15, 39–46.
[214] Hampden-Smith, M. J., and Kodas, T. T. (1995). Chemical vapor deposition of metals: Part 1. An overview of CVD processes. *Chem. Vapor Depos.* 1, 8–23.
[215] Koch, C. C., Scattergood, R. O., Darling, K. A., and Semones, J. E. (2008). Stabilization of nanocrystalline grain sizes by solute additions. *J. Mater. Sci.* 43, 7264–7272.
[216] Gu, C. D., You, Y. H., Yu, Y. L., Qu, S. X., and Tu, J. P. (2011). Microstructure, nanoindentation, and electrochemical properties of the

nanocrystalline nickel film electrodeposited from choline chloride-ethylene glycol. *Surf. Coat. Technol.* 205, 4928–4933.
[217] Afshari, V., and Dehghanian, C. (2009). Effects of grain size on the electrochemical corrosion behaviour of electrodeposited nanocrystalline Fe coatings in alkaline solution. *Corros. Sci.* 51, 1844–1849.
[218] Shtein, M., Mapel, J., Benziger, J. B., and Forrest, S. R. (2002). Effects of film morphology and gate dielectric surface preparation on the electrical characteristics of organic-vapor-phase-deposited pentacene thin-film transistors. *Appl. Phys. Lett.* 81, 268–270.
[219] Wiora, M. et al. (2009). Grain size dependent mechanical properties of nanocrystalline diamond films grown by hot-filament CVD. *Diam. Relat. Mater.* 18, 927–930.
[220] Michaelson, S., and Hoffman, A. (2006). Hydrogen bonding, content and thermal stability in nano-diamond films. *Diam. Relat. Mater.* 15, 486–497.
[221] Nistor, L. C., Van Landuyt, J., Ralchenko, V. G., Obraztsova, E. D., and Smolin, A. A. (1997). Nanocrystalline diamond films: transmission electron microscopy and Raman spectroscopy characterization. *Diam. Relat. Mater.* 6, 159–168.
[222] Ballutaud, D., Kociniewski, T., Vigneron, J., Simon, N., and Girard, H. (2008). Hydrogen incorporation, bonding and stability in nanocrystalline diamond films. *Diam. Relat. Mater.* 17, 1127–1131.
[223] Santner, E., and Czichos, H. (1989). Tribology of polymers. *Tribol. Int.* 22, 103–109.
[224] Liu, T., Li, B., Lively, B., Eyler, A., and Zhong, W. H. (2014). Enhanced wear resistance of high-density polyethylene composites reinforced by organosilane-graphitic nanoplatelets. *Wear* 309, 43–51.
[225] Yazdani, B. et al. (2015). Tribological performance of Graphene/Carbon nanotube hybrid reinforced Al_2O_3 composites. *Sci. Rep.* 5:11579.
[226] Moran, J., Sweetland, M., and Suh, N. P. (2004). "Low friction and wear on non-lubricated connector contact surfaces," in *Proceedings of the 50th IEEE Holm Conference on Electrical Contacts and the 22nd International Conference on Electrical Contacts Electrical Contacts,* Seattle, WA, 263–266.
[227] Lu, Z. P., and Friedrich, K. (1995). On sliding friction and wear of PEEK and its composites. *Wear* 181–183, 624–631.
[228] Leszczyñska, A., Njuguna, J., Pielichowski, K., and Banerjee, J. R. (2007). Polymer/montmorillonite nanocomposites with improved

thermal properties. Part I. Factors influencing thermal stability and mechanisms of thermal stability improvement. *Thermochim. Acta* 453, 75–96.

[229] Zhang, T., Wu, X., and Luo, T. (2014). Polymer nanofibers with outstanding thermal conductivity and thermal stability: fundamental linkage between molecular characteristics and macroscopic thermal properties. *J. Phys. Chem. C* 118, 21148–21159.

[230] Gulotty, R., Castellino, M., Jagdale, P., Tagliaferro, A., and Balandin, A. A. (2013). Effects of functionalization on thermal properties of single-wall and multi-wall carbon nanotube-polymer nanocomposites. *ACS Nano* 7, 5114–5121.

[231] Huang, X., Jiang, P., and Tanaka, T. (2011). A review of dielectric polymer composites with high thermal conductivity. *IEEE Electr. Insul. Mag.* 27, 8–16.

10

Computational Modeling and Theoretical Strategies for the Design of Chiral Recognition Sites Using Molecular Imprinting Technology

T. Sajini[1], Aravind Krishnan[1] and Beena Mathew[2]

[1]Research and Post Graduate Department of Chemistry,
St. Berchmans College, Changanassery, India
[2]School of Chemical Sciences, Mahatma Gandhi University,
Kottayam, India

10.1 Introduction

Chirality is an important universal phenomenon in nature. For the in-depth study of pharmacology and biology, efficient enantioselective technologies are essential [1]. Research on enantiomeric recognition of chiral compounds can provide us important information to understand the recognition process in biological systems [2]. Although some progress in chiral discrimination has been achieved during the past decades, the selective detection of individual enantiomer is still the most difficult analytical task owing to the similar physical and chemical properties and the similar molecular configurations of enantiomers [3]. Hence it is important to fabricate practical and rapidly available methods for chiral recognition and separation of enantiomers.

Molecular imprinting is a promising technique for the preparation of polymers with predetermined selectivity, specificity, and high affinity which involves arrangement of polymerizable functional monomers around a print molecule [4–7]. The combinations of quantum mechanics and molecular dynamics have produced considerable advancement in the designing and implementation of molecularly imprinted polymers (MIPs). Here in this article we describe a brief account of advancements in molecular imprinting technology for enantioselective sensing system (MIT) in accordance with the theoretical and computational aspects.

10.1.1 Enantiomeric Sensing System Tailored by Molecular Imprinting Technology

Many naturally occurring and synthetic chemicals exist in an optically active form. They therefore have pairs of enantiomers. Enantiomers of the same compound, indistinguishable on the basis of physical and chemical properties, differ, sometimes significantly so, in physiological effects. The differences may be subtle, but sometimes they have tremendous importance. They cannot be separated by commonly used methods such as fractional distillation or fractional crystallization, except when the solvent used is optically active. Currently used methods for analysis of optically active compounds are mainly high-performance separation methods such as gas chromatography (GC) and high-performance liquid chromatography (HPLC) using chiral stationary phases, chiral selectors in the mobile phase, or flow reactors for derivatization. Also, efficient electromigration techniques such as capillary electrophoresis (CE) employing chiral selectors are widely used for determination of enantiomers. Other methods are mass spectroscopy, nuclear magnetic resonance (NMR) for the study of molecular recognition, as well as some spectroscopic techniques such as circular dichroism. Those techniques require expensive equipment, and the analysis is often time-consuming.

Molecular recognition phenomena have been extensively studied due to the widespread importance of non-covalent interactions in complex formation, catalysis, biochemistry, and chemical sensors. These principles are realized by the application of host-guest chemistry and are based on intermolecular non-covalent forces. Molecular imprinting is a promising technique for the preparation of polymers with predetermined selectivity, specificity, and high affinity which involves arrangement of polymerizable functional monomers around a print or template molecule. MIPs have the ability to specifically distinguish and separate a particular molecule from other molecules of similar structures [8]. This property makes MIPs applicable in various fields such as in separation and purification of structurally related compounds, catalysis, biosensors, drug delivery, and in biotechnology. A schematic representation of the molecular imprinting process is shown in Scheme 10.1.

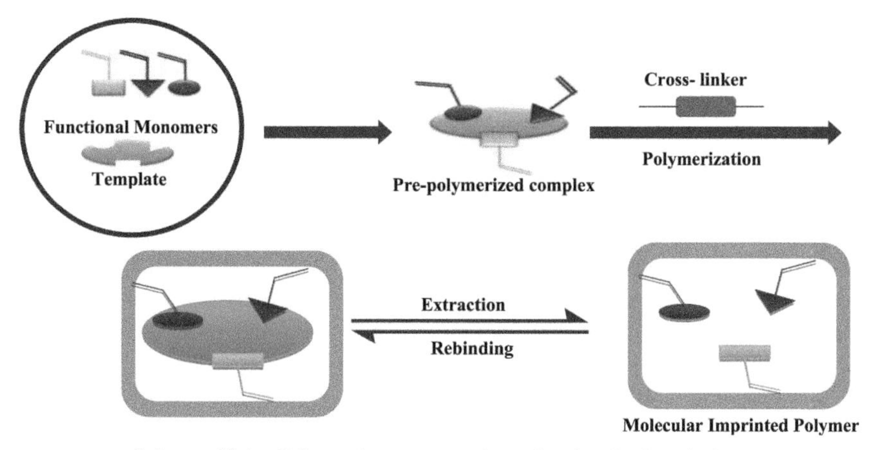

Scheme 10.1 Schematic representation of molecular imprinting.

Usually, there are two main methods to fabricate MIPs, based on covalent and non-covalent interactions between the template and the functional monomer. Covalent imprinting, being stoichiometric, ensures that functional monomer residues exist only in the imprinted cavities. It is a typical method and often uses readily reversible condensation reactions involving boronate esters, ketals/acetals, and Schiff's base. This covalent imprinting approach is regarded as a less flexible method since the reversible condensation reactions are limited. It is very difficult to reach thermodynamic equilibrium since the strong covalent interactions will result in slow binding and dissociation. Non-covalent imprinting can proceed by ionic interactions, hydrogen bonding, van der Waals forces and π–π interactions. Most commonly, the dominant interaction is hydrogen bonding, which often occurs between methacrylic acid (MAA) groups and primary amines in nonpolar solvents. In recent times, non-covalent imprinting has become the most popular and general synthesis strategy due to the simplicity of operation and rapidity of binding and removal.

Compared to other recognition systems, MIPs, which possess three major unique features of structure predictability, recognition specificity, and application universality, have received widespread attention and have become attractive in many fields, such as purification and separation, chemo/biosensing, artificial antibodies, drug delivery, and catalysis and degradation, owing to their high physical stability, straightforward preparation, remarkable robustness and low cost [9]. Among the various approaches used during synthesis of MIPs non-covalent interaction can be considered the best one due to easy removal of template, its applicability for a variety of molecules, and the economical and easy method [10].

10.1.2 Computational Modeling

The rapid improvements in computing power over the years and significant software developments have enhanced simulation capability for different areas including the complex strategies of the molecular imprinting process. Computational modeling has evolutionally shown a foremost role in MIP's development, especially as a powerful tool in providing preliminary prediction of the complex molecular interaction and the later performance of the MIP developed. The state of the art falls on using mathematical formulae to calculate the possible outcome based on hands-on information collected from past researchers' experiences. Adopting this idea, the wastage of consumables used in the laboratory can be reduced as the trial-and-error stage is carried out via software. For instance, matching and selection of the best monomer for a target template have been reported using software screening.

The support from computational modeling is still relevant in producing receptors of higher selectivity and reducing some initial preliminary lab work especially on establishing the interaction between the polymerizable ingredients. Basically, computational strategies can be divided into three major approaches, which are based on quantum chemical calculations, molecular dynamic consideration, and data processing. Quantum chemical calculation employs the approximation of electron distribution within a molecule as the basis to predict and describe the possible interaction between template and monomer. In the molecular dynamic approach, simulation is carried out by closely taking into account the experimental conditions. This method permits the qualitative evaluation of the types of interactions, the frequency of a species of complex formed, and the lifetime of the complex.

The computational approach is very useful to treat data generated out of an MIP system for the purpose of optimizing its usefulness for an intended application.

10.2 Theoretical and Computational Strategies in MIPs

Computational techniques are growing as a promising tool for the design of molecularly imprinted materials. Through the knowledge of the structures of the template and the monomer, it is possible to clearly evaluate the interaction between the two and hence to develop a fruitful combination depending on one's needs. Attempts have been made to formulate a library of monomers which can plausibly interact with a given template. By the availability of shared workstations and high-performance software, the "in silico" approaches are gaining predominant importance in this regard.

Quantum computation was the first method to be effectively used in this direction. In simple words, the approaches involve the optimization of the structures of the template and the monomers individually, followed by finding out their interaction, using energy calculations. It is now possible to determine accurately the nature of interaction existing between the two [11]. As these methods, in contrast to less sophisticated models and to other theoretical approaches, are able to describe the electronic structure, they generally yield considerably better representations of non-covalent interactions present in the system under study. In most studies these methods have been used to describe interactions present in pre-polymerization mixtures, in particular aiming to investigate the interaction between the template molecule and candidate functional monomers. In some studies, however, their use was extended beyond the pre-polymerization mixture, e.g., where these techniques have also been applied to the evaluation of recognition and rebinding of templates to MIPs. Various possibilities of the factors like positioning of the template and the monomer, the stoichiometric ratio between the two, etc. are generated and the system with the best energy gain is evaluated. Methodologies like ab initio, semiempirical, and DFT methods are commonly employed. It is to be remembered that the method and the basis sets involved are selected on the basis of the nature of the material under investigation. Solvent selection being a major factor in MIP interactions, computational approaches have been standardized to incorporate solvent-explicit models into the above strategies.

Another effective computational method is using molecular mechanics/dynamics simulation techniques. The method gains its significance in the pre-polymerization stage, where the arrangement of the various monomers (as in the polymer) around the template is evaluated. An example was illustrated in Figure 10.1. The solvent-explicit models are incorporated along with the monomer molecules to produce an exact replica of the experiment. Techniques like molecular mechanics (involving conformational analysis with time), simulated annealing process (cooling the entire system from a high to low temperature), etc. are employed to obtain the low energy configuration of the system of the interacting molecules. The cross-linker influence can also be evaluated in this step. The interaction energy is calculated for the best solvent-template-prepolymer system. CHARMM or AMBER force fields are widely used for the dynamics method.

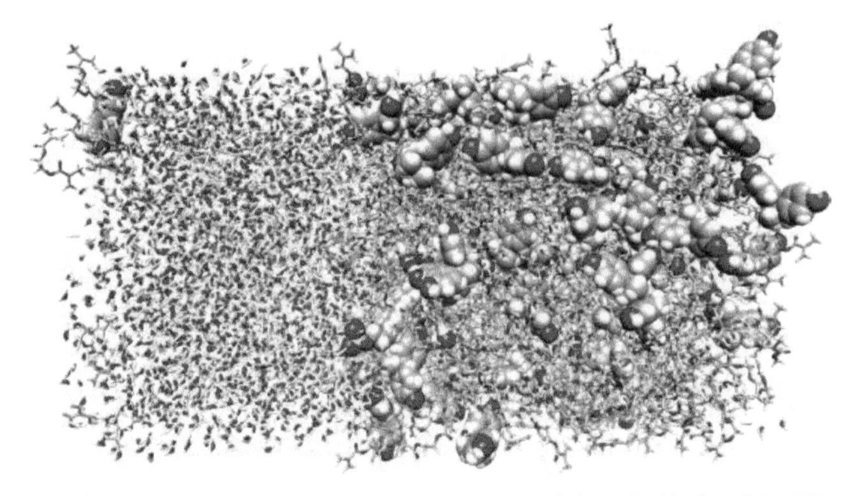

Figure 10.1 Snapshot from performed 2-phase MD simulation of a bisphenol-A MIP pre-polymerization mixture, applied in the production of MIP nanoparticles using the miniemulsion polymerization approach. (The template is presented using a van der Waal volume representation and all other pre-polymerization components are presented using a Licorice representation.) [12].

An example of the use of quantum-chemical calculations in molecular imprinting is the design of an ester hydrolysis catalyzing polymer based on PM3 calculations. These calculations provided support for the hypothesis that the template used in the MIP synthesis is a mimic of the transition state of the reaction to be catalyzed [13]. Finally, using a quite different approach, Voshell and Gagné [14] used HF/6–31G* and AM1 computational studies to determine the conformational rigidity of a dendritic system used in the imprinting of BINOL. Based on the results, a more rigid dendrimer structure was chosen to lower the binding-site heterogeneity and the enantioselectivity of the resulting MIPs was improved.

More examples of this kind include the use of the semi-empirical AM1 method for the calculation of a complex between (S)-nilvadipine and 4-vinylpyridine (4VP) [15], the optimization of a complex between 2,4,6-trichlorophenol and four molecules of 4VP [16], and the PM3 method for describing two complexes formed between (S)-naproxen and one or two molecules of acrylamide, respectively [17]. This strategy was employed in conjunction with density functional methods by Pietrzyk et al. [18] to model a complex between melamine and three functional monomer molecules on a B3LYP/3-21G(d) level of theory. The B3LYP functional was also used by Demircelik et al. [19] with a 6-31G(d,p) basis set, and by Riahi et al. [20], with a 6-311+G(d,p) basis set, for the modeling of template-monomer complexes; the latter study also included the effect of solvents using a polarizable continuum model (PCM). In PCM calculations, the effect of the solvent is approximated by placing the system in a cavity with a surface that is polarizable according to the dielectric constant of the modeled solvent.

The design of a polymer system imprinted with nicotinamide was undertaken using a broad range of calculations, e.g., different density functional methods and MP2 in combination with different basis sets, by Del Sole et al. [21]. This strategy was also used by Azenha et al. [22] in the design of a silicate-based polymer selective for β-damascenone, in this case using HF and B3LYP in combination with different basis sets. A growing number of other examples can be found in the literature, including the calculations of the structures of the complexes formed between a series of compounds and two molecules of MAA [23]. In this case, the structural parameters obtained by AM1 calculations could be correlated to the results of experimental data in the form of chromatographic studies. In a study by Lai and Feng [24], molecular geometries of buffer acids and bases optimized on the AM1-level were related to the metformin template binding in the respective media using an assumed mechanism for competitive binding. Moreover, Wu et al. [25] were able to demonstrate a relationship between experimentally determined capacity factors and the binding energies found by MP2/6-31G//HF/6-31G and PM3 calculations of complexes of various functional monomer–templates.

The self-consistent reaction field of Tomasi PCM was used by Junbo Liu et al. to simulate the interactions among Dicyandiamide (DCD), the template molecule, and MAA as well as DCD and different cross-linkers in ACN solution at 333 K [26]. The binding energy between DCD and the functional monomer (DEB1) was calculated using the Equation (10.1):

$$\Delta E_{B1} = E_{\text{template-monomer}} - E_{\text{template}} - \Sigma E_{\text{monomer}} \tag{10.1}$$

where $E_{\text{template-monomer}}$ is the energy of template and monomer, E_{template} represents the energy of template, and $\Sigma\, E_{\text{monomer}}$ means the sum of energy for monomer. The binding energy between the template and cross-linker (ΔE_{B2}) was calculated using the Equation (10.2):

$$\Delta E_{B2} = E_{\text{template-cross-linker}} - E_{\text{template}} - E_{\text{cross-linker}} \tag{10.2}$$

where $E_{\text{template-cross-linker}}$ is the energy of template and cross-linker, E_{template} represents the energy of template, and $E_{\text{cross-linker}}$ means the energy of cross-linker. The binding energy between monomer and cross-linker (ΔE_{B3}) was calculated using the Equation (10.3).

$$\Delta E_{B3} = E_{monomer-cross\text{-}linker} - E_{monomer} - E_{cross\text{-}linker} \quad (10.3)$$

where $E_{template-cross\text{-}linker}$ is the energy of template and cross-linker, $E_{monomer}$ represents the energy of monomer, and $E_{cross\text{-}linker}$ means the energy of cross-linker. Moreover, the basis set superposition error (BSSE) was taken into account through counterpoise correction method.

A significant limitation in the application of electronic structure methods to MIP systems is the difficulty in handling the large numbers of atoms necessary to provide a comprehensive picture of the pre-polymerization or polymer system. One particular complication appears to be the problems that can arise from the lack of an explicit solvent [27]. For electronic structure methods, the inclusion of a reasonable number of solvent molecules makes the calculation rather time-consuming; as a result, solvent effects are often omitted completely. However, methods such as PCM provide the possibility of including solvent effects without the inclusion of explicit solvent molecules. It can be concluded that many approaches have been used primarily for studying interactions in the pre-polymerization mixture, even for explaining recognition phenomena in the finished polymer. Nonetheless, most studies have been focused on single aspects of the process of designing, synthesizing or testing of a system, and the impact on MIP research has so far been rather limited. With the development of refined computational techniques and improved availability of computational power, however, the impact of electronic structure-based calculations can be expected to increase.

One of the first successes in using an MD-based approach for the screening of the best functional monomer for a given template was presented by Piletsky et al. (Figure 10.2) in 2001 [28]. In their approach, they created and utilized a virtual library consisting of a total number of 20 different functional monomers, which were screened against one enantiomer of their template molecule ephedrine. After initially running 30,000 MD steps via the Leapfrog algorithm and generating empirical binding scores, the top four functional monomer–template complexes were then selected and the corresponding MIPs were prepared and evaluated. Here, through a comparison of the final MIP performance using these top-ranked monomers and monomers displaying lower binding scores, they demonstrated the potential of their screening approach as a supporting tool in the development of MIPs.

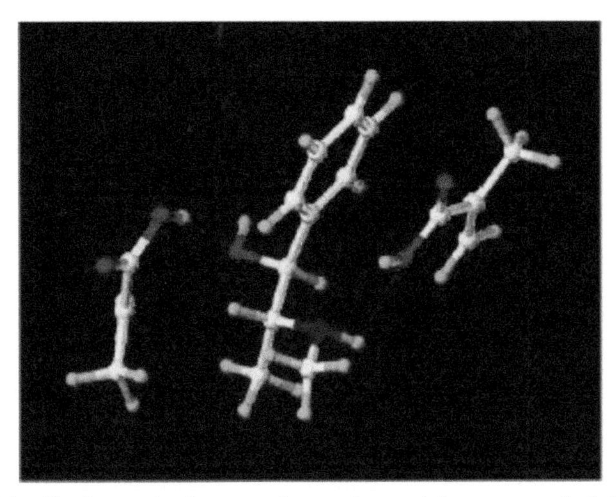

Figure 10.2 The interaction between the template and the monomer in 1:2 ratio [28].

It is commonly accepted that the origin of the predetermined recognition in MIPs can be correlated back to the nature and strength of functional monomer template complexes that are established at the pre-polymerization stage. One of the notable studies in this can be attributed to formulation of virtual libraries of monomers developed by many researchers. Wei et al. [29] imprinted 17β-estradiol after simulating either a monomer–template pair or a single template molecule solvated with eight functional monomers to incorporate also the effect of monomer dimerization on the degree of template complexation. In both these cases, each 17β-estradiol was surrounded by explicit non-polar porogen molecules (acetone or chloroform). The selection of the "best binding" monomer here was made from a virtual library comprised of a total of nine different functional monomers. Here, the use of an MD-based computational screening approach to find the optimal functional monomer suggested that MAA, methacrylamide, and 2-(diethylamino) ethyl methacrylate showed the strongest hydrogen bonding interactions to the selected template, with results in accordance with parallel experimental MIP-template batch rebinding studies. To investigate the impact of the growing polymer chain during polymerization on the stability of functional monomer–template complexation, studies have been performed in which homo- and co-polymeric chains of functional monomers have been used in the search for the optimal monomer.

Pavel and Lagowski [30] reported on a strategy where they computed potential energy differences for a series of functional monomer–template complexes. The binding energies were calculated after simulating an ensemble of functional monomers in the absence or presence of the template. Secondly, the effect of polymer chain growth on complex stability was investigated instead, using linear homo- and co-polymer chains. In this, using the target theophylline and a series of structural analogues, they demonstrated that the use of itaconic acid or ethylene dimethacrylate (EDMA) resulted in the best binding. Later, Pavel et al. [31] also reported on the successful virtual screening of a number of warfare agents. To study the stability of the complex formed between the template, 2,4-dichlorophenoxyacetic acid and the functional monomer 4VP which is believed to be stoichiometric of type 1:1. Molinelli et al. [32] reported a series of studies inserting a pre-minimized complex that was explicitly solvated by either chloroform or water molecules. From using different starting geometries in the different solvents, hydrogen bond interaction in chloroform and π–π stacking interaction in water, the authors proposed mechanisms to explain the nature of the interactions involved during the pre-polymerization stage as well as during MIP rebinding in aqueous solution.

To approach a more realistic representation of the "actual" pre-polymerization mixture, a series of efforts has been made towards mimicking the actual pre-polymerization mixture where large multiples of components are present in the stoichiometric ratios representative of the ones being used in typical MIP preparation protocols. O'Mahony et al. [33] used a number of MD simulations to study functional monomer–naproxen complexation and especially the effect of template dimerization on MIP performance. Although these mixtures did not include the initiator, extracted data from simulations suggested not only a high degree of 4VP naproxen complexation, but also that the cross-linking agent EDMA was found to contribute to the high degree of selectivity demonstrated by the MIP. On the same, Nicholls et al. theme, the role of template dimerization on final MIP performance was also reported by O'Mahony et al. for imprinting of quercetin using 4VP as functional monomer [34]. In this example, results revealed the formation of sheet-like structures of quercetin-4VP complexes with stabilities typically independent of the concentration of EDMA used.

The above system in Figure 10.3a is constructed using all the components used in the MIP synthesis and in the same stoichiometry as the synthesis. For small molecule templates, at least 10 template structures are generally used. Randomized initially packing geometries are used [36]. In Figure 10.3b, after a series of energy minimization steps and equilibration, first at NVT and then at NPT, the collection of production-phase data is undertaken (for small

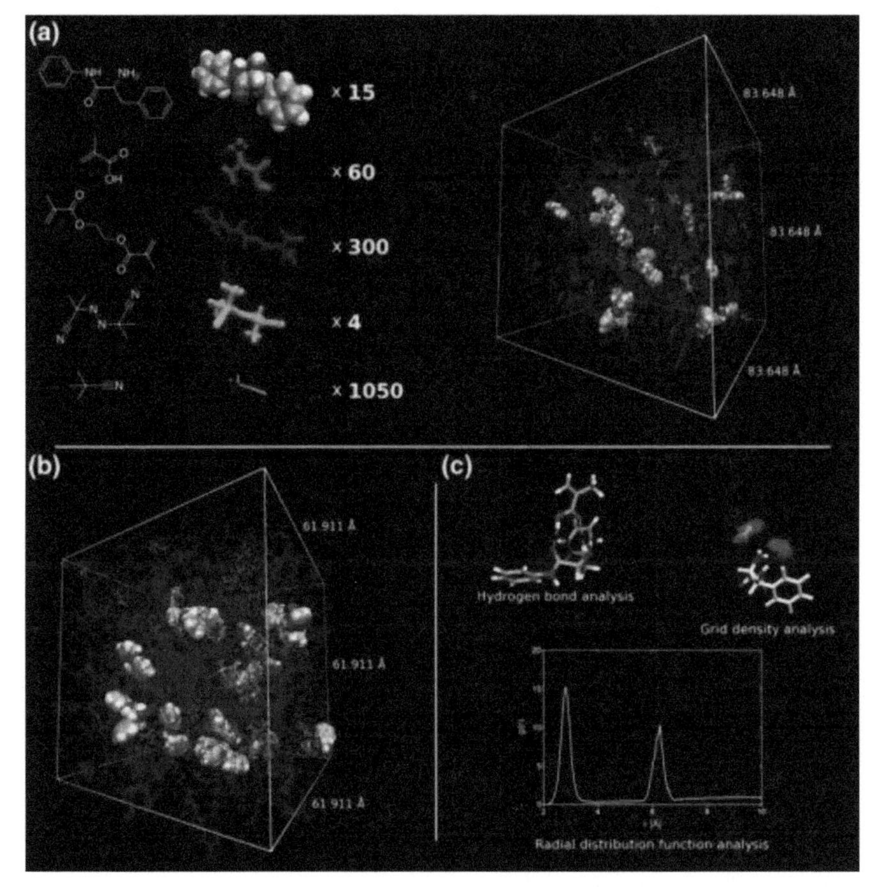

Figure 10.3 The strategy used for all-component MD simulations, here for a bupivacaine MIP pre-polymerization mixture [35].

organic templates, typically 5–10 ns production-phase at NVT, and generally >5 systems run in parallel). In Figure 10.3c, study of the interactions of the components in the pre-polymerization mixture is performed by examining the trajectories of all the components in the system over the production-phase, e.g., using radial distribution functions and hydrogen bond analyses. The prevalence of a given type of component around a particular element is illustrated using 3D grid density analyses. The use of MD simulation has further increased magnificently in recent years and is one of the promising methods available today for designing MIP materials.

10.3 Conclusions

Efforts to employ computational strategies for describing, predicting, and analyzing molecular imprinting systems are a rapidly expanding research area. Experimental work supported by proper theoretical calculation can surely be expected to extend the scope of application of MIPs, including in challenging areas where high specificity sensing is the need of the hour. We can expect that the computational tools become effective enough to generate molecular imprinting polymers for serving as personalized enantiomers or drug carriers, in the near future.

References

[1] Cuilan, C., Wang, X., Bai, Y., and Huwei, L. (2012). Applications of nanomaterials in enantioseparation and related techniques. *Trends Anal. Chem.* 39, 195–206.

[2] Samuel, K. T., Lorenz, H., and Morgenstern, A. S. (2010). Solubility of mandelic acid enantiomers and their mixtures in three chiral solvents. *J. Chem. Eng. Data.* 55, 5196–5200.

[3] Jayakrishnan, S. S., and Litta, E. G. (2012). *Intl. J. Biomed. Pharma. Sci.* 3.

[4] Andersson, L., Sellergren, B., and Mosbach, K. (1984). Imprinting of amino acid derivatives in macroporous polymers. *Tetrahedron Lett.* 25, 5211–5214.

[5] Piletsky, S. H., Chianella, I., and Whitcombe, M. J. (2013). Molecularly imprinted polymers. *Encycl. Phys.* 2013, 1596–1599.

[6] Takeuchi, T., and Matsui, J. (1996). Molecular imprinting: an approach to "tailor-made" synthetic polymers with biomimetic functions. *Acta Polym.* 47, 471–480.

[7] Ramstro, O., and Ansell, R. J. (1998). Molecular imprinting technology: challenges and prospects for the future. *Chirality* 10, 195–209.

[8] Cheong, W. J., Yang, S. H., and Ali, F. (2013). Molecular imprinted polymers for separation science: a review of reviews. *J. Sep. Sci.* 36, 609–628.

[9] Chen, L., Wang, X., Lu, W., Wu, X., and Li, J. (2016). Molecular imprinting: perspectives and applications. *Chem. Soc. Rev.* 45, 2137–2211.

[10] Nasrullah, S., Mazhar, U., Haneef, M., and Park, J. K. (2012). A brief overview of molecularly imprinted polymers: from basics to applications. *J. Pharm. Res.* 5, 3309–3317.

[11] Chen, L., Wang, X., Lu, W., Wu, X., and Li, J. (2016). Molecular imprinting: perspectives and applications. *Chem. Soc. Rev.* 45, 2137–2211.

[12] Nicholls, I. A., Karlsson, C. G., Olsson, G. D., and Rosengren, A. M. (2013). Computational strategies for the design and study of molecularly imprinted materials. *Indus. Chem. Eng. Res.* 371.

[13] Sagawa, T., Togo, K., Miyahara, C., Ihara, H., and Ohkubo, K. (2004). Rate-enhancement of hydrolysis of longchain amino acid ester by cross-linked polymers imprinted with a transition-state analogue: evaluation of imprinting effect in kinetic analysis. *Anal. Chim. Acta* 504, 37–41.

[14] Voshell, S., and Gagné, M. (2005). Rigidified dendritic structures for imprinting chiral information. *Organometallics* 24, 6338–6350.

[15] Fu, Q., Sanbe, H., Kagawa, C., Kunimoto, K. K., and Hagina, J. (2003). Uniformly sized molecularly imprinted polymer for (S)-nilvadipine. Comparison of chiral recognition ability with HPLC chiral stationary phases based on a protein. *Anal. Chem.* 75, 191–198.

[16] Schwarz, L., Holdsworth, C. I., Cluskey, A. M., and Bowyer, M. C. (2004). Synthesis and evaluation of a molecularly imprinted polymer selective to 2,4,6-trichlorophenol. *Aust. J. Chem.* 57, 759–764.

[17] Li, P., Rong, F., Xie, Y. B., Hu, V., and Yuan, C. W. (2004). Study on the binding characteristic of snaproxen imprinted polymer and the interactions between templates and monomers *J. Anal. Chem.* 59, 939–944.

[18] Pietrzyk, A., Kutner, W., Chitta, R., Zandler, M. E., D'Souza, F., Sannicolo, F., and Mussini, P. R. (2009). Melamine acoustic chemosensor based on molecularly imprinted polymer film. *Anal. Chem.* 81, 10061–10070.

[19] Demircelik, A. H., Andac, M., Andac, C. A., Say, R., and Denizli, A. (2009). Molecular recognition-based detoxification of aluminum in human plasma. *J. Biomater. Sci. Polym. Ed.* 20, 1235–1258.

[20] Riahi, S., Edris-Tabrizi, F., Javanbakht, M. M., Ganjali, R. and Norouzi, P. (2009). A computational approach to studying monomer selectivity towards the template in an imprinted polymer. *J. Mol. Model.* 15, 829–836.

[21] Dei Sole, R., Lazzoi, M. R., Arnone, M., Dei Sala, F., Cannoletta, D., and Vasapollo, G. (2009). Experimental and computational studies on non-covalent imprinted microspheres as recognition system for nicotinamide molecules. *Molecules* 14, 2632–2649.
[22] Azenha, M., Kathirvel, P., Nogueira, P., and Fernando-Silva, A. (2008). The requisite level of theory for the computational design of molecularly imprinted silica xerogels. *Biosens. Bioelectron.* 23, 1843–1849.
[23] Ogawa, T., Hoshina, K., Haginaka, J., Honda, C. Moto, T. T., and Uchida, T. (2005). Screening of bitterness-suppressing agents for quinine: the use of molecularly imprinted polymers. *J. Pharm. Sci.* 94, 353–362.
[24] Lai, E. P. C., and Feng, S. Y. (2003). Molecularly imprinted solid phase extraction for rapid screening of metformin. *Microchem. J.* 75, 159–168.
[25] Wu, L. Q., Sun, B. W., Li, Y. Z., and Chang, W. B. (2003). Study properties of molecular imprinting polymer using a computational approach. *Analyst* 128, 944–949.
[26] Liu, J., Wang, Y., Tang, S., Tang, S., Gao, Q., and Jin, R. (2017). Theoretical guidance for experimental research of the dicyandiamide and methacrylic acid molecular imprinted polymer. *New J. Chem.* 41, 13370–13376.
[27] Karlsson, B. C. G., O'Mahony, J., Karlsson, J. G. Bengtsson, H., Eriksson, L. A., and Nicholls, I. A. (2009). Structure and dynamics of monomer–template complexation: an explanation for molecularly imprinted polymer recognition site heterogeneity. *J. Am. Chem. Soc.* 131, 13297–13304.
[28] Piletsky, S. A., Karim, K., Piletska, E. V., Day, C. J., Freebairn, K. W, Legge, C., et al. (2001). Recognition of ephedrine enantiomers by molecularly imprinted polymers designed using a computational approach. *Analyst* 126, 1826–1830.
[29] Wei, S., Jakusch, M., and Mizaikoff, B. (2007). Investigating the mechanisms of 17β-estradiol imprinting by computational prediction and spectroscopic analysis. *Anal. Bioanal. Chem.* 389, 423–431.
[30] Pavel, D., and Lagowski, J. (2005). Computationally designed monomers for molecular imprinting of chemical warfare agents–Part V. *Polymer* 46, 7543–7556.
[31] Pavel, D., Lagowski, J., and Lepage, C. J. (2006). Computationally designed monomers for molecular imprinting of chemical warfare agents–Part V. *Polymer* 47, 8389–8399.

[32] Molinelli, A., O'Mahony, J., Nolan, K., Smyth, M. R., Jakusch, M., and Mizaikoff, B. (2005). Anatomy of a successful imprint: analysing the recognition mechanisms of a molecularly imprinted polymer for quercetin. *Anal. Chem.* 77, 5196–5204.

[33] O'Mahony, J., Karlsson, B. C. G., Mizaikoff, B., and Nicholls, I. A. (2007). Correlated theoretical, spectroscopic and X-ray crystallographic studies of a non-covalent molecularly imprinted polymerisation system. *Analyst* 132, 1161–1168.

[34] O'Mahony, J., Wei, S., Molinelli, A., and Mizaikoff, B. (2006). Imprinted polymeric materials. Insight into the nature of prepolymerization complexes of quercetin imprinted polymers. *Anal. Chem.* 78, 6187–6190.

[35] O'Mahony, J., Moloney, M., McCormack, M., Nicholls, I. A., Mizaikoff, B., and Danaher, M. (2013). Design and Implementation of an Imprinted Material for the Extraction of the endocrine disruptor bisphenol A from milk. *J. Chromatogr. B.* 931, 164–169.

[36] Olsson, G. D., Karlsson, B. C. G., Shoravi, S., Wiklander, J. G., and Nicholls, I. A. (2012). Mechanisms underlying molecularly imprinted polymer molecular memory and the role of crosslinker: resolving debate on the nature of template recognition in phenylalanine anilide imprinted polymers. *J. Mol. Recognit.* 25, 69–73.

11

Ultrafast Characterization 2D Semiconducting TMDC for Nanoelectronics Application

Yadu Nath V. K.[1], Raghvendra Kumar Mishra[1,2,3], Neelakandan M. S.[1], Bilahari Aryat[1], Parvathy Prasad[1] and Sabu Thomas[4]

[1]International Interuniversity centre for Nanoscience and Nanotechnology, Kerala, India
[2]Director, BSM Solar and Environmental Solution, A-348, Awas Vikas Colony, Unnao, UP, 261001, India
[3]Indian Institute of Space Science and Technology, India
[4]School of Chemical Sciences, Mahatma Gandhi University, Kottayam, India

Abstract

After the invention of graphene, nanoelectronics fields such as in transistors, detectors, sensors, etc., have been grown rapidly and which have been enhanced more and more even after the invention of the transition metal dichalcogenide (TMDC). In this review, we have discussed the properties and specialties of TMDC materials and how ultrafast process could help to characterize them, which will be directed to the evolution of nanoelectronics to another level and futuristic applications.

11.1 Introduction

Advancements in electronics in the last two decades were exceptionally heightened as a branch of science dealing with the flow and control of electrons. The giant leap of nanotechnology on the past 20 years has led to many developments in all scientific fields including electronics. Nanotechnology has influenced and deeply affected the rest of the scientific world so that the face of the science divisions changed so fast in the past 10 years. Today, the entire electronic engineer faces a serious trouble which is the size of the devices and circuits. They have tried to make the circuit tinier and tinier since this area is explored. Introducing nanotechnology to the electronics division could revolutionize the whole department by reducing the size to nanodimension. Nanoelectronics deals with the use of nanotechnology in electronic components to reduce the size to the minimum but no compromise in the efficiency of the device [1–7]. As an example, the newly developed semiconductors are manufactured with a feature size measured in nanometers with distinct electrical characteristics and many of them are crystalline or amorphous in nature. The cost, only fabricating a nano-regime semiconductor simply exceeds millions of dollars, which limits many of our well-known researchers on this field forcing them to work in advance developed semiconductors technology. After all, advanced semiconductor processing is already taking place well below 100 nm, the threshold for nanotechnology. The primary interest is that consumer products are becoming more feature rich with consumers unwilling to pay substantially more for these features and more advanced technologies combining in one resulting in a better product that has the ability to do multifunctions at the same time. The main challenge to the device designer is to put more feature into the same smaller device without significantly increasing the costs in order to get more functionality devices must shrink even further [8–13].

11.2 Ultrafast Characterization Process

When studying about the properties in atoms and molecules, there arises a phenomenal blockade; the fact is that the process occurs very fast, so these process is defined as any technology that relies on properties of a molecule/atom that are only extent up to a very short time period, or lesser than 10^{-9} s (nanoseconds to atto seconds). Many methods are evolved to study such fast dynamics, because using these techniques, we could measure the perturbed state dynamics of materials easily, and also many other non-equilibrium process are developed spanning change in time regime or energy ranges [14–17]. Atto seconds to picoseconds spectroscopy technique is considered on studying about the dynamics in time range of atto seconds to femtosecond; this kind of process is too hard to record electronically because they are so much faster than the equipment could respond. Even though arranging an ultrashort light pulses to perturb the energy states and initiate the process [18–21]. We widely use a titanium-sapphire laser as light source which emits light of wavelength ranging from 700 nm to 1100 nm. In this laser, titanium-doped sapphire is acting as a gain medium. The precision of the spectroscopical measurement lies within these pulse properties such as pulse duration, pulse energy, spectral phase, and spectral shape [22–24]. High harmonic generation refers to the non-linear process where conversion of a fixed frequency to the higher harmonics of that frequency of that intense laser radiation by takes place ionization and recollision of electrons. The three-step model is used to understand the high harmonic generation in atoms.

1. Ionization: The coulomb potential of the atom is modified by the laser field, and ionization occurs when electrons tunnel through the barrier.
2. Propagation: The free electron uses the laser field to gain momentum or accelerate.
3. Recombination: Emission of a photon with very high energy occurs when the field is reversed as the electron is accelerated back to the ionic parent [25–27].

For different spectroscopic technique requires different wavelength, and energy range for this has to be done, for this equipment like optical parametric oscillator and optical parametric amplifier is used [28–31].

11.3 Ultrafast Characterization Techniques

11.3.1 Ultrafast Transient Absorption

This method is generally a pulse probe experiment; in this, we use a pulsed laser to excite the molecules electrons to higher energy excited state from their ground state. A xenon arc lamp is used as probing light (using beam splitters could avoid the use of this lamp); as a result, we get an absorption spectrum of the compound corresponding to its excitations; after excitation, further excitation occurs and excitation to higher energy state is possible. After the absorption is over, the unabsorbed light is continued to a photodiode and the data are processed, to generate a spectrum [32–34].

11.3.2 Time-resolved Photo Electron Spectroscopy

Both of these spectroscopical techniques are combined as a pump-probe method with an angle-resolved photo emission. In this approach, a laser pulse is used to initiate excitation in a material and the second pulse is used to ionize the system, the kinetic energy is recorded, and the method is repeated many times by changing the time delays; this method builds up the picture and shows how the material relaxes. So we could study the relaxation dynamics using these techniques. There are many kinds of methods evolved to evaluate ultrafast process efficiently and precisely; some of them are multidimensional approach, by using the same principle established by two-dimensional (2D) nuclear magnetic resonance experiments, ultrafast imaging field which include electron diffraction imaging, kerr gated microscopy [27]. Photo dissociation and femtosecond probing–photo dissociation is a reaction of breaking a chemical compound by photons, termed as the contact of one or more photons with one target molecule. Photons with adequate energy can distress the bonds such as visible light, X-rays, gamma rays, and ultraviolet light. This field opens the possibility to study the bimolecular reaction at the individual collision level, which is complicated [35].

11.4 Graphene

Carbon got its designation from Latin expression carbo that denotes the charcoal component is so exceptional and that their electronic assembly permits for hybridization to form sp^3, sp^2, and sp networks, forming more stable allotropes. One of the common allotrope is graphite, which consists of sp^2 hybridized carbon atomic layers which are arranged together by weak van der Waals forces. Those monolayers of carbon atoms tightly stacked into 2D honeycomb crystal lattice are called graphene. Graphene is the thinnest material known and is also the strongest among them, and a superb conductor of both electricity and heat, used to create high-speed electronic devices such as sensors effective at detecting explosives, transistors that operate at higher frequency compared to others, as helping to store hydrogen in fuel cells, and also used in the production of low-cost display screens for mobiles. Widely investigational research on graphene concentrates on the electronic characteristics; an effective fact about the initial efforts on graphene transistors was the ability to continuously tune the charge carriers from holes to electrons. Graphene has high electron mobility at room temperature with stated values in excess of 15,000 $cm^2\ V^{-1}s^{-1}$; the hole and electron mobilities were predicted to be equal. Being a great conductor of electricity, graphene does not have a band gap, which is one of its disadvantages (means it cannot be switched off). This was the application of transition metal dichalgogenide risen up [36–38].

11.5 Two-Dimensional Semiconductors

General definition for the so-called 2D semiconductors basically is a type of natural semiconductor with the thickness on the atomic scale, and 2D materials display various electronic properties, such as stretching from insulating materials like hexagonal boron nitride and semiconducting transition metal dichalgogenide like molybdenum disulfide to semimetalic graphene. The fabrication of mechanically exfoliated graphene in 2004 ignited research on 2D materials that is growing at massive proportion. It is often claimed that the achievement of this research space is often sustained by the potential of these materials to have an explosive impact in the nanoelectronics and optoelectronics; till today, more than 10 different 2D semiconductors (with a band gap ranging from a few mill electron volts up to some electron volts) have been investigated and fabricated [39–41].

11.6 Direct and Indirect Band Gaps

The band gap represents the minimum energy difference between the top of the valence band and the bottom of the conduction band; however, the top of the valence band and the bottom of the conduction band are not generally at the same value of the electron momentum. In a direct band-gap semiconductor, the top of the valence band and the bottom of the conduction band occur at the same value of momentum. In semiconductors, band gaps are of two types—direct and indirect band gaps (Figures 11.1). Brillouin zone—if both K vectors are the same (momentum of electrons and momentum of holes are the same in both energy bands), then its direct band gap and if it is not its indirect band gap (Figures 11.1) [42–44]. MX_2 model (M=Mo,W; X=S, Se,) transition metal dichalcogenide (TMDC) multiple layers show that an indirect band gap in the samples is observed in the transition from direct to indirect band gap when the layer number exceeds 2 or 3. TMDC monolayers MoS_2, $MoSe_2$, WS_2, WSe_2, and $MoTe_2$ have direct band gap, used as transistors and detectors.

11.7 Transition Metal Dichalgogenide

Transition metal dichalgogenides are inorganic layered materials which exhibit a large variety of electronic nature such as semiconductivity, superconductivity, or charge density waves. The chalcogens are the name for the Periodic Table group 16 (or V1). The group consists of the elements: oxygen, sulfur, selenium, tellurium, and polonium. Semiconducting dichalgogenides like MoS_2 are promising materials for electronic application and their mechanical properties make them candidates for the fabrication of field-effect transistor (FET). Substituting the commonly used organic-based FETs which have a very low charge carrier mobility, indeed dichalgogenide FETs have very great charge carrier mobility equivalent to the one of Si FETs;

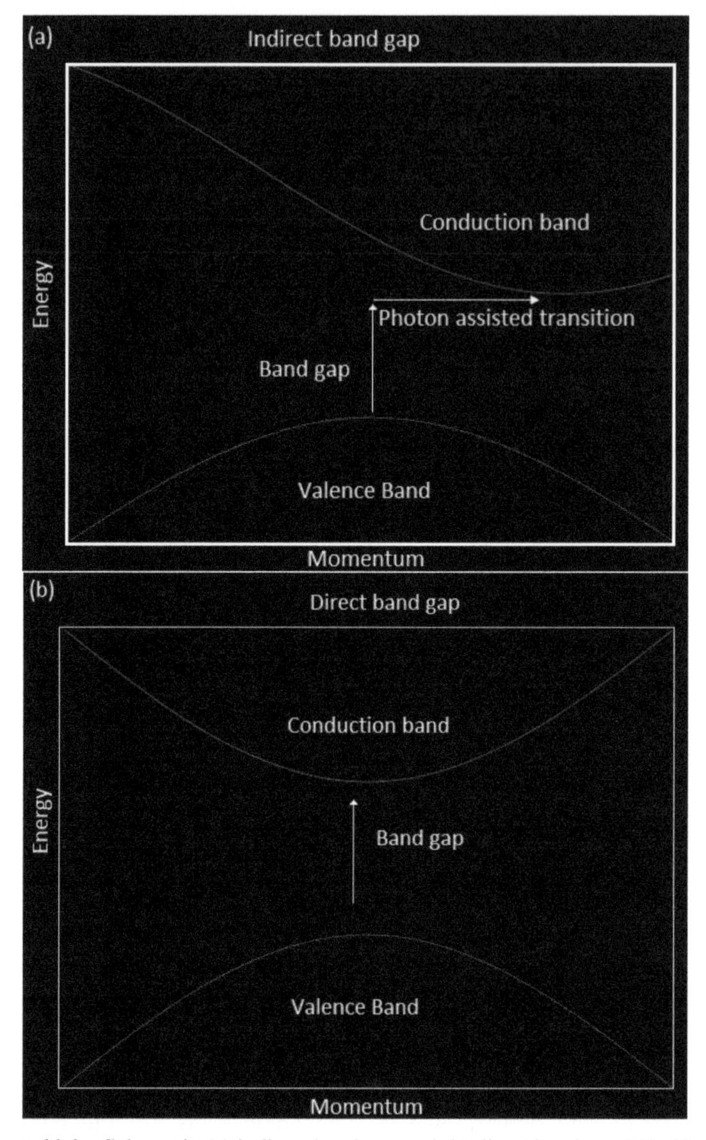

Figure 11.1 Schematic (a) indirect band gap and (b) direct band gap [42, 43, 45].

in addition, optical properties of these materials can be exploited and used for solar cell, photo electrochemical cell, and photo catalytic applications [48–51]. They are atomically thin (monolayer thickness) semiconductors using a general formula MX_2 (M as a transition metal and X as a chalcogen that is S, Se, and Te); this family of 2D materials is often studied after graphene due to their properties. These materials have intrinsic band gap with in the visible part of the spectrum range. The band gap intensely varies with the number of layers it had and a change in direct band gap to the indirect band is seen when the layer number changes from single layer to multilayers [52–54]. Other interesting criteria about these compounds are that having the large spin–orbit interaction in these compounds due to the heavy transition metals leads to splitting of valance band that strongly affects their spectra. In particular, molybdenum- and tungsten-based dichalgogenide exhibits optical band gap in the range of 1–2 eV which makes them appropriate for near infrared absorption. These materials have an indirect band gap in the bulk and little layer form; they become direct gap semiconductor with strong photoluminescence (PL) at the monolayer level [55–57]. The 2D TMDC materials are covalently bonded layers that are stacked or laminated together by the weak van der Waals forces which are ideal sources in 2D crystals [52, 58, 59]. There has been a rise of interest in the properties of these materials, specifically considering their 2D crystalline form, for nanoscale electronics and photonics applications [60–62]. For example, monolayer MoS_2 has been applied successfully in the fabrication of low-power FETs, logic circuits, and phototransistors. In their bulk states, MoX_2 and WX_2 (X = S, Se, and Te) LTMDs are indirect-gap semiconductors. It is well documented and found out that in the case of MoS_2, through both theory and experiment, that the material remains an indirect-gap semiconductor until samples are thinned down to a monolayer, at which point the gap becomes direct [42, 43, 49, 54, 63, 64].

11.8 Preparation of Transition Metal Dichalcogenides

11.8.1 Exfoliation Method for 2D Transition Metal Dichalcogenide

Graphene's successful exfoliation from bulk graphite paves the way for the fabrication of other graphene-like 2D materials through the simple even though much efficient "Scotch tape method" [65]. Due to high quality of monolayers occurring from mechanical exfoliation, this method is popularly used for intrinsic sheet production and fundamental research. Nevertheless, this method is not suitable for practical applications on a large scale due to its low yield and disadvantages in controlling sheet size and layer number [66]. In 2012, Karim Gacem's group proposed a general technique for fabricating

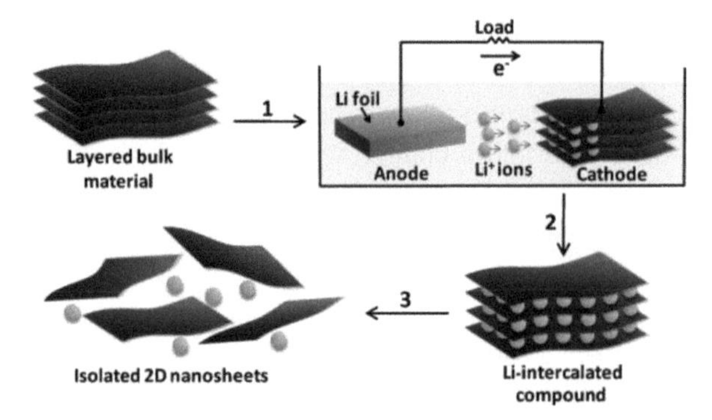

Figure 11.2 Electrochemical lithiation process for the fabrication of 2D nanosheets from the layered bulk material [77].

high-quality 2D layered materials, which was called anodic bonding. Sizes of few-layer MoS_2 obtained were relatively controllable and larger, which vary from 10 mm to several hundred microns [67, 68] Another class of exfoliation method is through chemical approach, including ion intercalation and solvent-based exfoliation [69–71]. In 2011, Hua Zhang's group investigated a faster and highly controllable method to exfoliate a series of semiconducting nanosheets. Through an electrochemical lithiation discharge process, bulk MoS_2 could be realized by lithium intercalation [72]. Then after subsequent ultrasonication, a high yield (92%) single-layer MoS_2 was achieved, as shown in Figure 11.2. Metallic 1T-MoS_2 accounted for a large proportion. Fabricated through the above way is called solvent-based exfoliation, also called Coleman method [73], which could obtain mostly semiconducting 2HeMoS_2 from exfoliating suspended bulk MoS_2 flakes in organic solvents. O'Neill et al. further optimized this method by carefully controlling and limiting the sonication time, resulting in higher reported flake concentration of about 40 mg/mL and relatively increasing flake size [74–76].

11.8.2 Chemical Vapor Deposition for Two-dimensional Transition Metal Dichalcogenide

Controllable synthesis of 2D TMDC with large area uniformity has still a big challenge. Chemical vapor deposition (CVD) approach has widely attracted attention because it could synthesis 2D TMDCs on a wafer scale which is very efficient and shows a great potential toward practical application like large-scale integrated electronics. Certain thickness of a thin film could be prepared using these and also avoids the contamination through the process. This part

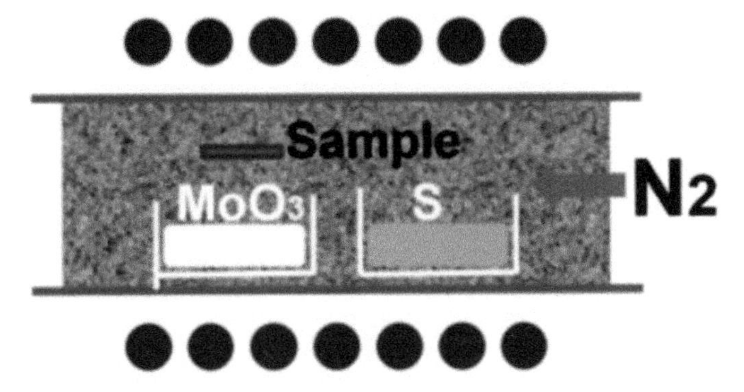

Figure 11.3 Schematic illustration for the experimental setup [82].

gives the systematical presentation of CVD synthesis of monolayer or few layers of MoS_2. Typically, the following precursors are used to prepare MoS_2 film including Mo-based compound powder deposited molybdenum-based film, ammonium thiomolybdates $[(NH_4)2MoS_4]$ film, and MoS_2 powder, as shown in Figure 11.3 [54, 78–81].

11.8.3 Characterization of Transition Metal Dichalcogenide

11.8.3.1 Optical properties

The photon energy which is equal or larger to the band gap of the TMDC material gets absorbed or light with shorter wavelength. In case of direct band gaps, they show to be great emitter if the minimum of the conduction band energy is at the same position of valance band energy, so comparing monolayers with bulk, up to two layers, it is so indirect that the efficiency in emission is lower compared to the single layer [83–85].

11.8.3.2 Raman spectra

Raman spectra is a non-destructive characterization technique which is generally used to measure specific heat, atomic bond, phase transition, and chemical composition. Figure 11.4 shows the spectra of MoS_2, WS_2, and MoS_2/WS_2 [86–89].

Considering the characterization of layered materials, the progress of changes in structural factors from the 3D bulk blocks to the 2D van der Waals bonded constructions could be illustrated by the Raman characterization method; for decades this has been widely used to study the quality and layer number of graphene. Correspondingly, Changgu Lee's group had been

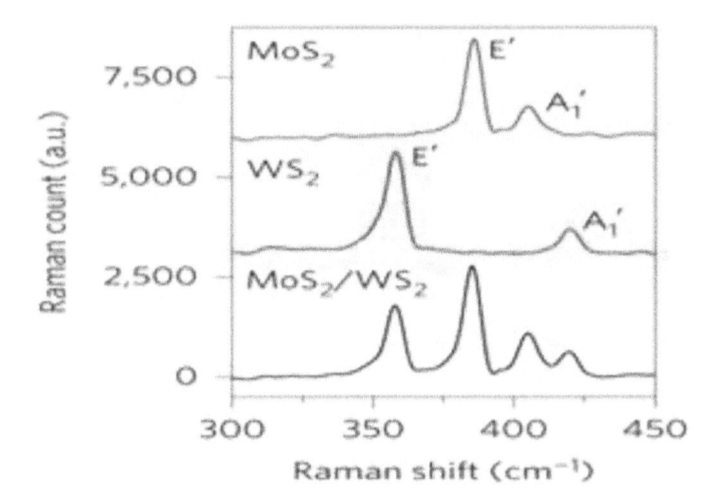

Figure 11.4 Raman spectra of an isolated MoS_2 monolayer (blue trace), an isolated WS_2 monolayer (red trace), and a MoS_2/WS_2 heterostructure (black trace) [90].

systematically characterized single-layer MoS_2 and few-layer MoS_2 in the early 2010. There are two typical Raman peaks observed $E^1{}_{2g}$ and $A^1{}_g$. These two peaks give information about the crystal structure of MoS_2. These two displays an in-plane and out-of-plane vibrations as present in the case of S atoms. There are three changes are strikes first when the transition from bulk to monolayer. First, $E^1{}_{2g}$ exhibits a regularly blue-shifted while $A^1{}_g$ shows an opposite red-shifted. $E^1{}_{2g}$ and $A^1{}_g$ locate at ~384 cm^{-1} and 405 cm^{-1} for single-layer MoS_2 [Figures 11.5(a,b)]. Second, the peak frequency difference between $E^1{}_{2g}$ and $A^1{}_g$ shows a clear decreasing trend as a function of layer number (Figure 11.5).

The frequency spacing is about 25 cm^{-1} and 19 cm^{-1} for bulk and monolayer MoS_2, respectively. From the data collected by making the layer thickness with the frequencies of $E^1{}_{2g}$ and $A^1{}_g$, we could find that two peak intensities almost increase linearly up to four layers with increasing layer thickness, while they decrease for thicker MoS_2. Yongjie Zhan's group has also reported the intensity ratio between $E^1{}_{2g}$ and silicon (Si) substrate is associated with layer thickness, ~0.05 and 0.09 for single- and double-layer samples.

It has been noted that the temperature-dependent Raman spectroscopy studies of 2D WS_2 were not explored till date. Expanded Raman spectrum indicates the difference between single-layer and bulk samples. Difference in

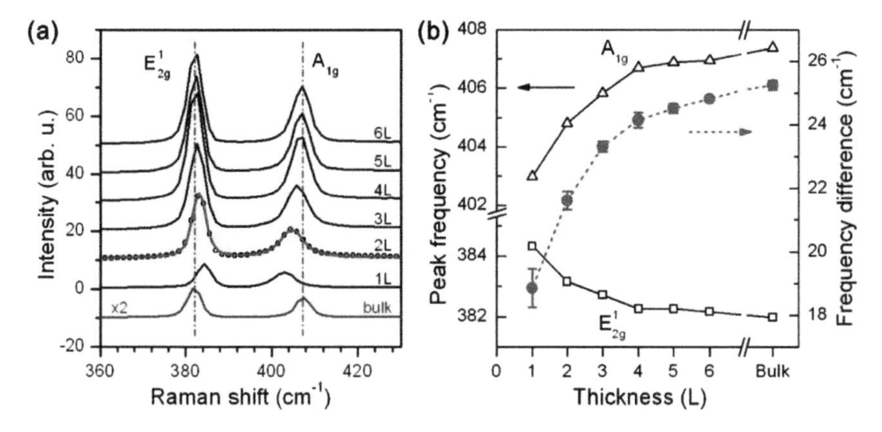

Figure 11.5 (a) Raman spectra of thin and bulk MoS_2 films. The solid line for the 2L spectrum is a double voigt fit through data. (b) Frequencies of $E^1{}_{2g}$ and $A^1{}_g$ Raman models [91].

peak position, width, and intensity can be clearly seen; the intensity of the Raman band escalates with a reduction in the number of layers.

11.8.3.3 Photoluminescence (PL) evaluation

PL spectra are found to be closely related to the number of layers in MoS_2 [92] shown in Figures 11.6(a,b). Andrea Splendiani' group has reported the distinct PL difference between monolayer (1L), bilayer (2L), quadrilayer (4L), and hexalayer (6L) samples. Two evident absorption peaks at 670 nm and 627 nm, identified as A1 and B1 excitons, can be observed in the spectrum [Figure 11.6(a)] for 1L MoS_2, while they both disappear in bulk MoS_2 [Figure 11.6(b)]. These two excitons are related with the energy split from valence band spin–orbital coupling. Prominent resonances in 1L sample indicate the direct excitonic transitions at the K point of the Brillouin zone, which is also consistent with the theoretical prediction of indirect (1.2 eV) band gap to direct (1.8 eV) band gap transition in changing from bulk to single-layer MoS_2. Thus, MoS_2 with its layer number varying from multilayer to monolayer will lead to qualitatively change in its band structure, further explaining the prominent PL effect in 1L sample [93].

Determine optical transition in multiple-layer MoS_2 samples using reflection spectroscopy and reflectivity difference between the silicon or quartz substrate and regions with ultrathin layers of MoS_2 measured. For ultrathin layers, the absorption coefficient is directly proportional to the difference in reflection. Two projecting absorption peaks can be identified at 670 nm and

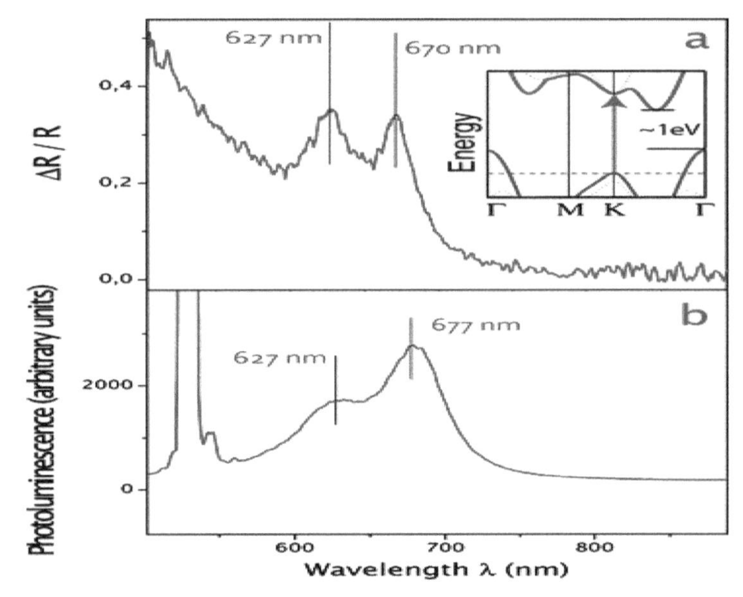

Figure 11.6 Reflection and photoluminescence (PL) spectra of ultrathin MoS_2 layers. (a) Reflection difference due to an ultrathin MoS_2 layer on a quartz substrate, which is proportional to the MoS_2 absorption constant. The observed absorption peaks at 1.85 eV (670 nm) and 1.98 eV (627 nm) correspond to the A1 and B1 direct excitonic transitions with the energy split from valence band spin–orbital coupling. The inset shows the bulk MoS_2 band structure neglecting the relatively weak spin–orbital coupling, which has an indirect bandgap around 1 eV and a single higher energy direct excitonic transition at the K point denoted by an arrow. (b) A strong PL is observed at the direct excitonic transitions energies in a monolayer MoS_2. Such luminescence is absent in the indirect bandgap bulk MoS_2 sample [93].

627 nm in the spectrum. Figure 11.6(b) displays the PL of monolayer MoS_2. This PL emission in monolayer is in striking contrast to its absence in bulk MoS_2, a consequence of bulk MoS_2 being an indirect band gap semiconductor like silicon substrate, as Figure 11.7 show differential luminescence for mono, bi, and hexa layers [93].

11.8.3.4 Electrical property

Comparing the bulk with monolayer samples, the bulk samples of TMDC have indirect band gap, while monolayers of TMDC have indirect band gap making them efficient for electronic applications. Many TMDCs have band structures that are comparable in their general features; in general, MoX_2 and WX_2 compounds are semiconducting whereas NbX_2 and TaX_2 are metallic [94–97]. Band structure is as revealed in Figure 11.8. We can observe

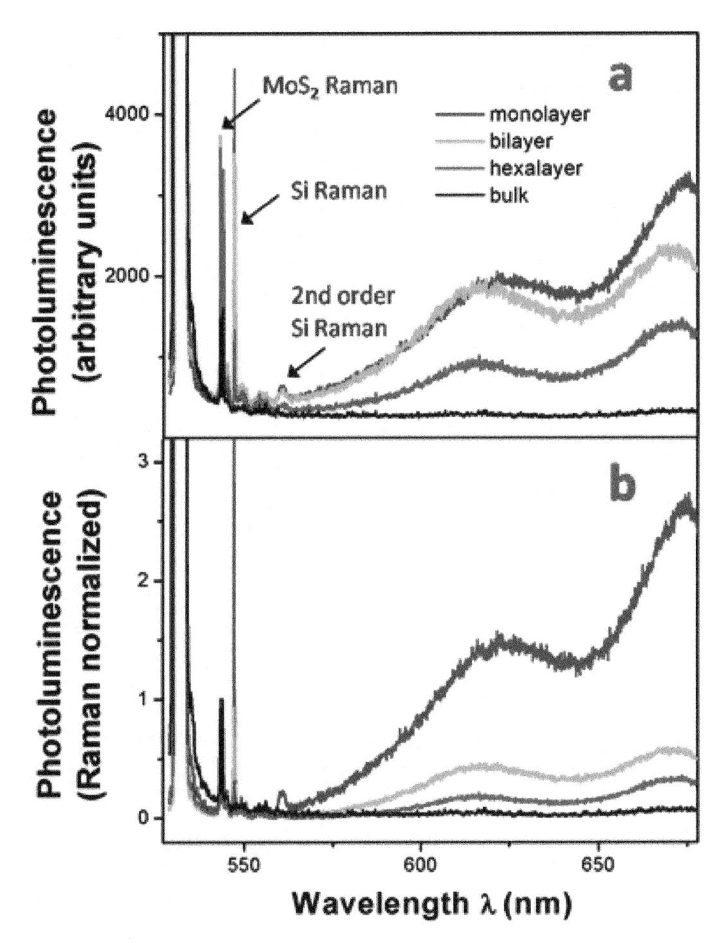

Figure 11.7 Layer dependence of photoluminescence (PL) efficiency in MoS_2. (a) PL and Raman spectra of MoS_2 monolayer, bilayer, hexalayer, and bulk sample. Different Raman peaks can be assigned to the MoS_2 and silicon vibration modes. For MoS_2 thin layers, monolayer MoS_2 Raman signal is relatively weak because less material is being excited. However, PL is the strongest in monolayer MoS_2 in spite of reduced material. (b) PL spectra normalized by Raman intensity for MoS_2 layers with different thickness, showing a dramatic increase of luminescence efficiency in MoS_2 monolayer [93].

a change in the Γ point where the band gap becomes indirect to direct as the material changes from bulk to monolayer [98]. The direct excitonic transitions at the K point remain relatively unaffected with the layer number [99, 100]. The quantum confinement and the difference occurred at hybridization between the P_z orbital on sulfur atoms and d orbital on molybdenum

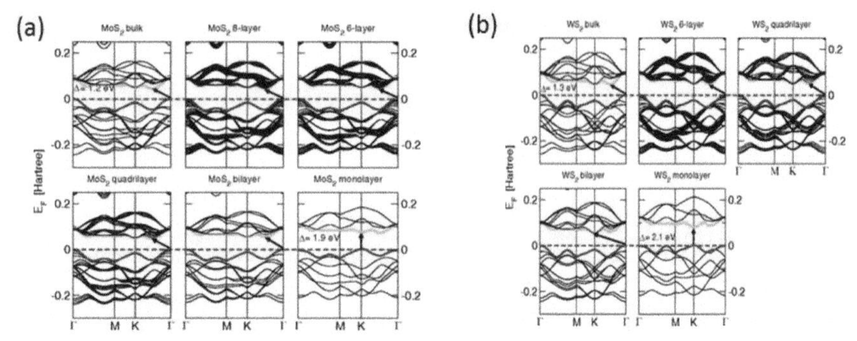

Figure 11.8 (a) Band structures of bulk MoS_2, its monolayer, as well as polylayers calculated at the density functional theory (DFT)/PBE level. The horizontal dashed lines indicate the Fermi level. The arrows indicate the fundamental band gap (direct or indirect) for a given system. The top of valence band (blue/dark gray) and bottom of conduction band (green/light gray) are highlighted. (b) Band structures of bulk WS_2, its monolayer, as well as polylayers calculated at the DFT/PBE level. The horizontal dashed lines indicate the Fermi level. The arrows indicate the fundamental band gap (direct or indirect) for a given system. The top of valence band (blue/dark gray) and the bottom of conduction band (green/light gray) are highlighted [98].

atoms cause a change in the band structure with layer number [101, 102]. For MoS_2, density functional theory (DFT) calculations efficiently evaluated that the localized d orbitals in the molybdenum atom cause the conduction-band states on the K-point which located in the central of the S–Mo–S layer sandwiches and relatively of the antibonding P_Z orbitals on the sulfur atoms and the d-orbitals on molybdenum atoms, and have a higher interlayer coupling effect [103, 104]. Therefore, when a change in layer number occurs, the direct excitonic effect near to the K point has no change, but on the other hand, changing from indirect band gap to direct band gap produces a visible change in the transition at the Γ-point. All MoX_2 and WX_2 samples are expected to go through a similar indirect to direct band gap transformation with decreasing layer numbers, covering the band gap energy range 1.1–1.9 eV. The band gaps in most semiconducting TMDCs are comparable to the 1.1-eV band gap of silicone as making them favorable digital transistors' fabrication. WS_2 is another competitor among the family of semiconducting TMDCs; according to optical calculations and measurements, the monolayer form displays a direct band gap of at least 2 eV; also the theoretical model of WS_2 suggests that due to reduced effective mass, it has the highest electron mobility, while valance band spitting of MoS_2 is 150 meV; because of the larger mass of W atoms, it is almost 3 times larger (426 meV).

11.8.3.5 Electrical transport property

One of the important applications of semiconductor is for fabricating transistors. Currently, electronic devices have silicon-based semiconductor FETs with a featured length of 22 nm. But the subsequent reduction in scale will soon approach restrictions due to quantum and statistical influence and struggle with heat dissipation [105–107]. Two-dimensional semiconductors such as MoS_2 and others TMDC propose an important advantage when equated to traditional 3D electric material. They have subnanometer width coupled with band gap naturally in the 1–2 eV regime which can result in a greater on/off ratio, their extreme thinness lets more effective mechanism above switching, and can assist to diminish short-channel effects and power dissipation, which are the main limiting factors to transistors' miniaturization. In 2D TMDC layers, the mobility of the carriers is affected by the following main scattering mechanism—acoustic and optical phonon scattering, Coulomb scattering at charged impurities, surface interface phonon scattering, and Roughness scattering. The degree which this scattering mechanism affects the carrier mobility is also affected by a layer thickness, temperature, carrier density, effective carrier mass electronic band structure, and photonic band structure [42, 43, 49, 54, 64, 108].

11.8.3.6 Electrical performance

Single-layer MoS_2 have a large direct band gap of 1.8 eV, being suitable acting as switching nanodevices. In 2011, Radisavljevic's group fabricated a single-layer MoS_2 transistor adopting a hafnium oxide as the gate dielectric material, in which the mobility of MoS_2 could be up to 200 $cm^2/(V\ s)$ at room temperature with a current on/off ratio of 1×10^8 [43, 49, 50, 109]. Generally, MoS_2-based transistors show the n-type behavior. Later in 2012, Zongyou Yin's group produced a phototransistor created on the mechanically exfoliated monolayer MoS_2 for the first time [49, 110, 111]. The switching character of this device is highly efficient, with photocurrent generation and annihilation within only 50 ms [112–115]. Under incident light, the photoresponsivity could reach 7.5 mA/W at a gate voltage of $\sim$50 V, much greater than that in graphene-corporated devices ($\sim$1 mA/W at a gate voltage of 60 V) [116–118]. This advantage gives MoS_2 for future applications in many fields, including transistors, photodetectors, and memory devices, broadening 2D graphene and graphene-like flexible materials into areas like transparent conducting and semiconducting areas.

11.8.4 Different Types of TMDC Materials

TMDCs, whose generalized formula is MX2 [M—transition metals (Ti, Zr, Hf, Nb, Ta, Mo, W, Tc, Co, Rh, Ir, Ni, Pd, Pt) and X—chalcogen (S, Se, Te)], have a large family of materials and their electronic characters could be semiconducting, metallic, and superconducting [119–121]. In the single layer of MoS_2 films, Mo(+4) and S (–2) are arranged to a sandwich structure by covalent bonds in a sequence of S–Mo–S [42, 122], whereas the sandwich layers are interacted by relatively weak van der Waals forces. Generally, the individual layer has a thickness of ~0.65 nm. Monolayer MoS_2 with triagonal prismatic polytype is found to be semiconducting (referred to as 2H), while that with octahedral crystal symmetry configuration (referred to as 1T) is metallic [54, 123]. Very similarly to graphene, MoS_2 is mechanically flexible with a Young's modulus of 0.33 ± 0.07 TPa [124–126].

11.8.4.1 Ultrafast process in MoS_2

Recently, genuine 2D semiconductors like MoS_2 have been isolated; in this kind of material, low dimensionality contributes to the presence of strongly bound excitons with remarkable properties [49, 54, 127]. Generally, PL sand differential absorption studies are able to help in understanding the optical properties and excited carrier dynamics in the monolayer and few layers of MoS_2. The electron–hole pair dynamics has been studied through the measured exciton lines, which are the most prominent advantage on such experiments [128–130]. Time-resolved studies have been enlightening valuable insight into exciton life times, including relaxation dynamics and defects. Using time-resolved approach, we could unravel the carrier dynamics near Brillouin zone has only recently become possible using a high photon energy laser via high harmonic generation [131–133].

Conduction band extreme and valance band extreme of monolayer MoS_2 are shown in Figure 11.9. The sample is first excited by an ultrashort pump pulse, optical pump achieved, and the system is then probed, while in the excited state, by photoemission using a time-delayed 25-eV probe pulse, both pump and pulses are in 30-fs duration. Electronic features of the MoS_2 sample are characterized by sharp dispersing VB features of monolayer MoS_2 [42, 49, 134]. The non-equilibrium carrier distribution is displayed in Figures 11.9(f–h), which show the intensity difference between the spectrum immediately after the optical excitation and spectrum taken before the arrival of the pump pulse. Red and blue indicates the increase and decrease in intensity. When system is pumped with 1.6 eV, it indicates an

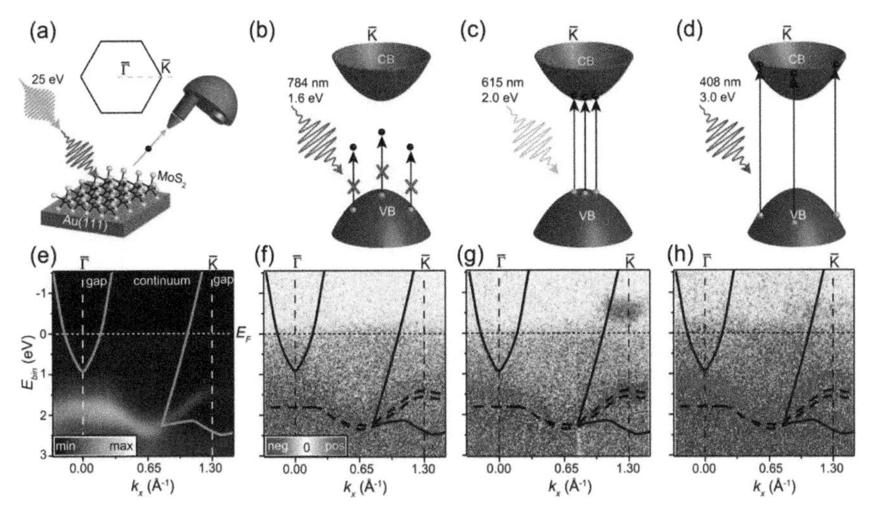

Figure 11.9 Tuning pump–probe ARPES experiments near the direct band gap of SL MoS_2. (a) Schematic of our photoemission experiment. Electrons (black spheres) excited by a pump pulse tuned across the visible to infrared range are photoemitted from epitaxial SL MoS_2 on Au(11) using a 25-eV probe pulse and detected along the high symmetry direction $\overline{\Gamma} - \overline{K}$ in the hexagonal BZ. (b–d) Diagrams of electron–hole pair excitations around $\overline{K}$ for the studied regimes of optical excitation: (b) below the CB onset, (c) at the band gap energy, and (d) above the CB onset. Holes are sketched as orange spheres. The pump pulse energies and corresponding wavelengths used in the experiments are stated. (e) Occupied band structure at equilibrium measured by synchrotron radiation. (f–h) Intensity difference obtained by subtracting the spectra at the peak of the optical excitation from a spectrum obtained before the arrival of the pump pulse. The orange lines in (e) and full black lines in (f–h) outline the boundaries between the projected bulk band continuum and gaps of Au(111). The dashed black curves in (f–h) are the fitted dispersion from (E). The optical excitation energies in (f–h) correspond to those given directly above in (b–d) [138].

energy well below the band gap [Figure 11.9(f)]. When the excitation energy increased to 2 eV, we observe a clear population of the conduction band at K [Figure 11.9(g)], indicating the pump exceeds the band gap and excited free carriers into the CB. The population strongly decreases at 3eV [135–137].

11.8.4.2 Ultrafast process in WSe_2

Another TMDC semiconductor with hexagonal crystalline structure is Wolfram diselinide (WSe_2). WSe_2 is flexible and has a low surface trap density. This draws attention toward making flexible electronics. The carrier mobilities similar to silicon are achieved by the WSe_2 FETs at room temperature [139–141]. Furthermore, current retrification has been seen in heterojunction diodes of InAs/WSe [142, 143]. The discovery of graphene is made a revolution in this field of electronics; there have been a tremendous efforts taken to fabricate 2D crystals other than graphene with the help of transition metal dichalgogenides [144, 145]. WSe_2 had been studied carefully throughout the decades even they mechanically exfoliated two layers of WSe_2 ultrathin in 1970 and recorded the influence of thickness of the layers on exciton transistors. More recently, WSe_2 2D crystals have been produced by modulated elemental reaction, solution phase synthesis, rapid selenization process, and mechanical exfoliation [146–148]. Photoluminescence measurements have been recorded an indirect to direct band gap transition in monolayer. The same result has reported as in case of the other TMDCs such as MoS_2 and $MoSe_2$ to measure the thickness; in order to study the lattice dynamics of the 2D WSe_2, Raman spectroscopy is employed; p-type and n-type field FETs with a hole mobility of 250 $cm^2/(Vs)$, comparable electron motilities, and on/off ratio about 10^6 at room temperature are fabricated using single-layer WSe_2 at room temperature. These specialities of single-layer WSe_2 on comparing with the properties of the single layers of MoS_2 found to be comparable or in some cases even more better, and it is also one of the four stable non-carbon compounds in the above conditions, and found to be efficient for the electronics applications. By using the transient absorption microscopy approach, we could learn about the exciton dynamics of both monolayer and bulk WSe_2 which is a time-resolved study that enlightens the data corresponding to the life time and the diffusion coefficient of the excitons in both the monolayer and the bulk WSe_2, and from this information, we could analyze the other parameters like diffusion length, mean free time, mean free path, and exciton mobility from which we could understand the dynamics even more precisely. The differential reflection of the probe pulse is used to analyze these results; for that, first, we need to use a pump pulse to inject excitons which is in 100 fs range. These experiments are carried out in a favorable condition and there will be no changing in sample during the study. This has to be done by measuring the differential reflection as a function of the probe wavelength and the probe delay [42, 54, 140].

11.8.5 Application

Synthesizing ultrathin layers of graphene by mechanical exfoliation has become an inspiration for the other 2D materials with diversity in electrical characteristic as well as the semiconducting MoS_2 and insulating hexagonal boron nitride. Especially, the semiconducting 2D materials such as MoS_2

WS_2 are potentially effective in electric devices due to their extreme thinness. Some examples are as follows.

11.8.5.1 Digital electronic devices

Digital electronics circuits are made up of a large number of logic gates and every logic gate is a combination of transistors. This is where transistor field of electronics brightens up, which is in fact lays on building the storage and reading of distinct voltage states. So the transistors become the essential constituent of current digital electronics. In today's semiconductors, the most commonly used transistors are metal–oxide–semiconductor FETs. The basic requirements for the transistors are high on/off current ratio which is easier to achieve for TMDC materials which have band gap more than 1 eV.

11.8.5.2 TMDC transistors

One-dimensional semiconducting transition metal dichalgogenides have sole features like lack of the dangling bond make them to use in the fabrication of FET as channel medium and structural stability and mobility comparable to silicon. WSe_2 is used as the earliest TMDC in FETs in 2004, which revealed mobility similar to silicon FET (up to 500 $cm^2\ v^{-1}s^{-1}$ for p-type conductivity at room temperature) ambipolar behavior and a 10^4 on/off ratio at a temperature of 60 K, the same properties achieved by the MoS_2 with a back-gated configuration resulting in a range of mobility between 0.1 and 10 $cm^2\ v^{-1}s^{-1}$. The first fabricated top-gated transistors based on monolayers of MoS_2 were reported by Kis et al.; as shown above, this device is efficient in case of on/off ratio (10^8), n-type conduction, and a room temperature mobility of $>$200 $cm^2\ v^{-1}s^{-1}$. Top-gate geometry allows reduction in the voltage necessary to switch the device while allowing the integration of multiple devices on the same substrate. Figure 11.10 shows the image of the transistor.

11.8.5.3 Optoelectronics

These devices are used for the purpose to interact, detect, control, or generate light, which are used in applications like lasers, solar cells, optical switches, photodetectors, and displays. Optoelectronics devices are flexible and transparent which are more important in solar arrays, wearable electronics, and transparent displays. The electronic band structure of semiconductors directly influences their ability to absorb and emit light. For direct-band semiconductors, photons with energy greater than the band gap energy can readily absorb or emit light. For indirect semiconductors, an additional

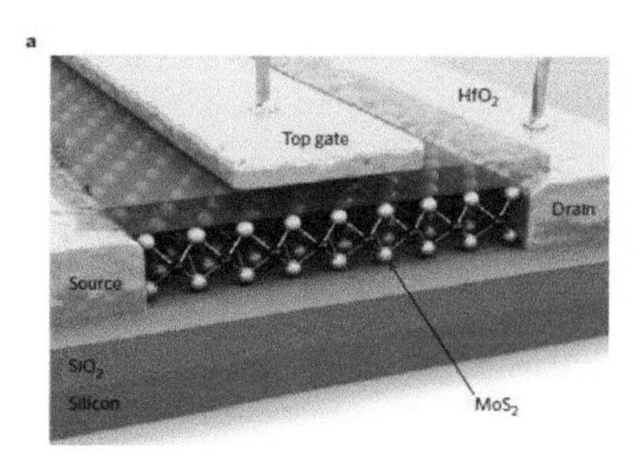

Figure 11.10 Schematic diagram for the transistor.

phonon must be absorbed or emitted to supply the difference in momentum, making the photon absorption or emission process much less efficient because the monolayer MoS_2 primarily has a direct semiconducting band gap which comes in handy for the application in optoelectronics, due to the fact that they are atomically thin and easier to process.

11.9 Conclusion

Using the ultrafast approach characterization, we could easily demonstrate that the unique variation in the optical property 2D TMDC crystal so hence to isolate for different purpose such as in the building block for heterostructures as optoelectronics and photocatalytic, etc. functions. In this review, we had discussed about the features of the four TMDC materials. We have evaluated the change in reflection with the time delay in the above sample to understand the dynamics of the materials MoS_2, $MoSe_2$, WS_2, and WSe_2.

References

[1] Mishra, R. K., Cherusseri, J., Allahyari, E., Kalarikkal, N., and Thomas, S. (2017). "Small-angle light and X-ray scattering in nanosciences and nanotechnology," in *Thermal and Rheological Measurement Techniques for Nanomaterials Characterization,* eds S. Thomas, R. Thomas, ans A. Zachariah (Amsterdam: Elsevier Science Publishing Co Inc), 233–269. doi: 10.1016/B978-0-323-46139-9.00010-4

[2] Mishra, R. K., Zachariah, A. K., and Thomas, S. (2017). "Energy-dispersive X-ray spectroscopy techniques for nanomaterial," in

Microscopy Methods in Nanomaterials Characterization, eds S. Thomas, R. Thomas, A. Zachariah, and R. Mishra (Amsterdam: Elsevier Science Publishing Co Inc), 383–405. doi: 10.1016/B978-0-323-46141-2.00012-2

[3] Thomas, S., Thomas, R., Zachariah, A., and Mishra, R. K. (2017). *Microscopy Methods in Nanomaterials Characterization.* Amsterdam: Elsevier Science Publishing Co Inc.

[4] Mishra, R. K., Cherusseri, J., Bishnoi, A., and Thomas, S. (2017). "Nuclear magnetic resonance spectroscopy," in *Spectroscopic Methods for Nanomaterials Characterizations*, eds S. Thomas, R. Thomas, A. Zachariah, and R. K. Mishra (Amsterdam: Elsevier Science Publishing Co Inc), 369–415. doi: 10.1016/B978-0-323-46140-5.00013-3

[5] Chirayil, C. J., Abraham, J., Mishra, R. K., George, S. C., and Thomas, S. (2017). "Chapter 1 – instrumental techniques for the characterization of nanoparticles," in *Thermal and Rheological Measurement Techniques for Nanomaterials Characterization,* eds S. Thomas, R. Thomas, A. Zachariah, and R. K. Mishra (Amsterdam: Elsevier Science Publishing Co Inc), 1–36.

[6] Loganathan, S., Valapa, R. B., Mishra, R. K., Pugazhenthi, G., and Thomas, S. (2017). in *Thermal and Rheological Measurement Techniques for Nanomaterials Characterization,* eds S. Thomas, R. Thomas, A. Zachariah, and R. K. Mishra (Amsterdam: Elsevier Science Publishing Co Inc), 67–108. doi: 10.1016/B978-0-323-46139-9.00004-9

[7] Thomas, S., Thomas, R., Zachariah, A. K., and Mishra, R. K. (2017). *Thermal and Rheological Measurement Techniques for Nanomaterials Characterization.* Amsterdam: Elsevier Inc.

[8] Park, J. S., Maeng, W. J., Kim, H. S., and Park, J. S. (2012). Review of recent developments in amorphous oxide semiconductor thin-film transistor devices. *Thin Solid Films* 520, 1679–1693.

[9] Forman, G. (2007). Feature selection for text classification. *Comput. Methods Featur. Sel.* 16, 257–274.

[10] Ohno, H. (1998). Making Nonmagnetic Semiconductors Ferromagnetic. *Science* 281, 951–956.

[11] Joo, S., et al. (2013). Magnetic-field-controlled reconfigurable semiconductor logic. *Nature* 494, 72–76.

[12] Farhan (2011). "Semiconductor heterostructures," in *Theory of Modern Electronic Semiconductor Devices,* eds K. F. Brennan and A. S. Brown (Ithaca, NY: Cornell Univ), 1–29. doi:10.1002/0471224618.ch2

[13] Diao, Y., Shaw, L., and Mannsfeld, S. C. B. (2014). Morphology control strategies for solution-processed organic semiconductor thin fi lms. *Energy Environ. Sci.* 7, 2145–2159.

[14] Nida-Rümelin, M. (2007). *Phenomenal Concepts and Phenomenal Knowledge: New Essays on Consciousness and Physicalism.* Oxford University Press: Oxford, doi:10.1093/acprof:oso/9780195171655.003.0013

[15] Jones, K. M., Tiesinga, E., Lett, P. D., and Julienne, P. S. (2006). Ultracold photoassociation spectroscopy: long-range molecules and atomic scattering. *Rev. Mod. Phys.* 78, 483–535.

[16] Bader, R. F. W. and Bayles, D. (2000). Properties of Atoms in Molecules: group additivity. *J. Phys. Chem. A* 104, 5579–5589.

[17] Kawai, M., and Kim, Y. (2009). Single-molecule chemistry. *AIP Conf. Proc.*1119, 49–51.

[18] Pabst, S. (2013). Atomic and molecular dynamics triggered by ultra-short light pulses on the atto- to picosecond time scale. *Eur. Phys. J. Spec. Top.* 221, 1–71.

[19] Morales, F., Pérez-Torres, J. F., Sanz-Vicario, J. L., and Martín, F. (2009). Probing H2 quantum autoionization dynamics with xuv atto and femtosecond laser pulses. *Chem. Phys.* 366, 58–63.

[20] Agostini, P., and DiMauro, L. F. (2004). The physics of attosecond light pulses. *Rep. Prog. Phys.* 67:1563.

[21] Helml, W., et al. (2014). Measuring the temporal structure of few-femtosecond free-electron laser X-ray pulses directly in the time domain. *Nat. Photonics* 8, 950–957.

[22] Oksenhendler, T., and Forget, N. (2010). *Advances in Solid State Lasers Development and Applications.* Rijeka: InTech, 347–386.

[23] Villeneuve, D. (2009). Attosecond light sources. *Phys. Canada* 65, 63–66.

[24] Kalachnikov, M. P., Karpov, V., Schönnagel, H., and Sandner, W. (2002). 100-Terawatt Titanium – Sapphire Laser System. *Laser Phys.* 12, 368–374.

[25] Ghimire, S., et al. (2011). Observation of high-order harmonic generation in a bulk crystal. *Nat. Phys.*7, 138–141.

[26] Corkum, P. B. (1993). Plasma perspective on strong-field multiphoton ionization. *Phys. Rev. Lett.* 71, 1994–1997.

[27] Shafir, D., et al. (2012). Resolving the time when an electron exits a tunnelling barrier. *Nature* 485, 343–346.

[28] Cerullo, G., and De Silvestri, S. (2003). Ultrafast optical parametric amplifiers. *Rev. Sci. Instrum.* 74, 1–18.

[29] Hansryd, J., Andrekson, P. A., Westlund, M., Li, J., and Hedekvist, P. O. (2002). Fiber-based optical parametric amplifiers and their applications. *IEEE J. Sel. Top. Quantum Electron.*8, 506–520.

[30] Werner, C. S., Beckmann, T., Buse, K., and Breunig, I. (2012). Blue-pumped whispering gallery optical parametric oscillator. *Opt. Lett.* 37, 4224–4226.

[31] Foster, M. A., et al. (2006). Broad-band optical parametric gain on a silicon photonic chip. *Nature* 441, 960–963.

[32] Link, S., Burda, C., Nikoobakht, B., and El-Sayed, M. A. (2000). Laser-induced shape changes of colloidal gold nanorods using femtosecond and nanosecond laser pulses. *J. Phys. Chem. B* 104, 6152–6163.

[33] Juban, E. A., and McCusker, J. K. (2005). Ultrafast dynamics of 2E state formation in Cr(acac)3. *J. Am. Chem. Soc.*127, 6857–6865.

[34] Zhang, X., Xia, Y., Friend, R. H., and Silva, C. (2006). Sequential absorption processes in two-photon-excitation transient absorption spectroscopy in a semiconductor polymer. *Phys. Rev. B Condens. Matter Mater. Phys.* 73:245201.

[35] Pound, R. V., and Rebka, G. A. (1960). Apparent weight of photons. *Phys. Rev. Lett.* 4, 337–341.

[36] Georgakilas, V., et al. (2012). Functionalization of graphene: covalent and non-covalent approaches, derivatives and applications. *Chem. Rev.* 112, 6156–6214.

[37] Compton, O. C., and Nguyen, S. T. (2010). Graphene oxide, highly reduced graphene oxide, and graphene: versatile building blocks for carbon-based materials. *Small* 6, 711–723.

[38] Lui, C. H., Liu, L., Mak, K. F., Flynn, G. W., and Heinz, T. F. (2009). Ultraflat graphene. *Nature* 462, 339–341.

[39] Shavanova, K., et al. (2016). Application of 2D non-graphene materials and 2D oxide nanostructures for biosensing technology. *Sensors* 16:223.

[40] Haigh, S. J., et al. (2012). Cross-sectional imaging of individual layers and buried interfaces of graphene-based heterostructures and superlattices. *Nat. Mater.* 11, 764–767.

[41] Mueller, T., Xia, F., and Avouris, P. (2010). Graphene photodetectors for high-speed optical communications. *Nat. Photonics* 4, 297–301.

[42] Mak, K. F., Lee, C., Hone, J., Shan, J., and Heinz, T. F. (2010). Atomically thin MoS_2: a new direct-gap semiconductor. *Phys. Rev. Lett.* 105:136805.

[43] Mak, K., Lee, C., Hone, J., Shan, J., and Heinz, T. (2010). Atomically thin MoS_2: a new direct-gap semiconductor. *Phys. Rev. Lett.* 105:136805.

[44] Park, J. S., et al. (2015). Ordering-induced direct-to-indirect band gap transition in multication semiconductor compounds. *Phys. Rev. B* 91, 1–5.

[45] Yablonovitch, E. (1993). Photonic band-gap structures. *J. Opt. Soc. Am. B* 10:283.

[46] Kim, E., Jiang, Z.-T., and No, K. (2000). Measurement and calculation of optical band gap of chromium aluminum oxide films. *Jpn. J. Appl. Phys.* 39, 4820–4825.

[47] Akamatsu, B., Hénoc, J., and Hénoc, P. (1981). Electron beam-induced current in direct band-gap semiconductors. *J. Appl. Phys.* 52, 7245–7250.

[48] Fujimoto, T., and Awaga, K. (2013). Electric-double-layer field-effect transistors with ionic liquids. *Phys. Chem. Chem. Phys.* 15:8983.

[49] Mak, K. F., et al. (2012). Tightly bound trions in monolayer MoS_2. *Nat. Mater.* 12, 207–211.

[50] Lembke, D., Bertolazzi, S., and Kis, A. (2015). Single-layer MoS_2 electronics. *Acc. Chem. Res.* 48, 100–110.

[51] Tiwari, S., and Greenham, N. C. (2009). Charge mobility measurement techniques in organic semiconductors. *Opt. Quantum Electron.* 41, 69–89.

[52] Duan, X., Wang, C., Pan, A., Yu, R., and Duan, X. (2015). Two-dimensional transition metal dichalcogenides as atomically thin semiconductors: opportunities and challenges. *Chem. Soc. Rev.* 44, 8859–8876.

[53] Zhao, W., et al. (2013). Evolution of electronic structure in atomically thin sheets of WS_2 and WSe_2. *ACS Nano* 7, 791–797.

[54] Radisavljevic, B., Radenovic, A., Brivio, J., Giacometti, V., and Kis, A. (2011). Single-layer MoS_2 transistors. *Nat. Nanotechnol.* 6, 147–150.

[55] Segets, D., et al. (2012). Determination of the quantum dot band gap dependence on particle size from optical absorbance and transmission electron microscopy measurements. *ACS Nano* 6, 9021–9032.

[56] Kim, K., Lee, J. U., Nam, D., and Cheong, H. (2016). Davydov splitting and excitonic resonance effects in Raman spectra of few-layer $MoSe_2$. *ACS Nano* 10, 8113–8120.

[57] Coehoorn, R., Haas, C., and De Groot, R. A. (1987). Electronic structure of MoSe2, MoS_2, and WSe_2. II. The nature of the optical band gaps. *Phys. Rev. B* 35, 6203–6206.

[58] Bondi, A. (1964). van der waals volumes and radii. *J. Phys. Chem.* 68, 441–451.

[59] Grimme, S. (2004). Accurate description of van der Waals complexes by density functional theory including empirical corrections. *J. Comput. Chem.* 25, 1463–1473.

[60] Kaul, A. B. (2013). Nano-electro-mechanical-systems (NEMS) and energy-efficient electronics and the emergence of two-dimensional layered materials beyond graphene. *Micro Nanotechnol. Sensors Syst. Appl. V* 8725, 1–6.

[61] Kaul, A. B. (2014). Two-dimensional layered materials: structure, properties, and prospects for device applications. *J. Mater. Res.* 29, 348–361.

[62] Mieszawska, A. J., Jalilian, R., Sumanasekera, G. U., and Zamborini, F. P. (2007). The synthesis and fabrication of one-dimensional nanoscale heterojunctions. *Small* 3, 722–756.

[63] Chakraborty, B., Matte, H. S. S. R., Sood, A. K., and Rao, C. N. R. (2013). Layer-dependent resonant Raman scattering of a few layer MoS_2. *J. Raman Spectrosc.* 44, 92–96.

[64] Terrones, H., López-Urías, F., and Terrones, M. (2013). Novel hetero-layered materials with tunable direct band gaps by sandwiching different metal disulfides and diselenides. *Sci. Rep.* 3:1549.

[65] Kelly, K. F., and Billups, W. E. (2013). Synthesis of soluble graphite and graphene. *Acc. Chem. Res.* 46, 4–13.

[66] Yi, M., and Shen, Z. (2015). A review on mechanical exfoliation for the scalable production of graphene. *J. Mater. Chem. A* 3, 11700–11715.

[67] Lee, C., et al. (2010). Anomalous lattice vibrations of single- and few-layer MoS_2. *ACS Nano* 4, 2695–2700.

[68] Yu, Y., et al. (2013). Controlled scalable synthesis of uniform, high-quality monolayer and few-layer MoS_2 films. *Sci. Rep.* 3:1866.

[69] Geng, J., Kong, B.-S., Yang, S. B., and Jung, H.-T. (2010). Preparation of graphene relying on porphyrin exfoliation of graphite. *Chem. Commun.* 46:5091.

[70] Hernandez, Y., et al. (2008). High-yield production of graphene by liquid-phase exfoliation of graphite. *Nat. Nanotechnol.* 3, 563–568.
[71] Nicolosi, V., Chhowalla, M., Kanatzidis, M. G., Strano, M. S., and Coleman, J. N. (2013). Liquid exfoliation of layered materials. *Science* 340, 1226419–1226419.
[72] Wang, L., Xu, Z., Wang, W., and Bai, X. (2014). Atomic mechanism of dynamic electrochemical lithiation processes of MoS_2 nanosheets. *J. Am. Chem. Soc.* 136, 6693–6697.
[73] Khan, U., O'Neill, A., Lotya, M., De, S., and Coleman, J. N. (2010). High-concentration solvent exfoliation of graphene. *Small* 6, 864–871.
[74] O'Neill, A., Khan, U., and Coleman, J. N. (2012). Preparation of high concentration dispersions of exfoliated MoS_2 with increased flake size. *Chem. Mater.* 24, 2414–2421.
[75] Show, K. Y., Mao, T., and Lee, D. J. (2007). Optimisation of sludge disruption by sonication. *Water Res.* 41, 4741–4747.
[76] Behabtu, N., et al. (2010). Spontaneous high-concentration dispersions and liquid crystals of graphene. *Nat. Nanotechnol.* 5, 406–411.
[77] Zeng, Z., et al. (2012). An effective method for the fabrication of few-layer-thick inorganic nanosheets. *Angew. Chem. Int. Ed.* 51, 9052–9056.
[78] Liu, K. K., et al. (2012). Growth of large-area and highly crystalline MoS_2 thin layers on insulating substrates. *Nano Lett.* 12, 1538–1544.
[79] Li, M.-Y., et al. (2015). Epitaxial growth of a monolayer WSe_2-MoS_2 lateral p-n junction with an atomically sharp interface. *Science* 349, 524–528.
[80] Tarasov, A. et al. (2014). Highly uniform trilayer molybdenum disulfide for wafer-scale device fabrication. *Adv. Funct. Mater.* 24, 6389–6400.
[81] Ji, Q., et al. (2013). Epitaxial monolayer MoS_2 on mica with novel photoluminescence. *Nano Lett.* 13, 3870–3877.
[82] Lee, Y. H., et al. (2012). Synthesis of large-area MoS_2 atomic layers with chemical vapor deposition. *Adv. Mater.* 24, 2320–2325.
[83] Kane, E. O. (1957). Band structure of indium antimonide. *J. Phys. Chem. Solids* 1, 249–261.
[84] López, R., and Gómez, R. (2012). Band-gap energy estimation from diffuse reflectance measurements on sol-gel and commercial TiO_2: a comparative study. *J. Sol Gel Sci. Technol.* 61, 1–7.
[85] Kao, K. H., et al. (2012). Direct and indirect band-to-band tunneling in germanium-based TFETs. *IEEE Trans. Electron Devices* 59, 292–301.

[86] Chen, S. Y., Zheng, C., Fuhrer, M. S., and Yan, J. (2015). Helicity-resolved Raman scattering of MoS_2, MoSe2, WS_2, and WSe_2 atomic layers. *Nano Lett.* 15, 2526–2532.
[87] Liu, Y., et al. (2013). Layer-by-layer thinning of MoS_2 by plasma. *ACS Nano* 7, 4202–4209.
[88] Stacy, A. M., and Hodul, D. T. (1985). Raman spectra of IVB and VIB transition metal disulfides using laser energies near the absorption edges. *J. Phys. Chem. Solids* 46, 405–409.
[89] Zhao, W., et al. (2013). Lattice dynamics in mono- and few-layer sheets of WS_2 and WSe_2. *Nanoscale* 5:9677.
[90] Hong, X., et al. (2014). Ultrafast charge transfer in atomically thin MoS_2/WS_2 heterostructures. *Nat. Nanotechnol.* 9, 682–686.
[91] Lee, C., et al. (2010). Anomalous lattice vibrations of single-and few-layer MoS_2. *ACS Nano* 4, 2695–700.
[92] Xiao, S., et al. (2016). Atomic-layer soft plasma etching of MoS_2. *Sci. Rep.* 6:19945.
[93] Splendiani, A., et al. (2010). Emerging photoluminescence in monolayer MoS_2. *Nano Lett.* 10, 1271–1275.
[94] Terrones, H., and Terrones, M. (2014). Electronic and vibrational properties of defective transition metal dichalcogenide Haeckelites: new 2D semi-metallic systems. *2D Mater.* 1:11003.
[95] Kumara, A., and Ahluwalia, P. K. (2012). Electronic structure of transition metal dichalcogenides monolayers 1H-MX2 (M = Mo, W; X = S, Se, Te) from ab-initio theory: new direct band gap semiconductors. *Eur. Phys. J. B* 85:186.
[96] Shemella, P., and Nayak, S. K. (2009). Electronic structure and band-gap modulation of graphene via substrate surface chemistry. *Appl. Phys. Lett.* 94:032101.
[97] Koåmider, K., and Fernández-Rossier, J. (2013). Electronic properties of the MoS_2-WS_2 heterojunction. *Phys. Rev. B* 87:075451.
[98] Kuc, A., Zibouche, N., and Heine, T. (2011). Influence of quantum confinement on the electronic structure of the transition metal sulfide TS_{2}. *Phys. Rev. B* 83:245213.
[99] Mak, K. F., Shan, J., and Heinz, T. F. (2011). Seeing many-body effects in single- and few-layer graphene: observation of two-dimensional saddle-point excitons. *Phys. Rev. Lett.* 106:046401.
[100] Chichibu, S. F., et al. (2006). Improvements in quantum efficiency of excitonic emissions in ZnO epilayers by the elimination of point defects. *J. Appl. Phys.* 99:093505.

[101] Gao, J., Amara, P., Alhambra, C., and Field, M. J. (1998). A generalized hybrid orbital (GHO) method for the treatment of boundary atoms in combined QM/MM calculations. *J. Phys. Chem. A* 102, 4714–4721.

[102] King, P. D. C., et al. (2014). Quasiparticle dynamics and spin–orbital texture of the SrTiO3 two-dimensional electron gas. *Nat. Commun.* 5:3414.

[103] Sun, M. (2004). On the incorporation of nickel and cobalt into MoS_2-edge structures. *J. Catal.* 226, 32–40.

[104] Feng, J., et al. (2015). Electrochemical reaction in single layer MoS<inf>2</inf>: nanopores opened atom by atom. *Nano Lett.* 15, 3431–3438.

[105] Auth, C., et al. (2012). "A 22nm high performance and low-power CMOS technology featuring fully-depleted tri-gate transistors, self-aligned contacts and high density MIM capacitors," in *Proceedings of the Digest of Technical Papers – Symposium on VLSI Technology,* Honolulu, HI, 131–132. doi: 10.1109/VLSIT.2012.6242496

[106] Knoch, J., and Appenzeller, J. (2002). Impact of the channel thickness on the performance of Schottky barrier metal-oxide-semiconductor field-effect transistors. *Appl. Phys. Lett.* 81, 3082–3084.

[107] Fuhrer, A., Füchsle, M., Reusch, T. C. G., Weber, B., and Simmons, M. Y. (2009). Atomic-scale, all epitaxial in-plane gated donor quantum dot in silicon. *Nano Lett.* 9, 707–710.

[108] Liu, X., et al. (2013). Top–down fabrication of sub-nanometre semiconducting nanoribbons derived from molybdenum disulfide sheets. *Nat. Commun.* 4:1776.

[109] Zhang, C., Johnson, A., Hsu, C. L., Li, L. J., and Shih, C. K. (2014). Direct imaging of band profile in single layer MoS_2 on graphite: quasiparticle energy gap, metallic edge states, and edge band bending. *Nano Lett.* 14, 2443–2447.

[110] Wu, J., et al. (2013). Layer thinning and etching of mechanically exfoliated MoS_2 nanosheets by thermal annealing in air. *Small* 9, 3314–3319.

[111] Zhang, W., et al. (2013). High-gain phototransistors based on a CVD MoS_2 monolayer. *Adv. Mater.* 25, 3456–3461.

[112] Yang, F., and Forrest, S. R. (2008). Photocurrent generation in nanostructured organic solar cells. *ACS Nano* 2, 1022–1032.

[113] Liu, C. C., Chang, S. J., Huang, G. Y., and Lin, Y. Z. (2010). A 10-bit 50-MS/s SAR ADC with a monotonic capacitor switching procedure. *IEEE J. Solid State Circ.* 45, 731–740.

[114] Pandey, A. K. (2015). Highly efficient spin-conversion effect leading to energy up-converted electroluminescence in singlet fission photovoltaics. *Sci. Rep.*5:7787.

[115] Uji, H., Tanaka, K., and Kimura, S. (2016). O_2-triggered directional switching of photocurrent in self-assembled monolayer composed of porphyrin- and fullerene-terminated helical peptides on gold. *J. Phys. Chem. C* 120, 3684–3689.

[116] Hung, Y.-J., et al. (2011). Polarization dependent solar cell conversion efficiency at oblique incident angles and the corresponding improvement using surface nanoparticle coating. *Nanotechnology* 22:485202.

[117] Lee, M.-K., Chu, C.-H., Wang, Y.-H., and Sze, S. M. (2001). 155-μm and infrared-band photoresponsivity of a Schottky barrier porous silicon photodetector. *Opt. Lett.* 26:160.

[118] Yao, B., et al. (2012). Photoresponsivity enhancement of pentacene organic phototransistors by introducing C60 buffer layer under source/drain electrodes. *Appl. Phys. Lett.* 101:163301.

[119] Lu, N., et al. (2014). MoS_2/MX2 heterobilayers: bandgap engineering via tensile strain or external electrical field. *Nanoscale* 6:2879.

[120] Guo, H., Lu, N., Wang, L., Wu, X., and Zeng, X. C. (2014). Tuning electronic and magnetic properties of early transition-metal dichalcogenides via tensile strain. *J. Phys. Chem. C* 118, 7242–7249.

[121] Çakır, D., Peeters, F. M., and Sevik, C. (2014). Mechanical and thermal properties of h-MX2 (M = Cr, Mo, W; X = O, S, Se, Te) monolayers: a comparative study. *Appl. Phys. Lett.* 104:203110.

[122] Byskov, L. S., Hammer, B., Nørskov, J. K., Clausen, B. S., and Topsøe, H. (1997). Sulfur bonding in MoS_2 and Co-Mo-S structures. *Catal. Lett.* 47, 177–182.

[123] Mattheiss, L. F. (1973). Band structures of transition-metal-dichalcogenide layer compounds. *Phys. Rev. B* 8, 3719–3740.

[124] Jiang, J.-W., Wang, J.-S., and Li, B. (2009). Young's modulus of graphene: a molecular dynamics study. *Phys. Rev. B* 80:113405.

[125] Castellanos-Gomez, A., et al. (2012). Mechanical properties of freely suspended semiconducting graphene-like layers based on MoS_2. *Nanoscale Res. Lett.* 7:233.

[126] Lee, C., Wei, X., Kysar, J. W., and Hone, J. (2008). Measurement of the elastic properties and intrinsic strength of monolayer graphene. *Science,* 321, 385–388.

[127] Yu, H., Cui, X., Xu, X., and Yao, W. (2015). Valley excitons in two-dimensional semiconductors. *Natl. Sci. Rev.* 2, 57–70.

[128] Korn, T., Heydrich, S., Hirmer, M., Schmutzler, J., and Schller, C. (2011). Low-temperature photocarrier dynamics in monolayer MoS_2. *Appl. Phys. Lett.* 99:102109.

[129] Najmaei, S., et al. (2014). Plasmonic pumping of excitonic photoluminescence in hybrid MoS_2-Au nanostructures. *ACS Nano* 8, 12682–12689.

[130] Mouri, S., Miyauchi, Y., and Matsuda, K. (2013). Tunable photoluminescence of monolayer MoS_2 via chemical doping. *Nano Lett.* 13, 5944–5948.

[131] Eich, S., et al. (2014). Time- and angle-resolved photoemission spectroscopy with optimized high-harmonic pulses using frequency-doubled Ti:Sapphire lasers. *J. Electron Spectros. Relat. Phenomena* 195, 231–236.

[132] Li, W., et al. (2008). Time-resolved dynamics in N_2O_4 probed using high harmonic generation. *Science* 322, 1207–1211.

[133] Berntsen, M. H., Götberg, O., and Tjernberg, O. (2011). An experimental setup for high resolution 10.5 eV laser-based angle-resolved photoelectron spectroscopy using a time-of-flight electron analyzer. *Rev. Sci. Instrum.* 82:095113.

[134] Jin, W., et al. (2015). Tuning the electronic structure of monolayer graphene/MoS_2 van der Waals heterostructures via interlayer twist. *Phys. Rev. B* 92:201409.

[135] Baxter, J. B., and Schmuttenmaer, C. A. (2009). Carrier dynamics in bulk ZnO. II. Transient photoconductivity measured by time-resolved terahertz spectroscopy. *Phys. Rev. B* 80:235206.

[136] Zheng, K., et al. (2016). High excitation intensity opens a new trapping channel in organic–inorganic hybrid perovskite nanoparticles. *ACS Energy Lett.* 1, 1154–1161.

[137] Manjón, F. J., et al. (2004). Band structure of indium selenide investigated by intrinsic photoluminescence under high pressure. *Phys. Rev. B* 70:125201.

[138] Grubišić Čabo, A., et al. (2015). Observation of ultrafast free carrier dynamics in single layer MoS_2. *Nano Lett.* 15, 5883–5887.

[139] Huang, J. K., et al. (2014). Large-area synthesis of highly crystalline WSe_2 monolayers and device applications. *ACS Nano* 8, 923–930.

[140] Huang, C., et al. (2014). Lateral heterojunctions within monolayer MoSe2–WSe_2 semiconductors. *Nat. Mater.* 13, 1096–1101.

[141] Chuang, H. J., et al. (2014). High mobility WSe_2 p – And n – Field-effect transistors contacted by highly doped graphene for low-resistance contacts. *Nano Lett.* 14, 3594–3601.

[142] Htay, M. T., et al. (2011). A cadmium-free Cu 2 ZnSnS 4/ZnO het-rojunction solar cell prepared by practicable processes. *Jpn. J. Appl. Phys.*50:32301.

[143] Sharma, B. K., Khare, N., and Ahmad, S. (2009). A ZnO/PEDOT:PSS based inorganic/organic hetrojunction. *Solid State Commun.* 149, 771–774.

[144] Geim, A. K. (2009). Graphene: status and prospects. *Science* 324, 1530–1534.

[145] Hicks, J., et al. (2012). A wide-bandgap metal–semiconductor–metal nanostructure made entirely from graphene. *Nat. Phys.* 9, 49–54.

[146] Poellmann, C., et al. (2015). Resonant internal quantum transitions and femtosecond radiative decay of excitons in monolayer WSe_2. *Nat. Mater.* 14, 889–893.

[147] Cui, Q., Ceballos, F., Kumar, N., and Zhao, H. (2014). Transient absorption microscopy of monolayer and bulk WSe_2. *ACS Nano* 8, 2970–2976.

[148] Zhou, H., et al. (2015). Large area growth and electrical properties of p-type WSe_2 atomic layers. *Nano Lett.* 15, 709–713.

Index

About the Editors

Didier Rouxel was born in Lyon, France, on March 5, 1965. He received an Engineer degree (five-year degree) in materials in 1989 from the Ecole Supérieure des Sciences et Techniques de l'Ingénieur de Nancy, Vandoeuvre, France. He received his Ph.D. in Material Sciences and Engineering from the University of Nancy I, France in 1993. He is now full professor in the Université de Lorraine and he leads the team « Micro and Nano Systems » in Institut Jean Lamour in Nancy, France. His research has been mainly focused on physical chemistry of surfaces and thin films, piezoelectric materials, elastic properties of materials studied by Brillouin spectroscopy, and development of micro-devices based on polymer-nanoparticles nanocomposite materials, in particular for biomedical applications. He was expert for the french agency ANSES for the topic "Nanomaterials and Health" and Member of the Year of the French Society of Nanomedicine in 2014.

Professor Sabu Thomas is the Director of the International and Interuniversity Centre for Nanoscience and Nanotechnology and full professor of Polymer Science and Engineering at the School of Chemical Sciences of Mahatma Gandhi University, Kottayam, Kerala, India. He is an outstanding leader with sustained international acclaims for his work in Polymer Science and Engineering, Polymer Nanocomposites, Elastomers, Polymer Blends, Interpenetrating Polymer Networks, Polymer Membranes, Green Composites and Nanocomposites, Nanomedicine and Green Nanotechnology. Dr. Thomas's ground breaking inventions in polymer nanocomposites, polymer blends, green nano technological and nano-biomedical sciences, have made transformative differences in the development of new materials for automotive, space, housing and biomedical fields. Professor Thomas has received a number of national and international awards which including, Fellowship of the Royal Society of Chemistry, London, MRSI medal, Nano Tech Medal, CRSI medal, Distinguished Faculty Award, and Sukumar Maithy Award for the best polymer researcher in the country. He is in the list of most productive researchers in India and holds a position of No. 5. Thomas has been conferred Honoris Causa (D.Sc) by the University of South Brittany, Lorient, France in 2015 and in May 2016 he was awarded Loyalty Award in "International Materials Technology Conference Exhibition. Professor Thomas has published over 700 peer reviewed research papers, reviews and book chapters. He has co-edited nearly 58 books and is the inventor of 5 patents. He has supervised 79 Ph.D. theses and his H index is 78 with nearly 27184 citations. Prof. Thomas has delivered over 300 Plenary/Inaugural and Invited lectures in national/international meetings over 30 countries. He has established a state of the art laboratory at Mahatma Gandhi University in the area of Polymer Science and Engineering and Nanoscience and Nanotechnology.

Nandakumar Kalarikkal, Ph.D., is an Associate Professor of Physics at the School of Pure and Applied Physics, as well as the Director of the International and Inter University Centre for Nanoscience and Nanotechnology, Mahatma Gandhi University, India. Dr. Kalarikkal's research group focuses on the specialized areas of nanomultiferroics, nanosemicondustors and nanophosphors, nanocomposites, nanoferroelectrics, nanoferrites, nanomedicine, nanosenors, ion beam radiation effects and phase transitions etc. The research group has extensive exchange programs with different industries and various research and academic institutions all over world and is performing world-class collaborative research in various fields. The Dr. Nandakumar Kalarikkal Centre is equipped with various sophisticated instruments and has established state-of-art experimental facilities that cater to the needs of researchers within the country and abroad. He has more than 100 International publications He has received several award and also edited several books with colleagues.

Sajith T. Abdulrahman, M.Tech, is a research scholar at the International and Inter University Center for Nanoscience and Nanotechnology, Mahatma Gandhi University, Kottayam, India. He was formerly an Assistant Professor at the School of Engineering at the Cochin University of Science and Technology, Cochin, India. He holds an M.Tech degree in nanotechnology and a B.Tech degree in mechanical engineering. His areas of research include nanotechnology, mechanical engineering, and polymer nanocomposites. He has written several papers presented at international conferences.